UNIVERSAL OBJECTIVE FORESTRY

(2nd Revised Edition)

(UPSC, PSCs, SRF/JRF/AFO, State PG & Ph.D. Entrance examinations and interviews of all Forest services)

Mohit Husain
M.Sc. (Forestry)
(UG-Goldmedalist)

Published by:

SCIENTIFIC PUBLISHERS (INDIA)
5A, New Pali Road, P.O. Box 91
Jodhpur 342 001, India
E-mail: info@scientificpub.com
Website: www.scientificpub.com

ISBN: 978-93-89184-26-6
eISBN: 978-93-89184-27-3

Printed in India

UNIVERSAL OBJECTIVE FORESTRY

(2nd Revised Edition)

Related Books

Title	*Author*
A Text Book of Environmental Science	Vidya Thakur
Agroforestry Theory and Practices	S.B. Lal
Biodiversity Traditional Knowledge and Intellectual Property Rights	S. Ram Reddy
Comprehensive Forestry (English-Gujarati)	R. L. Meena
Essentials of Plant Nursery Management *2nd Ed*	P.K. Ray
FAQ's on Forestry for IFS Aspirants	K.T. Parthiban
Forest Ecology	S.S. Sagwal
Forest Fires and their Control	E.S. Artsybashev
Forest Mensuration Handbook	G.J. Hamilton
Forest Nursery: How to Raise and Manage	S.S. Sagwal
Forest Pathology: Principles and Practice in Forestry	B.K.Bakshi
Forest Tree Seeds: Handbook	S.S. Sagwal
Forest Tribology and Anthropology	Vinod M. Mhaiske
Forestry Practicals: A Complete Practical Solution for Students	K.K. Chandra
Forestry Principles and Applications	A.J. Raj
Indian Manual of Plant Ecology	R. Misra
Indiras Objective Agriculture *2nd Ed.*	R.L. Arya
Introduction to Forestry & Agroforestry	K.T. Parthiban
Nursery and Plantation Practices in Forestry	Vinod Kumar
Objective Forestry: For All Competitive Examination *3rd Ed*	K.T. Parthiban
Operations Research	P.K. Mohanty
Practical Manual on Plantation Forestry	Pankaj Panwar
Research Methodology in Plant Science *2nd Ed*	P.S. Narayana
Textbook on Wood Products and Utilization	M.P. Divya
Tree Breeding and Improvement: Theory and Techniques	K.T. Parthiban
Universal Forestry 2nd Ed. (UPSC, PCS, ARS/SRF/JRF/AFO Exams)	Mohit Husain
Vaniki Aur Krishi Vaniki Se Parichiye (Hindi)	K.T. Parthiban
Wildlife Ecology and Conservation	M. Balakrishnan

Preface

Forestry is a broad subject. It covers all the basic subject of science. Therefore students have to pass through all the forestry related subject with elementary courses during under graduation. After passing graduation, higher degrees are done in specialized field of forestry, so there is no need to read all forestry subject in M.Sc. and Ph.D. programme.

But for the preparation of various forestry competition examinations students have to read all the basic books of forestry to cover the syllabus. And at that time most of the students do not have all the necessary books and ample of time to read them.

Therefore, to overcome these problems the author have tried his best to write this book through reading various books, states forest reports and other sources of forestry, to covers brief information of all the subjects in one book.

Covering all the subjects of forestry in one books is a very hard task so some errors may be present, therefore your suggestions are invited (mohit.husainchf@gmail.com) for further improvement of this book.

I express my appreciation for the hard work and affinity of my sincere and respected professors who have guided me for bring all the content in this single book.

The forbearance shown by my mother, Mrs. Hanipha Khatoon, father Mr. Amisher Khan, brothers Mr. Kamil Hussain & Mr. Suhaiv Akhtar and sister Ms. Nargis Khatoon, is worth placing high on record who always favoured me in collecting and bearing the assignment.

I hope this book would be very helpful for those appearing in various competition exams of forestry.

Mohit Husain

About the Author

Mr. Mohit Husain was born in Dhoondli (Bijnor), a small village in the state of Uttar Pradesh. His father, Amisher Khan is a businessman and had motivated him to pursue an under graduation studies in forestry science.

Mr. Mohit Husain, did his B.Sc. (Forestry) from college of Horticulture and Forestry, Agricultural University Kota, Rajasthan in 2016. Currently, he is pursuing his Master degree in forestry science from Sher-e-Kashmir University of Agricultural Science and Technology of Kashmir. His current research focuses on Vegetation analysis, Biomass production and carbon stock in the grasslands of Gulmarg. He has qualified ICAR-UG and ICAR-PG examinations and has received ICAR scholarship.

He was awarded gold medal in B.Sc. And was also awarded as *Best student of the year-2016* during the same.

Mr. Mohit Husain has attended various National and International conferences/ seminars and gave oral presentations as well as poster presentation. He has done various trainings in both agricultural and Forestry field. He has published 23 research papers and articles in National and International journals.

He has also written a practical manual on tree physiology, forest nursery and also edited a book, entitled *objective forestry*.

Dedicated to

my beloved parents

Mr. Amisher Khan
&
Mrs. Hanipha Khatoon

NET Syllabus for Agroforestry

UNIT 1. National Forest Policy 1894, 1952 and 1988; Indian Forest Act, 1927; Forest Conservation Act, 1980 and Wildlife Protection Act, 1972; Forests-extent, basis for classification and distribution in India; Geographical distribution and salient features of major world forest types; Phytogeographical regions and vegetation of India; Role of forests in national economy - productive, protective and ameliorative, tribal and rural livelihoods; Forest types of India: distribution and types; Succession, climax and retrogression; Concepts of biomass, productivity, energy flow and nutrient cycling in forest ecosystem; Migration and dispersal mechanism.

UNIT 2. Concept and definition of agroforestry, social forestry, community forestry and farm forestry; Benefits and constraints of agroforestry; Historical development of agroforestry and overview of global agroforestry systems. Classification of agroforestry systems: structural, functional, socio-economic and ecological; Diagnosis and design of agroforestry system; Land capability classification and land use; Criteria of an ideal agroforestry design, productivity, sustainability and adoptability; multipurpose tree species and their characteristics suitable for agroforestry.

UNIT 3. Plant management practices in agroforestry; Tree-crop interactions: ecological and economic; Concept of complementarity, supplementarity and competition; Productivity, nutrient cycling and light, water and nutrient competition in agroforestry; Concept of allelopathy and its impact on agroforestry; Energy plantations - choice of species and management; Lopping of top-feed species such as frequency and intensity of lopping; Organic farming; Financial analysis and economic evaluation of agroforestry systems: cost benefit analysis and land equivalent ratio; Agroforestry practices and systems in different agro - ecological zones of India.

UNIT 4. Extent and causes of land denudation; Effects of deforestation on soil erosion, land degradation, environment and rural economy; Wastelands: their extent, characteristics and reclamation; Watershed management and its role in social, economic and ecological development; Biomass production for fuel wood, small timber, raw material for plant-based cottage industries, non-wood forest products such as gums, resins & tannins, medicinal plants, essential oils, edible fruits, spices, bamboo and canes; Wood quality and wood preservation; Plywood and pulp industries.

UNIT 5. Forest mensuration - definition, object and scope; Measurement of diameter, girth, height, stem form, bark thickness, crown width and crown length; Measurement methods and their principles. Measurement and computation of volume of logs and felled/standing trees; Construction and application of volume tables; Biomass measurement; Growth and increment; Measurement of crops; Forest inventory: kinds of enumeration, sampling methods, sample plots and photo interpretation; Geographic information systems and remote sensing - concept and scope.

UNIT 6. Definition, object and scope of silviculture; Site factors - climatic, edaphic, physiographic, biotic and their influence on forest vegetation; Forest regeneration: natural and artificial; Silvicultural systems - high forest and coppice systems; Silviculture of important tree species - Populus, Eucalyptus, Dalbergia, Acacia, Tectona, Shorea, Prosopis, Casurina, Pinus, Gmelina, Azadirachta, Diospyros, Pterocarpus, Anogeissus, Santalum, Quercus and Albizia.

UNIT 7. Seed collection, processing, storage, viability and pre-treatment; Seed dormancy and methods for breaking dormancy; Seed testing and germination tests; Seed certification and ISTA Rules; Forest nursery - need, selection and preparation of site, layout and design of nursery beds; Types of containers; Root trainers; Growing media and sowing methods; Management of nursery-shading, watering, manuring, fertilizer application, weed control, insect pest and diseases control; Planting techniques: site selection, evaluation and protection; Soil working techniques for various edaphic and climatic conditions; Planting patterns; Plant spacing, manure and fertilizer application, irrigation/moisture conservation techniques; Choice of species. Afforestation on difficult sites: saline-alkaline soils, coastal sands, lateritic soils, wetlands, ravines and sand dunes, dry and rocky areas, cold desert; Tending operations - weeding, cleaning, climber cutting, thinning - mechanical, ordinary, crown and selection thinning, improvement felling, pruning and girdling; Forest fires: causes, types, impacts and control measures; Major forest pests and weeds.

UNIT 8. Forest management: definition and scope; Concept of sustained yield and normal forest; Rotation; Estimation of growing stock, density and site quality; Management of even aged and uneven aged forest; Regulation of yield in regular and irregular forests by area, volume, increment and number of trees; Working plan; Joint forest management; Conservation and management of natural resources including wildlife; Forest evaluation; Internal rate of return, present net worth and cost benefit analysis.

JRF Syllabus for Forestry

UNIT I: Importance of Agriculture/Forestry/Livestock in national economy. Basic principles of crop production. Important rural development programmes in India Elementary principles of economics and agri-extension. Organizational set up of Agricultural Research, education and extension in India. Major diseases and pests of crops. Elements of statistics.

UNIT II: Forest-importance, types, classification, ecosystem, biotic and abiotic components, ecological succession and climax, nursery and planting technique, social forestry, farm forestry, urban forestry, community forestry, forest management, silvicultural practices, forest mensuration, natural regeneration, man-made plantations, shifting cultivation, taungya, dendrology, hardwoods, softwoods, pulp woods, fuel woods, multipurpose tree species, wasteland management. Agro forestry – importance and land use systems, forest soils, classification and conservation, watershed management, forest genetics and biotechnology and tree improvement, tree seed technology, rangelands, wildlife – importance, abuse, depletion, management, major and minor forest products including medicinal and aromatic plants, forest inventory, aerial photo interpretation and remote sensing, forest depletion and degradation – importance and impact on environment, global warming, role of forests and trees in climate mitigation, tree diseases, wood decay and discolouration, tree pests, integrated pest and disease management, biological and chemical wood preservation, forest conservation, Indian forest policies, Indian forest act, forest engineering, forest economics, joint forest management and tribology.

Contents

UNIT 9. Tree improvement: nature and extent of variations in natural population; Natural selection; Concept of seed source/ provenance; Selection of superior trees; Seed production areas, exotic trees, land races; Collection, evaluation and maintenance of germplasm; Provenance testing. Genetic gains; Tree breeding: general principles, mode of pollination and floral structure; Basics of forest genetics - inheritance, Hardyweinburg Law, genetic drift; Aims and methods of tree breeding. Seed orchard: types, establishment, planning and management, progeny test and designs; Clonal forestry - merits and demerits; Techniques of vegetative propagation, tissue culture, mist chamber; Role of growth substances in vegetative propagation.

UNIT 10. Forestry in bio-economic productivity of different agro-eco-systems and environmental management; Global overview and classification of agroforestry systems; Tree-crop interaction in agroforestry; Biomass production for fuel' wood, small timber, raw material for plants-based cottage industries, non-wood forest products such as gums, resins, tannins, medicinal plants, essential oils, edible fruits, bamboos and canes; Principle and criteria of plant selection in agroforestry; Resource use-efficiency in agroforestry.

UNIT 11. Measurement of trees and stand – diameter, girth, height, form and crown characteristics; Measurement methods and their principles; Volume/biomass estimation, volume tables; Measurement of rangeland productivity; Forest enumeration: sampling methods, sample plots, surveys and photo interpretation; Concept and application of GIS and remote sensing; Introduction to internal rate of return, present net worth, cost benefit analysis and land equivalent ratio; Agroforestry and environmental conservation; Role of green revolution in forest conservation in India.

UNIT 12. Climate change: greenhouse effect, sources and sinks of green-house gases, major greenhouses gases; Global climate change – its history and future predictions; Impact of climate change on agriculture, forestry, water resources, sea level; Livestock, fishery and coastal ecosystems; International conventions on climate change; Global warming: effect of enhanced CO_2 on productivity; Ozone layer depletion; Disaster management, floods, droughts, earthquakes; Tsunami, cyclones and landslides; Agroforestry and carbon sequestration.

UNIT 13. Statistics: definition, object and scope; Frequency distribution; Mean, median, mode and standard deviation, introduction to correlation and regression; Experimental designs: basic principles, completely randomized, randomized block, Latin square and split plot designs.

CHAPTER 1

General Forestry

1. Term forest is derived from-
   a. Greek word　　b. **Latin word**
   c. English word　　d. French word
2. Total forest cover as percentage of geographical area according, FSI-2015-
   a. **21.34**　　b. 23.14
   c. 24.08　　d. 19.75
3. Total tree cover as percentage of geographical area according, FSI-2015-
   a. 1.80　　b. **2.82**
   c. 3.85　　d. 4.55
4. State having largest total forest cover in India is-
   a. Jammu & Kashmir　　b. Uttarakhand
   c. **Madhya Pradesh**　　d. Assam
5. State having highest forest cover as % of its geographical area-
   a. **Mizoram**　　b. Uttarakhand
   c. Madhya Pradesh　　d. Assam
6. State having highest mangrove forest cover in India is-
   a. Mizoram　　b. Uttarakhand
   c. Madhya Pradesh　　d. **West Bengal**
7. Currently Mangrove cover in India is-
   a. 1740 km²　　b. 2740 km²
   c. 3740 km²　　d. **4740 km²**

Answers

1. b　2. a　3. b　4. c　5. a　6. d
7. d

8. The total carbon stock in the country's forest is around 7, 044 million tonnes.
   a. 4,044 mil tonnes b. 5,044 mil tonnes
   c. **7,044 mil tonnes** d. 9,044 mil tonnes
9. Percent forest area of total geographic area in Andhra Pradesh according FSI, 2015-
   a. **2.48 percent** b. 0.91 percent
   c. 2.06 percent d. 2.32 percent
10. Number of total protected area (national park plus wildlife sanctuary) in Andhra Pradesh-
    a. 12 b. 30
    c. **16** d. 11
11. State animal of Andhra Pradesh-
    a. Deer b. Leopard
    c. **Blackbuck** d. Tiger
12. Percent forest area of total geographic area in Arunchal Pradesh according FSI, 2015-
    a. 2.48 percent b. **0.91 percent**
    c. 2.06 percent d. 2.32 percent
13. Number of total protected area (national park plus wildlife sanctuary) in Arunachal Pradesh-
    a. 10 b. **13**
    c. 23 d. 30
14. State animal of Arunchal Pradesh-
    a. Sangai b. Rhinoceros
    c. Gaur d. **Mithun**
15. Percent forest area of total geographic area in Assam according FSI, 2015-
    a. 2.48 percent b. 0.91 percent
    c. **2.06 percent** d. 2.32 percent
16. Number of total protected area (national park plus wildlife sanctuary) in Assam -
    a. **23** b. 30
    c. 32 d. 35

Answers

8. c	9. a	10. c	11. c	12. b	13. b
14. d	15. c	16. a			

17. State animal of Assam-
    a. **Rhinoceros**
    b. Blackbuck
    c. Hangul
    d. Asiatic lion

18. Percent forest area of total geographic area in Bihar according FSI, 2015-
    a. 2.48 percent
    b. 0.91 percent
    c. 2.06 percent
    d. **2.32 percent**

19. Number of total protected area (national park plus wildlife sanctuary) in Bihar-
    a. 10
    b. **13**
    c. 18
    d. 25

20. State animal of Bihar-
    a. **Ox**
    b. Leopard
    c. Blackbuck
    d. Tiger

21. Percent forest area of total geographic area in Chhattisgarh according FSI, 2015-
    a. 2.48 percent
    b. 0.91 percent
    c. **2.68 percent**
    d. 2.32 percent

22. Number of total protected area (national park plus wildlife sanctuary) in Chhattisgarh-
    a. 10
    b. **14**
    c. 20
    d. 22

23. State animal of Chhattisgarh-
    a. Sangai
    b. Rhinoceros
    c. Gaur
    d. **Wild Buffalo**

24. Percent forest area of total geographic area in Goa according FSI, 2015-
    a. 2.48 percent
    b. **08.78 percent**
    c. 2.06 percent
    d. 2.32 percent

25. Number of total protected area (national park plus wildlife sanctuary) in Goa-
    a. 5
    b. **7**
    c. 10
    d. 13

Answers

17. a	18. d	19. b	20. a	21. c	22. b
23. d	24. b	25. b			

26. State animal of Goa-
    a. **Gaur**
    b. Blackbuck
    c. Hangul
    d. Asiatic lion

27. Percent forest area of total geographic area in Gujarat according FSI, 2015-
    a. **4.04 percent**
    b. 0.91 percent
    c. 2.06 percent
    d. 2.32 percent

28. Number of total protected area (national park plus wildlife sanctuary) in Gujarat-
    a. **27**
    b. 30
    c. 35
    d. 39

29. State animal of Gujarat-
    a. Tiger
    b. **Asiatic lion**
    c. Deer
    d. Leopard

30. Percent forest area of total geographic area in Haryana according FSI, 2015-
    a. 2.48 percent
    b. 0.91 percent
    c. **3.07 percent**
    d. 2.32 percent

31. Number of total protected area (national park plus wildlife sanctuary) in Haryana-
    a. 5
    b. 7
    c. **10**
    d. 15

32. State animal of Haryana-
    a. Goat
    b. **Blackbuck**
    c. Tiger
    d. Hangul

33. Percent forest area of total geographic area in Himanchal Pradesh according FSI, 2015-
    a. 2.48 percent
    b. **1.36 percent**
    c. 2.06 percent
    d. 2.32 percent

34. Number of total protected area (national park plus wildlife sanctuary) in Himanchal Pradesh-
    a. 15
    b. 23
    c. 29
    d. **33**

Answers

26. a	27. a	28. a	29. b	30. c	31. c
32. b	33. b	34. d			

35. State animal of Himanchal Pradesh-
a. Rhinoceros b. Blackbuck
c. **Snow leopard** d. Hangul

36. Percent forest area of total geographic area Jammu & Kashmir according FSI, 2015-
a. 2.48 percent b. **3.76 percent**
c. 2.06 percent d. 2.32 percent

37. Number of total protected area (national park plus wildlife sanctuary) in Jammu & Kashmir-
a. 12 b. 15
c. **19** d. 25

38. State animal of Jammu & Kashmir-
a. Rhinoceros b. Blackbuck
c. Snow leopard d. **Hangul**

39. Percent forest area of total geographic area in West Bengal according FSI, 2015-
a. **2.35 percent** b. 0.91 percent
c. 2.06 percent d. 2.32 percent

40. Number of total protected area (national park plus wildlife sanctuary) in West Bengal-
a. 16 b. **21**
c. 26 d. 29

41. State animal of West Bengal-
a. Hangul b. **Fishing cat**
c. Elephant d. Rhinoceros

42. Percent forest area of total geographic area in Jharkhand according FSI, 2015-
a. 2.48 percent b. 0.91 percent
c. **3.49 percent** d. 2.32 percent

43. Number of total protected area (national park plus wildlife sanctuary) in Jharkhand-
a. 10 b. **12**
c. 18 d. 29

Answers

35. c	36. b	37. c	38. d	39. a	40. b
41. b	42. c	43. b			

44. State animal of Jharkhand-

a. Sangai | b. Rhinoceros
c. Gaur | d. **Elephant**

45. Percent forest area of total geographic area in Karnataka according FSI, 2015-

a. **2.89 percent** | b. 0.91 percent
c. 2.06 percent | d. 2.32 percent

46. Number of total protected area (national park plus wildlife sanctuary) in Karnataka-

a. 24 | b. 28
c. **32** | d. 37

47. State animal of Karnataka-

a. Swamp deer | b. Blackbuck
c. Nilgai Thar | d. **Elephant**

48. Percent forest area of total geographic area in Kerala according FSI, 2015-

a. 2.48 percent | b. 0.91 percent
c. 2.06 percent | d. **7.59 percent**

49. Number of total protected area (national park plus wildlife sanctuary) in Kerala-

a. 14 | b. 16
c. **23** | d. 28

50. State animal of Kerala-

a. Swamp deer | b. Blackbuck
c. Nilgai Thar | d. **Elephant**

51. Percent forest area of total geographic area in Madhya Pradesh according FSI, 2015-

a. **2.52 percent** | b. 0.91 percent
c. 2.06 percent | d. 2.32 percent

52. Number of total protected area (national park plus wildlife sanctuary) in Madhya Pradesh-

a. 15 | b. 24
c. 28 | d. **34**

Answers

44. d 45. a 46. c 47. d 48. d 49. c
50. d 51. a 52. d

53. State animal of Madhya Pradesh-
a. **Swamp deer** b. Blackbuck
c. Snow leopard d. Hangul

54. Percent forest area of total geographic area in Maharashtra according FSI, 2015-
a. 2.48 percent b. 0.91 percent
c. 2.06 percent d. **3.11 percent**

55. Number of total protected area (national park plus wildlife sanctuary) in Maharashtra-
a. 21 b. 28
c. 35 d. **47**

56. State animal of Maharashtra-
a. Deer b. Tiger
c. **Giant squirrel** d. Blackbuck

57. Percent forest area of total geographic area in Manipur according FSI, 2015-
a. 2.48 percent b. **1.09 percent**
c. 2.06 percent d. 2.32 percent

58. Number of total protected area (national park plus wildlife sanctuary) in Manipur-
a. 2 b. **3**
c. 7 d. 9

59. State animal of Manipur-
a. **Sangai** b. Rhinoceros
c. Gaur d. Elephant

60. Percent forest area of total geographic area in Meghalaya according FSI, 2015-
a. 2.48 percent b. 0.91 percent
c. 2.06 percent d. **3.17 percent**

61. Number of total protected area (national park plus wildlife sanctuary) in Meghalaya-
a. **5** b. 12
c. 16 d. 23

Answers

53. a	54. d	55. d	56. c	57. b	58. b
59. a	60. d	61. a			

62. State animal of Meghalaya-
    a. Deer
    b. Tiger
    c. **Clouded leopard**
    d. Blackbuck
63. Percent forest area of total geographic area in Mizoram according FSI, 2015-
    a. 2.48 percent
    b. 0.91 percent
    c. **2.54 percent**
    d. 2.32 percent
64. Number of total protected area (national park plus wildlife sanctuary) in Mizoram-
    a. 7
    b. **10**
    c. 15
    d. 21
65. State animal of Mizoram-
    a. Rhinoceros
    b. Blackbuck
    c. Snow leopard
    d. **Serow**
66. Percent forest area of total geographic area in Nagaland according FSI, 2015-
    a. **2.30 percent**
    b. 0.91 percent
    c. 2.06 percent
    d. 2.32 percent
67. Number of total protected area (national park plus wildlife sanctuary) in Nagaland-
    a. 2
    b. **4**
    c. 8
    d. 10
68. State animal of Nagaland-
    a. Deer
    b. **Mithun**
    c. Clouded leopard
    d. Blackbuck
69. Percent forest area of total geographic area in Odisha according FSI, 2015-
    a. 2.48 percent
    b. 0.91 percent
    c. **2.56 percent**
    d. 2.32 percent
70. Number of total protected area (national park plus wildlife sanctuary) in Odisha-
    a. 15
    b. 19
    c. **21**
    d. 27

Answers

62. c	63. c	64. b	65. d	66. a	67. b
68. b	69. c	70. c			

71. State animal of Odisha-
a. Rhinoceros
b. Blackbuck
c. Snow leopard
d. **Sambar**

72. Percent forest area of total geographic area in Punjab according FSI, 2015-
a. 2.48 percent
b. 0.91 percent
c. **3.07 percent**
d. 2.32 percent

73. Number of total protected area (national park plus wildlife sanctuary) in Punjab-
a. 8
b. 10
c. **13**
d. 15

74. State animal of Punjab-
a. Deer
b. Tiger
c. Giant squirrel
d. **Blackbuck**

75. Percent forest area of total geographic area in Rajasthan according FSI, 2015-
a. 2.48 percent
b. 0.91 percent
c. 2.06 percent
d. **2.42 percent**

76. Number of total protected area (national park plus wildlife sanctuary) in Rajasthan-
a. 12
b. 19
c. 26
d. 30

77. State animal of Rajasthan-
a. Deer
b. Tiger
c. **Camel**
d. Blackbuck

78. Percent forest area of total geographic area in Sikkim according FSI, 2015-
a. 2.48 percent
b. **0.50 percent**
c. 2.06 percent
d. 2.32 percent

79. Number of total protected area (national park plus wildlife sanctuary) in Sikkim-
a. 5
b. **8**
c. 12
d. 19

Answers

71. d	72. c	73. c	74. d	75. d	76. d
77. c	78. b	79. b			

80. State animal of Sikkim-
   a. Rhinoceros
   b. Blackbuck
   c. Snow leopard
   d. **Red Panda**

81. Percent forest area of total geographic area in Tamil Nadu according FSI, 2015-
   a. **3.46 percent**
   b. 0.91 percent
   c. 2.06 percent
   d. 2.32 percent

82. Number of total protected area (national park plus wildlife sanctuary) in Tamil Nadu-
   a. 15
   b. 20
   c. 25
   d. **34**

83. State animal of Tamil Nadu-
   a. Swamp deer
   b. Blackbuck
   c. **Nilgai Thar**
   d. Elephant

84. Percent forest area of total geographic area in Telangana according FSI, 2015-
   a. **2.22 percent**
   b. 0.91 percent
   c. 2.06 percent
   d. 2.32 percent

85. Number of total protected area (national park plus wildlife sanctuary) in Telangana-
   a. 8
   b. **12**
   c. 18
   d. 23

86. State animal of Telangana-
   a. Phayre's langur
   b. Blackbuck
   c. Elephant
   d. **Deer**

87. Percent forest area of total geographic area in Tripura according FSI, 2015-
   a. **2.22 percent**
   b. 0.91 percent
   c. 2.06 percent
   d. 2.32 percent

88. Number of total protected area (national park plus wildlife sanctuary) in Tripura-
   a. **6**
   b. 9
   c. 11
   d. 13

Answers

80. d	81. a	82. d	83. c	84. a	85. b
86. d	87. a	88. a			

89. State animal of Tripura-
a. **Phayre's langur** b. Blackbuck
c. Elephant d. Deer

90. Percent forest area of total geographic area in Uttar Pradesh according FSI, 2015-
a. 2.48 percent b. 0.91 percent
c. 2.06 percent d. **2.92 percent**

91. Number of total protected area (national park plus wildlife sanctuary) in Uttar Pradesh-
a. 12 b. 18
c. **25** d. 35

92. State animal of Uttar Pradesh-
a. Tiger b. **Swamp deer**
c. Blackbuck d. Cow

93. Percent forest area of total geographic area in Uttarakhand according FSI, 2015-
a. 2.48 percent b. 0.91 percent
c. **1.41 percent** d. 2.32 percent

94. Number of total protected area (national park plus wildlife sanctuary) in Uttarakhand-
a. 10 b. **13**
c. 18 d. 24

95. State animal of Uttarakhand-
a. **Musk deer** b. Blackbuck
c. Nilgai Thar d. Elephant

96. Advanced Research Centre for Bamboo and Rattan is situated in-
a. **Aizawl** b. Bhopal
c. Kerala d. New Delhi

97. Agricultural and Processed Food Products Export Development Authority (APEDA)-
a. Kanpur b. Bangalore
c. Madhya Pradesh d. **New Delhi**

Answers

89. a	90. d	91. c	92. b	93. c	94. b
95. a	96. a	97. d			

98. Arid Forest Research Institute (AFRI) is situated-
a. Jaipur b. **Jodhpur**
c. New Delhi d. Bombay

99. Headquarter of Botanical Survey of India (BSI)-
a. New Delhi b. Dehra Dun
c. **Kolkata** d. Lucknow

100. Center for Ecological Sciences is situated-
a. Kanpur b. **Bangalore**
c. Madhya Pradesh d. New Delhi

101. Center for International Forestry Research (CIFOR) is situated in-
a. India b. Geneva
c. **Indonesia** d. Japan

102. Center for Tropical Forest Science is located in-
a. Morgis b. Geneva
c. Glans d. Panama

103. Central Drug Research Institute is located in-
a. New Delhi b. Dehra Dun
c. Kolkata d. **Lucknow**

104. Centre for Forest-based Livelihoods and Extension is situated in-
a. Morgis b. Geneva
c. Glans d. **Agartala**

105. Centre for Social Forestry and Eco-Rehabilitation is located in-
a. Jaipur b. **Allahabad**
c. New Delhi d. Bombay

106. Forest Research Institute is located in-
a. New Delhi b. **Dehra Dun**
c. Kolkata d. Lucknow

107. Headquarter of Forest Survey of India is situated in-
a. New Delhi b. Kolkata
c. Mumbai d. **Dehra Dun**

108. Foundation for Revitalizations of Local Health Traditions (FRLHT) is located in-
a. Kanpur b. **Bangalore**
c. Madhya Pradesh d. New Delhi

Answers

98. b	99. c	100. b	101. c	102. d	103. d
104. d	105. b	106. b	107. d	108. b	

109. Himalayan Forest Research Institute is situated in-
a. New Delhi b. Kolkata
c. Mumbai d. **Almora**

110. Indian Grassland and fodder research institute (IGFRI) located in-
a. **Uttar Pradesh** b. Rajasthan
c. Odisha d. Kerala

111. Indian Institute of Forest Management (IIFM) is located in-
a. Aizawl b. **Bhopal**
c. Kerala d. New Delhi

112. Indian Institute of Plantation Management (IIPM) is situated-
a. New Delhi b. Kolkata
c. Mumbai d. **Bangalore**

113. Indian institute of Soil science (IISS) is located in-
a. Aizawl b. **Bhopal**
c. Kolkata d. New Delhi

114. Indian Institute of Natural Resins and Gums is situated in-
a. Dehra Dun b. Bihar
c. **Ranchi** d. Lucknow

115. Indian Plywood Industries Research and Training Institute is located-
a. New Delhi b. Kolkata
c. Mumbai d. **Bangalore**

116. Indira Gandhi National Forest Academy (IGNFA) located-
a. **Dehra Dun** b. Bihar
c. Ranchi d. Lucknow

117. Institute of Forest Biodiversity (IFB) is located in-
a. New Delhi b. **Hyderabad**
c. Mumbai d. Bangalore

118. Institute of Forest Genetics and Tree Breeding is located-
a. **Coimbatore** b. Mumbai
c. Dehra Dun d. Ranchi

119. Institute of Forest Productivity is located-
a. Coimbatore b. Mumbai
c. Dehra Dun d. **Ranchi**

Answers

109. d	110. a	111. b	112. d	113. b	114. c
115. d	116. a	117. b	118. a	119. d	

120. Institute of Wood Science and Technology is located in-
    a. New Delhi
    b. Hyderabad
    c. Mumbai
    d. **Bangalore**

121. International Bureau of Plant Genetic Resources is located-
    a. Geneva
    b. **Rome**
    c. Washington
    d. Switzerland

122. International Union of Forest Research Organizations (IUFRO) is located-
    a. Switzerland
    b. Italy
    c. Paris
    d. **Austria**

123. Headquarter of National Bank for Agriculture and Rural Development (NABARD)-
    a. New Delhi
    b. Hyderabad
    c. **Mumbai**
    d. Bangalore

124. Headquarter of National Biodiversity Authority is located-
    a. Mumbai
    b. Dehra Dun
    c. New Delhi
    d. **Chennai**

125. Headquarter of National Biodiversity Board is located in-
    a. Mumbai
    b. Dehra Dun
    c. **New Delhi**
    d. Chennai

126. National Bureau of Plant Genetic Resources (NBPGR) located in-
    a. **New Delhi**
    b. Hyderabad
    c. Mumbai
    d. Bangalore

127. National Bureau of Soil Survey and Land Use Planning located in
    a. Faridabad
    b. New Delhi
    c. **Nagpur**
    d. Bhopal

128. National Museum of Natural History is located-
    a. Mumbai
    b. Dehra Dun
    c. **New Delhi**
    d. Chennai

129. Headquarter of National Tiger Conservation Authority (NTCA) is located-
    a. **New Delhi**
    b. Hyderabad
    c. Mumbai
    d. Dehra Dun

Answers

120. d	121. b	122. d	123. c	124. d	125. c
126. a	127. c	128. c	129. a		

130. Rain Forest Research Institute (RFRI) situated in-
   a. Faridabad b. New Delhi
   c. **Jorhat** d. Bhopal

131. Headquarter of CITES is situated in-
   a. Geneva b. Rome
   c. **Washington** d. Switzerland

132. Headquarter of IUCN is situated in-
   a. America b. Rome
   c. Washington d. **Switzerland**

133. Tropical Forest Research Institute is located in-
   a. **Jabalpur** b. New Delhi
   c. Jorhat d. Bhopal

134. Wildlife Institute of India is situated in-
   a. New Delhi b. Hyderabad
   c. Mumbai d. **Dehra Dun**

135. Headquarter of World Agroforestry Center is located in-
   a. Geneva b. Rome
   c. **Nairobi** d. Switzerland

136. Central Arid Zone Research Institute situated in-
   a. Jaipur b. **Jodhpur**
   c. New Delhi d. Bombay

137. Central Institute of Cotton Research located in-
   a. Faridabad b. New Delhi
   c. **Nagpur** d. Bhopal

138. Central Plantation Crops Research Institute situated in-
   a. Kochi b. **Kasargod**
   c. Cuttack d. Shimla

139. Central Sheep and Wool Research Institute located in-
   a. New Delhi b. Lucknow
   c. Jodhpur d. **Avikanagar**

140. Central Soil Salinity Research Institute is located-
   a. New Delhi b. Hyderabad
   c. Mumbai d. **Karnal**

Answers

130. c	131. c	132. d	133. a	134. d	135. c
136. b	137. c	138. b	139. d	140. d	

141. Central Tobacco Research Institute is located in-
   a. **Rajahmundry**   b. Nagpur
   c. Assam   d. Trivandrum

142. Directorate of Medicinal and Aromatic Plants Research is situated-
   a. **Anand**   b. Bhopal
   c. Lucknow   d. Dehra Dun

143. Directorate of Rapeseed & Mustard Research is located-
   a. Faridabad   b. New Delhi
   c. **Bharatpur**   d. Bhopal

144. Directorate of Seed Research is located-
   a. Gujarat   b. Rajahmundry
   c. **Mau**   d. Dehra Dun

145. Indian Agricultural Statistics Research Institute is situated-
   a. **New Delhi**   b. Hyderabad
   c. Mumbai   d. Dehra Dun

146. National Research Centre for Agroforestry is located-
   a. **Jhansi**   b. Debra Dun
   c. Coimbatore   d. New Delhi

147. National Research Centre on Camel is situated in-
   a. Anand   b. Bhopal
   c. Lucknow   d. **Bikaner**

148. National Research Centre on Orchids is situated-
   a. Coimbatore   b. Ajmer
   c. **Sikkim**   d. Almora

149. Sugarcane Breeding Research Institute is located-
   a. **Coimbatore**   b. Ajmer
   c. Kanpur   d. Almora

150. Per capita forest are available in India is-
   a. 0.03 hectare   b. **0.06 hectare**
   c. 0.64 hectare   d. 1.6 hectare

151. Per capita forest are available in world is-
   a. 0.03 hectare   b. 0.06 hectare
   c. **0.64 hectare**   d. 1.6 hectare

Answers

141. a	142. a	143. c	144. c	145. a	146. a
147. d	148. c	149. a	150. b	151. c	

152. Total volume of water in the hydrosphere is-
a. **1.4 billion cubic kilometres** b. 1.6 billion cubic kilometres
c. 1.8 billion cubic kilometres d. 2.4 billion cubic kilometres

153. How much percent fresh water is available on the earth-
a. **2.5%** b. 2.8%
c. 4.0% d. 4.5%

154. Maximum angiospermic plants found in India-
a. Uttar Pradesh b. **Tamil Nadu**
c. Andhra Pradesh d. Karnataka

155. Minimum gymnospermic plants found in India-
a. Uttar Pradesh b. Tamil Nadu
c. **Andhra Pradesh** d. Karnataka

156. 19th Commonwealth Forestry Conference was held on-
a. **3rd April at FRI, Dehra Dun** b. 3rd March at FRI, Dehra Dun
c. 6th May at FRI, Dehra Dun d. 6th June at FRI, Dehra Dun

157. The theme of 2017 International Day of Forest is-
a. Save the forest for future
b. Conserve natural resources for future
c. **Forest and Energy**
d. Forest and Climate change

158. The theme of 2017 International Day of Water is-
a. Save the water save life
b. Water conservation for life
c. Save environment and save the water
d. **Why Waste Water**

159. Chipko movement was started in the year-
a. 1956 b. **1974**
c. 1976 d. 1987

160. Apiko movement was started in the year-
a. 1956 b. 1974
c. 1976 d. **1983**

161. Indian forest service was started in the year-
a. 1893 b. 1948
c. **1966** d. 1983

Answers

152. a	153. a	154. b	155. c	156. a	157. c
158. d	159. b	160. d	161. c		

162. Indian Administrative Service was started in the year-
   a. **1893** b. 1948
   c. 1966 d. 1983

163. Indian Police Service was started in the year-
   a. 1893 b. **1948**
   c. 1966 d. 1983

164. Selvas in Brazil are-
   a. Big grassland in plain b. Highland pastureland
   c. **Dense equatorial forest** d. Open scrub forest

165. Golden Triangle in South East Asia is known in the world for-
   a. Mustard oil production b. **Opium production**
   c. Dense equatorial forest d. Gold mining

166. Chilgoza is obtained from-
   a. *Cycas revoluta* b. ***Pinus gerardiana***
   c. *Pinus roxburghii* d. *Pinus wallichiana*

167. First country in the World to ban deforestation-
   a. Canada b. Germany
   c. Holland d. **Norway**

168. Average rainfall of India is-
   a. 1000 mm b. **1194 mm**
   c. 1250 mm d. 1500 mm

169. Annual world precipitation is-
   a. **1000 mm** b. 1194 mm
   c. 1250 mm d. 1500 mm

170. Geographical area of India is-
   a. 128.84 m ha b. 228.94 m ha
   c. 318.24 m ha d. **328.74 m ha**

171. Regur refers to-
   a. Red soil b. **Black soil**
   c. Alluvial soil d. Sandy soil

172. Most predominant soil type in India is-
   a. Sandy loam b. Black soil
   c. **Red soil** d. Brown soil

Answers

162. a	163. b	164. c	165. b	166. b	167. d
168. b	169. a	170. d	171. b	172. c	

173. World wetlands day is celebrated on-
  a. **Feb-02** b. Mar-03
  c. Mar-21 d. Mar-22

174. World wildlife day is celebrated on-
  a. Feb-02 b. **Mar-03**
  c. Mar-21 d. Apr-22

175. Indian Penel Code was formulated in the year-
  a. **1860** b. 1875
  c. 1908 d. 1927

176. Indian forest Act-1927 contains-
  a. 10 chapters and 85 sections b. 15 chapters and 85 sections
  c. **13 chapters and 86 sections** d. 17 chapters and 91 sections

177. Indian forest Act was formulated in the year-
  a. 1860 b. 1886
  c. **1927** d. 1976

178. International day of forest is celebrated on-
  a. Mar-02 b. Mar-03
  c. **Mar-21** d. Apr-22

179. World water day is celebrated on-
  a. Feb-02 b. Mar-03
  c. Mar-21 d. **Mar-22**

180. Water (Prevention & control of pollution) Act was enacted in the year-
  a. 1927 b. **1974**
  c. 1972 d. 1976

181. Earth day is celebrated on-
  a. May-02 b. June-05
  c. Mar-21 d. **Apr-22**

182. Which chapter of Indian Forest Act (IFA-1927) deals with Preliminary definition-
  a. ***Chapter i*** b. *Chapter ii*
  c. *Chapter iii* d. *Chapter iv*

Answers

173. a	174. b	175. a	176. c	177. c	178. c
179. d	180. b	181. d	182. a		

183. Forest Conservation Act was enacted in the year-
   a. 1927 b. **1980**
   c. 1972 d. 1986

184. Which chapter of Indian Forest Act (IFA-1927) deals with reserved forests-
   a. *Chapter i* b. ***Chapter ii***
   c. *Chapter iii* d. *Chapter iv*

185. World biodiversity day is celebrated on-
   a. Mar-02 b. **May-22**
   c. Mar-21 d. Apr-22

186. World environment day is celebrated on-
   a. Jul-11 b. May-22
   c. Mar-21 d. **Jun-05**

187. Criminal Procedure code was enacted in the year-
   a. 1927 b. **1973**
   c. 1986 d. 1996

188. International year of desert and desertification-
   a. 2001 b. 2004
   c. **2006** d. 2009

189. Which chapter of Indian Forest Act (IFA-1927) deals with protected forests-
   a. *Chapter i* b. *Chapter ii*
   c. *Chapter iii* d. ***Chapter iv***

190. Air (Prevention & control of pollution) Act-
   a. 1927 b. 1973
   c. **1981** d. 1986

191. World International tiger day is celebrated on-
   a. **Jul-29** b. May-22
   c. Mar-21 d. Jun-29

192. International year of biodiversity-
   a. 2002 b. 2005
   c. 2006 d. **2010**

Answers

183. b	184. b	185. b	186. d	187. b	188. c
189. d	190. c	191. a	192. d		

193. Which chapter of Indian Forest Act (IFA-1927) deals with village forests-
a. *Chapter i* b. *Chapter ii*
c. ***Chapter iii*** d. *Chapter iv*

194. World lion day is celebrated on-
a. Jun-11 b. Mar-03
c. Sept-21 d. **Aug-10**

195. World elephant day is cclcbrated on-
a. May-23 b. **Aug-12**
c. Sept-23 d. Aug-10

196. International year of forests-
a. 2006 b. 2009
c. **2011** d. 2012

197. National Environment Protection Act was enacted in the year-
a. 1957 b. 1973
c. **1986** d. 1996

198. International ozone day is celebrated on-
a. Jun-11 b. May-22
c. **Sept-16** d. Aug-23

199. International year of soils-
a. 2009 b. 2013
c. 2014 d. **2015**

200. International year of pulses-
a. 2009 b. 2013
c. 2014 d. **2016**

201. Which chapter of Indian Forest Act (IFA-1927) deals with cattle tress pass-
a. *Chapter ii* b. *Chapter iv*
c. ***Chapter x*** d. *Chapter iv*

202. World rhino day is celebrated on-
a. May-23 b. Aug-12
c. **Sept-22** d. Aug-10

Answers

193. c	194. d	195. b	196. c	197. c	198. c
199. d	200. d	201. c	202. c		

203. Biosphere Reserve Schemes was enacted in the year-
a. 1957 b. 1973
c. **1986** d. 1996

204. Control over forests and lands not being the property government deals chapter of IFA-1927-
a. ***Chapter v*** b. *Chapter iv*
c. *Chapter ix* d. *Chapter xi*

205. World animal day is celebrated on-
a. Jun-23 b. Sept-22
c. **Oct-04** d. Dec-23

206. World soil day is celebrated on-
a. Jun-23 b. Sept-22
c. Oct-04 d. **Dec-05**

207. International mountain day is celebrated on-
a. Jun-11 b. **Dec-11**
c. Sept-27 d. Dec-22

208. Biodiversity Act was enacted in the year-
a. 1972 b. 1986
c. **2002** d. 2006

209. Which chapter of Indian Forest Act (IFA-1927) deals with forest officers -
a. *Chapter ii* b. *Chapter iv*
c. *Chapter x* d. ***Chapter xi***

210. Which chapter of Indian Forest Act (IFA-1927) deals with subsidiary rules -
a. *Chapter ix* b. *Chapter xi*
c. ***Chapter xii*** d. *Chapter xiii*

211. Wind speed is measure by-
a. Actinometer b. Wind vane
c. **Anemometer** d. Audiometer

212. Barometer is used measure-
a. Heat of combustion b. **Air pressure**
c. Rate of evaporation d. Light intensity

Answers

203. c	204. a	205. c	206. d	207. b	208. c
209. d	210. c	211. c	212. b		

213. Control over forests and lands not being the property government is deals in the chapter, IFA-
   a. *Chapter ix* b. ***Chapter v***
   c. ***Chapter xii*** d. *Chapter xiii*

214. Green revolution is related with-
   a. Milk production b. Fish production
   c. **Rice and wheat production** d. Onion production

215. Wind direction is measured-
   a. Actinometer b. **Wind vane**
   c. Anemometer d. Audiometer

216. The duty on timber and other forest-produce is deals in the chapter of IFA, 1927-
   a. ***Chapter vi*** b. *Chapter iv*
   c. *Chapter ix* d. *Chapter xi*

217. Bomb calorimeter is used to measure-
   a. **Heat of combustion** b. Air pressure
   c. Rate of evaporation d. Light intensity

218. Potometer is used to measure-
   a. Humidity b. Air pressure
   c. Rate of evaporation d. **Transpiration**

219. The collection of drift and stranded timber is deals in the chapter of IFA, 1927-
   a. *Chapter vi* b. *Chapter iv*
   c. ***Chapter viii*** d. *Chapter xi*

220. Hygrometer is used to measure-
   a. **Humidity** b. Air pressure
   c. Rate of evaporation d. Transpiration

221. Which chapter of Indian Forest Act (IFA-1927) deals with penalties and procedure -
   a. ***Chapter ix*** b. *Chapter xi*
   c. *Chapter xii* d. *Chapter xiii*

Answers

213. b	214. c	215. b	216. a	217. a	218. d
219. c	220. a	221. a			

222. The control of timber and other forest-produce in transit is deals in the chapter of IFA, 1927-
a. *Chapter vi* b. *Chapter v*
c. ***Chapter vii*** d. *Chapter xi*

223. Rain gauge is used to measure-
a. Humidity b. **Rainfall**
c. Rate of evaporation d. Transpiration

224. Conference on Human Environment was held in 1972 at-
a. **Stockholm** b. Montreal
c. Rio-de-Jenerio d. Johncsberg

225. Conference on Montreal Protocol was held in 1987 at-
a. Stockholm b. **Montreal**
c. Rio-de-Jenerio d. Johncsberg

226. White revolution is related with-
a. **Milk production** b. Fish production
c. Rice and wheat production d. Onion production

227. Blue revolution is related with-
a. Milk production b. **Fish production**
c. Rice and wheat production d. Onion production

228. Conference on Earth Summit was held in 1992 at-
a. Stockholm b. Montreal
c. **Rio-de-Jenerio** d. Nairobi

229. Yellow revolution is related with-
a. Milk production b. **Oilseeds production**
c. Rice and wheat production d. Onion production

230. Conference on Kyoto Protocol was held in 1997 at-
a. Stockholm b. Montreal
c. Rio-de-Jenerio d. **Kyoto**

231. Red revolution is related with-
a. Milk production b. **Tomato production**
c. Rice and wheat production d. Onion production

232. Conference on United Nation was held in 1997 at-
a. Stockholm b. Montreal
c. Rio-de-Jenerio d. **Nairobi**

Answers

222. c	223. b	224. a	225. b	226. a	227. b
228. c	229. b	230. d	231. b	232. d	

233. Golden revolution is related with-
a. Oilseeds production b. **Fruit production**
c. Rice and wheat production d. Onion production

234. Conference on World Summit was held in 2002 at-
a. Stockholm b. **Johncsberg**
c. Rio-de-Jenerio d. Nairobi

235. Agriculture is a-
a. Greek word b. **Latin word**
c. Arabic word d. French word

236. Round revolution is related with-
a. Onion production b. **Potato production**
c. Rice and wheat production d. Onion production

237. Extension is a-
a. Greek word b. **Latin word**
c. Arabic word d. French word

238. Black revolution is related with-
a. Onion production b. Potato production
c. Rice and wheat production d. **Biofuel/Jatropa revolution**

239. Monsoon is a-
a. Greek word b. Latin word
c. **Arabic word** d. French word

240. Pink revolution is related with-
a. **Onion production** b. Potato production
c. Rice and wheat production d. All sector of agriculture

241. Entomology is a-
a. **Greek word** b. Latin word
c. Arabic word d. French word

242. Rainbow revolution is related with-
a. Onion production b. Potato production
c. Rice and wheat production d. **All sector of agriculture**

243. Meteorology is a-
a. **Greek word** b. Latin word
c. Arabic word d. French word

Answers

233. b	234. b	235. b	236. b	237. b	238. d
239. c	240. a	241. a	242. d	243. a	

244. Forest mensuration is a-
a. Greek word
b. **Latin word**
c. Arabic word
d. French word

245. Agro climatic zones of India is-
a. 10
b. 12
c. **15**
d. 20

246. Silviculture is a-
a. Greek word
b. **Latin word**
c. Arabic word
d. French word

247. Forest Research Institute was established in the year-
a. 1901
b. **1906**
c. 1919
d. 1947

248. Dendrology is a-
a. **Greek word**
b. Latin word
c. Arabic word
d. French word

249. Young ones of antelope is called-
a. **Calf**
b. Fawn
c. Kid
d. Lamb

250. Forest Pathology is a-
a. **Greek word**
b. Latin word
c. Arabic word
d. French word

251. Ergonomics is a-
a. **Greek word**
b. Latin word
c. Arabic word
d. French word

252. Young ones of camel and cattle is called-
a. **Calf**
b. Fawn
c. Kid
d. Lamb

253. Communication is a-
a. Greek word
b. **Latin word**
c. Arabic word
d. French word

254. Young ones of deer is called-
a. Calf
b. **Fawn**
c. Kid
d. Lamb

Answers

244. b	245. c	246. b	247. b	248. a	249. a
250. a	251. a	252. a	253. b	254. b	

255. World environment day is celebrated on-
a. **June-5** b. May-22
c. Mar-21 d. Jun-22

256. Economics is a-
a. **Greek word** b. Latin word
c. Arabic word d. French word

257. Young ones of tiger is called-
a. Calf b. Fawn
c. **Cub** d. Lamb

258. Group of camel is called-
a. Herd b. Flock
c. **Caravan** d. Pride

259. Young ones of rhinoceros is called-
a. **Calf** b. Fawn
c. Cub d. Lamb

260. Meat of cattle is called-
a. Chevon b. Mutton
c. **Beef** d. Venison

261. Group of leopard is called-
a. Herd b. **Leap**
c. Pack d. Pride

262. Young ones of sheep is called-
a. Calf b. Fawn
c. Cub d. **Lamb**

263. Meat of deer is called-
a. Chevon b. Mutton
c. Beef d. **Venison**

264. Group of lion is called-
a. Herd b. Leap
c. Pack d. **Pride**

265. Meat of goat is called-
a. **Chevon** b. Mutton
c. Beef d. **Venison**

Answers

255. a	256. a	257. c	258. c	259. a	260. c
261. b	262. d	263. d	264. d	265. a	

266. Group of tiger is called-
a. **Ambush** b. Leap
c. Pack d. Pride

267. Group of red panda is called-
a. Ambush b. Leap
c. **Pack** d. Pride

268. ICAR was established in the year-
a. 1806 b. 1906
c. **1929** d. 1972

269. Headquarter of ICAR is situated in-
a. Dehra Dun b. **New Delhi**
c. Lucknow d. Mumbai

270. RBI was established in the year-
a. 1806 b. 1906
c. 1929 d. **1935**

271. Headquarter of RBI is situated in-
a. Dehra Dun b. New Delhi
c. Lucknow d. **Mumbai**

272. AGMARK was established in the year-
a. 1806 b. 1906
c. 1929 d. **1937**

273. Headquarter of AGMARK is situated in-
a. Dehra Dun b. New Delhi
c. **Nagpur** d. Mumbai

274. GATT was established in the year-
a. 1920 b. **1948**
c. 1979 d. 1987

275. Headquarter of GATT is situated in-
a. Dehra Dun b. New Delhi
c. **Geneva** d. Mumbai

276. IUCN was established in the year-
a. 1875 b. 1906
c. 1938 d. **1948**

Answers

266. a	267. c	268. c	269. b	270. d	271. d
272. d	273. c	274. b	275. c	276. d	

277. Headquarter of IUCN is situated in-
a. Gland b. **Morgis**
c. Geneva d. Nairobi

278. IMD was established in the year-
a. 1875 b. 1938
c. **1961** d. 1985

279. Headquarter of IMD is situated in-
a. Dehra Dun b. New Delhi
c. **Pune** d. Mumbai

280. ICRAF was established in the year-
a. 1875 b. 1956
c. 1961 d. **1977**

281. Headquarter of ICRAF is situated in-
a. Jhansi b. New Delhi
c. Geneva d. **Nairobi**

282. NABARD was established in the year-
a. 1835 b. 1948
c. **1982** d. 1991

283. Headquarter of NABARD is situated in-
a. Dehra Dun b. New Delhi
c. Lucknow d. **Mumbai**

284. ITTO was established in the year-
a. 1875 b. 1938
c. 1961 d. **1985**

285. Headquarter of ITTO is situated in-
a. Dehra Dun b. New Delhi
c. Geneva d. **Japan**

286. WTO was established in the year-
a. 1895 b. **1995**
c. 1999 d. 2000

287. Headquarter of WTO is situated in-
a. Washington b. New Delhi
c. **Geneva** d. Morgis

Answers

277. b	278. c	279. c	280. d	281. d	282. c
283. d	284. d	285. d	286. b	287. c	

288. WWF was established in the year-

a. 1875 b. 1938
c. **1961** d. 1985

289. Headquarter of WWF is situated in-

a. Dehra Dun b. New Delhi
c. Nairobi d. **Gland**

290. FCI was established in the year-

a. 1875 b. **1964**
c. 1982 d. 1985

291. Headquarter of FCI is situated in-

a. Dehra Dun b. **New Delhi**
c. Pune d. Mumbai

292. WIPO was established in the year-

a. 1925 b. **1967**
c. 1975 d. 1985

293. Headquarter of WIPO is situated in-

a. Washington b. New Delhi
c. **Geneva** d. Morgis

294. CITES was established in the year-

a. 1875 b. 1967
c. **1975** d. 1985

295. Headquarter of CITES is situated in-

a. **Washington** b. New Delhi
c. Geneva d. Morgis

296. Botanical name of axlewood is-

a. ***Anogeissus latifolia*** b. *Acrocarpus fraxinifolius*
c. *Bauhania variegata* d. None

297. Tree of heaven is botanically know as-

a. *Abies pindrew* b. *Adina cordifoia*
c. *Acacia auriculoformis* d. ***Ailanthus excelsa***

298. Devil's tree is botanically know as-

a. *Abies pindrew* b. ***Alostonia scholaris***
c. *Acacia auriculoformis* d. *Ashoka indica*

Answers

288. c	289. d	290. b	291. b	292. b	293. c
294. c	295. a	296. a	297. d	298. b	

299. God Fruit tree is botanically know as-
   a. *Abies pindrew* b. *Alostonia scholaris*
   c. ***Aegle marmelos*** d. *Ashoka indica*

300. Kapok tree is botanically know as-
   a. *Ceiba patendra* b. *Alostonia scholaris*
   c. *Aegle marmelos* d. ***Bombax ceiba***

301. Indian kapok tree is botanically know as-
   a. ***Ceiba patendru*** b. *Alostonia scholaris*
   c. *Butea monosperma* d. *Bombax ceiba*

302. Chironzi tree is botanically know as-
   a. *Ceiba patendra* b. ***Buchanania lanzan***
   c. *Bauhania variegata* d. *Ashoka indica_*

303. Forest flame tree is botanically know as-
   a. *Ceiba patendra* b. ***Butea monosperma***
   c. *Bauhania variegata* d. *Ashoka indica*

304. Indian laburnum wood botanically know as-
   a. *Ceiba patendra* b. *Acacia catechu*
   c. *Bauhania variegata* d. ***Cassia fistlua***

305. Beef wood botanically know as-
   a. *Ceiba patendra* b. *Butea monosperma*
   c. ***Casuarina equisitifolia*** d. *Cassia fistlua*

306. Sago palm botanically know as-
   a. *Pinus wallichiana* b. ***Cycas revoluta***
   c. *Pinus gerardiana* d. *Cedrus deodara*

307. Rose wood botanically know as-
   a. *Azadirachta indica* b. *Cupresses torulosa*
   c. ***Dalbergia latifolia*** d. *Cedrus deodara*

308. Malawar wood botanically know as-
   a. ***Hopea parviflora*** b. *Cupresses torulosa*
   c. *Dalbergia sissoo* d. *Cedrus deodara*

309. Indian elm botanically know as-
   a. *Pinus wallichiana* b. *Abies pindrew*
   c. *Ulmus wallichiana* d. ***Holoptelia integrifolia***

Answers

299. c	300. d	301. a	302. b	303. b	304. d
305. c	306. c	307. c	308. a	309. d	

310. White teak is botanically know as-
a. ***Gmelia arborea*** b. *Butea monosperma*
c. *Tecomella undulata* d. *Tectona grandis*

311. Marwar teak is botanically know as-
a. *Gmelia arborea* b. *Butea monosperma*
c. ***Tecomella undulata*** d. *Tectona grandis*

312. Butter tree is botanically know as-
a. *Gmelia arborea* b. *Butea monosperma*
c. *Tecomella undulata* d. ***Madhuca indica***

313. Drumstick is botanically know as-
a. *Ceiba patendra* b. *Buchanania lanzan*
c. *Bauhania variegata* d. ***Moringa oleifera***

314. Midnight horror is botanically know as-
a. ***Oroxylum indicum*** b. *Acacia nilotica*
c. *Holoptelia integrifolia* d. *Moringa oleifera*

315. Bengal Kino is botanically know as-
a. *Oroxylum indicum* b. *Acacia nilotica*
c. *Caryota urens* d. ***Pterocapus marsupium***

316. Ceylon oak botanically know as-
a. ***Schleria oleosa*** b. *Cupresses torulosa*
c. *Quercus rubra* d. *Callistimon viminalis*

317. White willow is botanically know as-
a. *Gmelia arborea* b. *Adina cordifoia*
c. *Tecomella undulata* d. ***Salix alba***

318. Red cedar botanically know as-
a. ***Toona ciliata*** b. *Cupresses torulosa*
c. *Dalbergia latifolia* d. *Cassia fistlua*

319. Teak belong to family-
a. Combretaceae b. Myrtaceae
c. **Verbenaceae** d. Meliaceae

320. Marwar teak belong to family-
a. Combretaceae b. Myrtaceae
c. Verbenaceae d. **Bignoniaceae**

Answers

310. a	311. c	312. d	313. d	314. a	315. d
316. a	317. d	318. a	319. c	320. d	

321. Sal belong to family-
   a. Combretaceae b. **Dipterocaarpaceae**
   c. Verbenaceae d. Mimoseae

322. Pink cedar belong to family-
   a. Combretaceae b. Dipterocaarpaceae
   c. Verbenaceae d. **Legumniosae**

323. *Crocus sativus* belong to family-
   a. Combretaceae b. Meliaceae
   c. Verbenaceae d. **Iridaceae**

324. *Eucalyptus globulus* belong to family-
   a. Combretaceae b. Dipterocaarpaceae
   c. **Myrtaceae** d. Iridaceae

325. Tendu belong to family-
   a. Combretaceae b. Dipterocaarpaceae
   c. Verbenaceae d. **Ebenaceae**

326. Kusum belong to family-
   a. Rubiaceae b. **Sapindaceae**
   c. Meliaceae d. Legumniosae

327. *Cochlosprmum religiosum* belong to family-
   a. Rutaceae b. Iridaceae
   c. Ebenaceae d. **Bixaceae**

328. Cycas belong to family-
   a. **Cycadaceae** b. Iridaceae
   c. Ebenaceae d. Bixaceae

329. Mysoore gum belong to family-
   a. **Myrtaceae** b. Iridaceae
   c. Ebenaceae d. Bixaceae

330. *Gmelia arborea* belong to family-
   a. Rubiaceae b. Sapindaceae
   c. Meliaceae d. **Lamiaceae**

331. *Grewia tilifolia* belong to family-
   a. Combretaceae b. **Malvaceae**
   c. Verbenaceae d. Legumniosae

Answers

321. b	322. d	323. d	324. c	325. d	326. b
327. d	328. a	329. a	330. d	331. b	

332. *Jatropha curcus* belong to family-
   a. **Euphorbiaceae** b. Sapindaceae
   c. Meliaceae d. Lamiaceae

333. *Lagerstomia parviflora* belong to family-
   a. Cycadaceae b. Iridaceae
   c. **Lytraceae** d. Bixaceae

334. *Madhuca indica* belong to family-
   a. **Sapotaceae** b. Iridaceae
   c. **Lytraceae** d. Bixaceae

335. *Picea smithiana* belong to family-
   a. Salicaceae b. Santalaceae
   c. Verbenaceae d. **Pineaceae**

336. *Populus ciliata* belong to family-
   a. **Salicaceae** b. Santalaceae
   c. Verbenaceae d. Pineaceae

337. Elm belong to family-
   a. Salicaceae b. Santalaceae
   c. **Ulmaceae** d. Pineaceae

338. *Wrightia tinctoria* belong to family-
   a. Combretaceae b. Myrtaceae
   c. Verbenaceae d. **Apocynaceae**

339. Salix belong to family-
   a. Combretaceae b. **Salicaceae**
   c. Verbenaceae d. Pineaceae

340. Which climatic factor play most important role in vegetation distribution on earth-
   a. Rainfall b. **Temperature**
   c. Topography d. Light

341. *Pterocapus marsupium* belong to family-
   a. Combretaceae b. **Dipterocaarpaceae**
   c. Verbenaceae d. Santalaceae

342. First organic state in India is-
   a. Mizoram b. Kerala
   c. **Sikkim** d. Meghalaya

Answers

332. a	333. c	334. a	335. d	336. a	337. c
338. d	339. b	340. b	341. b	342. c	

CHAPTER 2

Ecology

1. Term Ecology was given by-

   a. **E. Haeckel** c. Reiter

   b. E. P. Odum d. Mishra

2. Father of Ecology –

   a. E. Haeckel c. **Reiter**

   b. E. P. Odum d. Mishra

3. Study of the relationship of single species with its environment called-

   a. Morphology c. Synecology

   b. Ecology d. **Autecology**

4. Study of the relationship of the group of the different species with their environment called-

   a. **Synecology** c. Autecology

   b. Morphology d. Ecology

5. The smallest unit of ecological hierarchy and basic unit of ecological study-

   a. **Organism** c. Species

   b. Population d. All

6. The basic unit of classification-

   a. Population c. Organism

   b. **Species** d. Community

7. The group of living organism similar in structure, function and can interbreed under natural conditions and reproductively isolated from the other group of organism-

   a. Group c. **Species**

   b. Population d. Community

Answers

1. a	2. c	3. d	4. a	5. a	6. b
7. c					

8. A species which is found at a particular area is called-
   a. Endangered species
   b. Rare species
   c. **Endemic species**
   d. Extinct species

9. The species which have great influence on the community's characteristics relative to their low abundance and biomass are called-
   a. **key stone species**
   b. Apex species
   c. Dominant species
   d. None

10. The activities of which species determine the structure of the community-
    a. **Key species**
    b. Producer
    c. Apex species
    d. Decomposer

11. The behaviour of an individual to increase the chances of the survival of other individuals of same species called-
    a. Amensalism
    b. Commensalism
    c. Mutualism
    d. **Altruism**

12. Several members of a species may cover a definite area for searching food, mates and shelter, this is called-
    a. **Home range**
    b. Cover
    c. Habitat
    d. Territory

13. In home range particular geographic area/space marked by an individual or group of breeding individuals for breeding called-
    a. Habitat
    b. Cover
    c. Home range
    d. **Territory**

14. According need of light a deep lake has differentiated into following zone-
    a. **3**
    b. 5
    c. 4
    d. 6

15. The zone which is found at bank of lake and have maximum diversity (Rooted vegetation)-
    a. Profundal zone
    b. **Littoral zone**
    c. Limnetic zone
    d. Benthic zone

Answers

8. c 9. a 10. a 11. d 12. a 13. d
14. a 15. b

16. This zone receive sufficient amount of light to entire surface area. In this zone different types of floating plants, suspended and submerged plants are present-
    a. **Limnetic zone**
    b. Littoral zone
    c. Profundal zone
    d. Benthic zone

17. It is the deep area of lake where light does not reach up to the bottom. Only heterotrophs are present in this zone-
    a. **Profundal zone**
    b. Littoral zone
    c. Limnetic zone
    d. Benthic zone

18. Free floating small organisms can swim due to water current without locomotory organs called-
    a. Nekton
    b. Zooplankton
    c. Phytoplankton
    d. **Planktons**

19. Microscopic, inactive floating plants called-
    a. **Phytoplankton**
    b. Zooplankton
    c. Planktons
    d. Nekton

20. Microscopic, inactive floating animals called-
    a. Nekton
    b. Zooplankton
    c. Phytoplankton
    d. Planktons

21. Those aquatic plants or animals which are capable of swimming actively are called-
    a. Phytoplankton
    b. Zooplankton
    c. **Nektons**
    d. Planktons

22. The sedentary organisms of sea are called-
    a. Planktons
    b. Zooplankton
    c. Phytoplankton
    d. **Benthonic**

23. Winter sleep or period of dormancy called-
    a. **Hibernation**
    b. Diapause
    c. Aestivation
    d. Mitigation

24. Hibernation is shown by-
    a. **Polar bear**
    b. Star Fish
    c. Siberian crane
    d. Red Panda

Answers

16. a	17. a	18. d	19. a	20. b	21. c
22. d	23. a	24. a			

25. Summer sleep or escape from heat of sun called-
 a. **Aestivation** c. Hibernation
 b. Diapause d. Mitigation

26. Transition form between two ecotypes called-
 a. **Eco-cline** c. Ecotype
 b. Ecotone d. Co-ecotone

27. Organism of same trophic level called –
 a. **Guild** c. Herd
 b. Troops d. Flocks

28. Under unfavourable conditions many zooplankton species in lakes and ponds are known to enter –
 a. Mitigation c. Hibernation
 b. Aestivation d. **Diapause**

29. The phenotypes show variations due to difference in the environmental conditions within the local habitat such types of variations are known as –
 a. Adaptation plasticity c. Genotypic plasticity
 b. **Phenotypic plasticity** d. Ecological plasticity

30. Low lying area's covered with shallow water are called-
 a. **Wet land** c. Shallow land
 b. Grassland d. Wasteland

31. Member of same species living in different climatic conditions have some genetically variations called-
 a. Ecotypes c. Ecological races
 b. Breed d. **All**

32. Member of same species living in different climatic conditions have some external variations (Genetically they are similar) called-
 a. **Ecades** c. Ecophenes
 b. Breed d. a and b

33. Group of different species that live in common area which are interrelated and interdependent called-
 a. Population c. **Community**
 b. Species d. Group

Answers

25. a	26. a	27. a	28. d	29. b	30. a
31. d	32. a	33. c			

34. Lianas, epiphytes and epizones show-
a. **Commensalism** c. Amensalism
b. Mutualism d. Symbiosis

35. Association between two organism, when one behave as master and another as slave called-
a. Commensalism c. Symbiosis
b. Mutualism d. **Helotism**

36. Association between the algae and fungi is called-
a. **Lichen** c. Mycorrhiza
b. Symbiosis d. None

37. Association of fungi with the roots of higher plants is called-
a. **Mychorrhiza** c. Lichen
b. Symbiosis d. None

38. Development of plant community on barren area is called-
a. **Ecological succession** c. Botanical succession
b. Biotic succession d. both a and c

39. The first community to inhibit an area during the succession called-
a. Climax community c. **Pioneer community**
b. Seral community d. both a and b

40. The last and stable community in an area during the succession called-
a. **Climax community** c. Pioneer community
b. Seral community d. both a and b

41. In succession communities or stage which comes in between pioneer community and climax called-
a. Climax community c. Pioneer community
b. **Seral communities** d. both a and b

42. The entire series of communities during succession called-
a. Community c. **Sere**
b. Ecosystem d. Succession series

43. Succession in fresh water called-
a. **Hydrosere** c. Halosere
b. Oxalosere d. Xerosere

Answers

34. a	35. d	36. a	37. a	38. a	39. c
40. a	41. b	42. c	43. a		

44. Succession in salty water called-
a. Hydrosere
c. **Halosere**
b. Oxalosere
d. Xerosere

45. Succession in acidic water-
a. Hydrosere
c. Halosere
b. **Oxalosere**
d. Xerosere

46. Succession in dry region-
a. Lithosere
c. Psamosere
a. Mesosere
d. **Xerosere**

47. Succession on rock-
a. **Lithosere**
c. Psamosere
b. Mesosere
d. Serula

48. Succession in on sand-
a. Lithosere
c. **Psamosere**
b. Mesosere
d. Serula

49. Succession at moistened region-
a. Lithosere
c. Psamosere
b. **Mesosere**
d. Serula

50. Succession of microbes on decomposed matter-
a. Lithosere
c. Psamosere
b. Mesosere
d. **Serula**

51. Succession occurs in the barren area where there is no previously any type of living matter present-
a. **Primary succession**
c. Secondary succession
b. Tertiary succession
d. Zero succession

52. Afforestation is also a type of succession-
a. **Primary succession**
c. Secondary succession
b. Tertiary succession
d. Allogenic succession

53. Succession occurs where previously vegetation was present but lost due to natural or artificial causes called-
a. Primary succession
c. **Secondary succession**
b. Tertiary succession
d. Autogenic succession

Answers

44. c	45. b	46. d	47. a	48. c	49. b
50. d	51. a	52. a	53. c		

54. Reforestation is also a type of succession-
   a. Primary succession
   b. Tertiary succession
   c. **Secondary succession**
   d. Allogenic succession

55. Which grassland is a type of retrogressive succession (Forest to grass-
   a. **Savana**
   b. Veldts
   c. Pampas
   d. Down

56. Member of one or both the interacting species are benefitted but neither is harmed called-
   a. **Positive interaction**
   b. Neutral interaction
   c. Negative interaction
   d. None

57. Member of one or both the interacting species are harmed called-
   a. Positive interaction
   b. Neutral interaction
   c. **Negative interaction**
   d. None

58. Total root parasite is-
   a. **Rafflesia**
   b. Santalum
   c. Cuscuta
   d. Viscum

59. Total stem parasite-
   a. Viscum
   b. **Cuscuta**
   c. Rafflesia
   d. Santalum

60. Partial root parasite-
   a. **Santalum**
   a. Viscum
   c. Rafflesia
   d. Cuscuta

61. Partial stem parasite-
   a. Viscum
   b. Santalum
   c. Laranhus
   d. **both a and c**

62. Organism eaten by own species called-
   a. **Cannabilism**
   b. Commensalism
   c. Macrophagy
   d. Autophagy

63. Transition zone between two communities having greater number of species and diversity-
   a. **Ecotone**
   b. Estuaries
   c. Ecotype
   d. Eco cline

Answers

54. c	55. a	56. a	57. c	58. a	59. b
60. a	61. d	62. a	63. a		

64. Species which occur most abundantly and spend their time in ecotone are called-
   a. Key stone species
   b. **Edge species**
   c. Pioneer species
   d. Climax species
65. The tendency to increase variety and density of some organism at the community border is known as-
   a. **Edge effect**
   b. Edge spectrum
   c. Edge habitat
   d. None
66. Ecological Niche term was given by-
   a. Mohit Husain
   b. Haeckel
   c. **Grinnel**
   d. E. P. Odum
67. The functional role of any species in an ecosystem or community or its functional position or status in an ecosystem is called-
   a. **Niche**
   b. Strata
   c. Habitat.
   d. Trophic
68. Physical area where an organism lives called-
   a. Niche
   b. Strata
   c. **Habitat**
   d. Trophic
69. This principle state that two closely related species competing for the same resources cannot co-exist indefinitely and the competitively inferior are will be eliminated eventually-
   a. Allen Rule
   b. Mohit 's competitive exclusion principle
   c. **Gause's competitive exclusion principle**
   d. Grinnel 's competitive exclusion principle
70. Gause's competitive exclusion principle may be true if resources are-
   a. **Limited**
   b. Neither limited nor unlimited
   c. Unlimited
   d. None
71. Sub-division of habitat is called-
   a. **Micro climate**
   b. Mega climate
   c. Macro climate
   d. None
72. Term ecosystem was given by-
   a. **A. G. Tansley**
   b. S. G. Tansley
   c. H. G. Tansley
   d. G. A. Tansley

Answers

64. b	65. a	66. c	67. a	68. c	69. c
70. a	71. a	72. a			

73. Father of ecosystem ecology is-
a. Mishra
b. **E. P. Odum**
c. Reiter
d. Elton

74. Smallest structural and functional unit of the nature or environment-
a. Population
b. Species
c. Organism
d. **Ecosystem**

75. Every ecosystem is composed of following component-
a. Biotic component
b. Abiotic component
c. Living component
d. **All of above**

76. Biotic components of ecosystem are-
a. Plants
b. Microbes
c. Animals
d. **All of above**

77. Abiotic components of ecosystem are-
a. Light
b. Water
c. Air
d. **All of above**

78. All the autotrophs of an ecosystem called-
a. Producer
b. Converter
c. Transducer
d. **All of the above**

79. The ultimate source of energy on the earth-
a. **Solar radiation**
b. Ultra violet radiation
c. Electromagnetic radiation
d. Infra-red

80. Organism who are obtained their food directly from producers or plants-
a. **Primary consumer**
b. Tertiary consumer
c. Secondary consumer
d. Top consumer

81. Primary consumer also called as-
a. **Primary producers**
b. Tertiary producer
c. Secondary producers
d. Decomposer

82. Organism who are obtained their food directly from primary consumer-
a. Secondary consumers
b. Tertiary consumer
c. Primary carnivore
d. **Both a and c**

Answers

73. b	74. d	75. d	76. d	77. d	78. d
79. a	80. a	81. a	82. d		

83. Those animals which kill other animal and eat them, but they are not killed and eaten by other animal in the nature called-
  a. Tertiary consumer
  b. **Top consumers**
  c. Secondary consumers
  d. Decomposer

84. The main decomposers in an ecosystem are-
  a. Bacteria
  b. Earthworm
  c. Fungi
  d. **All of above**

85. In the process of decomposition some nutrients get tied up with the biomass of microbes and become temporarily unavailable to other organism, such incorporation of nutrient in living microbes (bacteria, fungi) is called-
  a. **Nutrient immobilization**
  b. Nutrient leaching
  c. Nutrient pumping
  d. Nutrient cycling

86. Plant parasites in ecosystem called as-
  a. **Primary consumers**
  b. Secondary consumers
  c. Primary producers
  d. Secondary producers

87. Animal parasites in ecosystem called as-
  a. Primary consumers
  b. **Secondary consumers**
  c. Primary producers
  d. Secondary producers

88. In an ecosystem every organism depends on other organism for food material and all organism are (herbivore to carnivore) arranged in a series in which food energy is transferred through repeated eating and being eaten called-
  a. **Food chain**
  b. Food Pyramid
  c. Food web
  d. Food Network

89. How many level are present in the ecosystem-
  a. 3
  b. 5
  c. **4**
  d. 6

90. In food chain energy flow is-
  a. **Unidirectional**
  b. Multidirectional
  c. Bidirectional
  d. Non-directional

91. How many types of food chains are present in nature-
  a. 2
  b. 4
  c. **3**
  d. 5

Answers

83. b	84. d	85. a	86. a	87. b	88. a
89. c	90. a	91. c			

92. Food chain which start from small organism and goes to big organism called-
    a. Grazing food chain
    b. Parasitic food chain
    c. Predatory food chain
    d. **Both a and c**

93. Food chain which starts from producers but in successive order it goes from big organism to small organism called-
    a. Grazing food chain
    b. **Parasitic food chain**
    c. Predatory food chain
    d. Detritus food chain

94. The food chain begins with decomposition of dead organic matter by decomposer called-
    a. Saprophytic food chain
    b. Parasitic food chain
    c. Detritus food chain
    d. **Both a and c**

95. Graphical representation of ecological parameters at different trophic level in ecosystem is called-
    a. Trophics
    b. **Pyramids**
    c. Food web
    d. Food chain

96. First of all ecological pyramid was formed by-
    a. A. G. Tansley
    b. Mohit Husain
    c. E. P. Odum
    d. **Charls Elton**

97. Which pyramid are mostly upright-
    a. **Number**
    b. Energy
    c. Biomass
    d. All of above

98. Pyramid of energy are always-
    a. **Upright**
    b. Irregular
    c. Inverted
    d. Linear

99. 10% law of energy was given by-
    a. **Lindeman**
    b. Odum
    c. Elton
    d. Clements

100. Total amount of living organic matter present in particular area in particular time in an ecosystem is known as-
    a. Standing form
    b. **Standing crop**
    c. Standing state
    d. Standing matter

Answers

92. d	93. b	94. d	95. b	96. d	97. a
98. a	99. a	100. b			

101. Total amount of inorganic matter present in particular area in particular time in an ecosystem is known as-

a. Standing crop
c. **Standing state**
b. Standing form
d. Standing matter

102. Biomass is calculated by the formula-

a. **Biomass = Volume *x* sp.gravity**
b. Biomass= Weight *x* sp.gravity
c. Biomass = Volume *x* gravity
d. Biomass = Volume *x* weight

103. Pyramid of biomass are mostly-

a. **Upright**
c. Inverted
b. Irregular
d. Linear

104. In which ecosystem the pyramid of number is inverted-

a. Pond ecosystem
c. Grassland ecosystem
b. Forest ecosystem
d. **Tree ecosystem**

105. Network of many food chain called-

a. Food stock
c. Food Chain
b. **Food web**
d. Energy Chain

106. In which ecosystem the pyramid of number is inverted-

a. Pond ecosystem
c. Grassland ecosystem
b. **Tree ecosystem**
d. Desert ecosystem

107. The rate of biomass production (g/m^2/yr.) called-

a. **Productivity**
c. Respiration
b. Photosynthesis
d. Transformation

108. The amount of biomass or organic matter produced per unit area over a time period by plants during photosynthesis -

a. **Primary productivity**
c. Secondary Productivity
b. Net Productivity
d. Gross Productivity

109. The total amount of energy fixed in an ecosystem in unit time including the organic matter used up in respiration during the measurement is called-

a. Primary productivity
c. Secondary Productivity
b. Net Primary Productivity
d. **Gross Primary Productivity**

Answers

101. c	102. a	103. a	104. d	105. b	106. b
107. a	108. a	109. d			

110. Primary productivity can be divided into-
   a. GPP c. NPP
   b. TPP d. **Both a and c**

111. The amount of stored organic matter in plants tissue after respiratory utilization is called-
   a. Primary productivity c. Secondary Productivity
   b. **Net Primary Productivity** d. Gross Primary Productivity

112. The rate of formation of new organic matter by consumers is called-
   a. Primary productivity c. Gross Primary Productivity
   b. **Secondary Productivity** d. Net Primary Productivity

113. Maximum productivity per unit area is found in-
   a. Temperate forest c. Dry deciduous forest
   b. **Tropical rain forest** d. Desert

114. Solar radiation before entering the atmosphere carries energy at a constant rate i.e., 2 Cal cm^{-2} min^{-1} called-
   a. **Solar constant** c. Energy constant
   b. Energy coefficient d. Solar power

115. NPP is equal to-
   a. **NPP = GPP – R** c. NPP = GPP + R
   b. NPP = GPP × R d. NPP = SP – R

116. Range of photosynthetically active radiation (PAR) is-
   a. 400-700 mm c. **400-700 nm**
   b. 400-700 um d. 400-700 cm

117. The tails, ears, limbs, snout of mammals are smaller in colder region and larger in warmer region. This is called-
   a. **Allens rule** c. Burger rule
   b. Jordan rule d. Bergman rule

118. The ability of a surface to reflect the incoming radiation is called-
   a. Emission c. Reflectivity
   b. Transmission d. **Albido**

119. Albido value for forest and agriculture crops are respectively-
   a. **5-10% & 33%** c. 20-30% & 80%
   b. 18% & 30% d. 50% & 30%

Answers

110. d	111. b	112. b	113. b	114. a	115. a
116. c	117. a	118. d	119. a		

120. Albido value for snow and sand are respectively-
 a. **80% & 20-30%** c. 50% & 30%
 b. 75% & 90% d. 90% & 10%

121. Every organism has minimum and maximum limit of tolerance with respect to environmental factors like temperature, sunlight or nutrient concentration in between those limits the central optimum range are found in which organisms are found in which organisms are abundant this is known as optimum zone of tolerance. This law called-
 a. Darwin law c. Allen law
 b. **Shelford law** d. Lamarckism

122. The plants growing in high temperature throughout the year called-
 a. **Megatherms** c. Mesotherms
 b. Microtherms d. Hekistotherms

123. The plants growing in alternate high and low temperature called-
 a. Hekistotherms c. Megatherms
 b. Microtherms d. **Mesotherms**

124. The plants growing in low temperature throughout the year called-
 a. Megatherms c. Mesotherms
 b. **Microtherms** d. Hekistotherms

125. The plants growing in very low temperature throughout the year-
 a. **Hekistotherms** c. Mesotherms
 b. Megatherms d. Microtherms

126. The region where river enter into the ocean are known as-
 a. **Estuaries** c. Gully
 b. Reef d. Canal

127. The deeper part of ocean where light do not reach called-
 a. Limnetic zone c. Sedimentation zone
 b. Profundal zone d. **Abyssal zone**

128. Height of a place from mean sea level called-
 a. **Altitude** c. Latitude
 b. Pole d. Isohyet

Answers

120. a	121. b	122. a	123. d	124. b	125. a
126. a	127. d	128. a			

129. Distance of any place from the equator called-
   a. Altitude   c. **Latitude**
   b. Isobar   d. Isohyet

130. A transparent gaseous envelope surrounding the earth known as-
   a. **Atmosphere**   c. Climate
   b. Environment   d. Weather

131. Atmosphere is divided into how many layer-
   a. 4   c. **5**
   b. 6   d. 7

132. Ozone layer is present in-
   a. Troposphere   c. **Stratosphere**
   b. Mesosphere   d. Thermosphere

133. Winds, clouds, spores, pollen & more than 90% atmospheric gases are found-
   a. **Troposphere**   c. Stratosphere
   b. Mesosphere   d. Thermosphere

134. Satellite are launched in which region-
   a. Stratosphere   c. Troposphere
   b. Mesosphere   d. **Thermosphere**

135. In troposphere temperature decrease with increase in altitude. The vertical temperature gradient over earth's surface is called-
   a. **Lapse rate**   c. Albido
   b. Pulse rate   d. Emission rate

136. Lapse rate is equal to-
   a. 5.5°C per 1000 meter   c. **6.5°C per 1000 meter**
   b. 7.5°C per 1000 meter   d. 8.5°C per 1000 meter

137. All the living and non-living things of the earth combine together to constitute a big ecosystem called-
   a. **Biosphere**   c. Atmosphere
   b. Ecosystem   d. Lithosphere

138. Biosphere also called as-
   a. Lithosphere   c. Hydrosphere
   b. **Ecosphere**   d. Atmosphere

Answers

129. c	130. a	131. c	132. c	133. a	134. d
135. a	136. c	137. a	138. b		

139. Biosphere is divide into how many spheres-
   a. 2 c. **3**
   b. 4 d. 5

140. Which is not a component of Biosphere-
   a. Hydrosphere c. Lithosphere
   b. Atmosphere d. **All of above**

141. Smoke plus fog collectively called as-
   a. **Smog** c. Fog
   b. Smoke d. Smoke-fog

142. Word smog was given by-
   a. Loss Angel c. Ringlmann
   b. **Desvoeux** d. Mohit Husain

143. London smog also known as-
   a. Nitrogen smog c. **Sulphur smog**
   b. Hydrogen smog d. Photochemical smog

144. Loss Angles smog also known as-
   a. **Photochemical smog** c. Sulphur smog
   b. Hydrogen smog d. Nitrogen smog

145. Due to inhalation of H_2SO_4 vapours with fog 4000 people died in 1952 in-
   a. India c. China
   b. **London** d. America

146. Term acid rain was given by-
   a. Michel August c. Meyer August
   b. Albert August d. **Robert August**

147. 147.NO_2 and SO_2 released from different sources in the form of smoke and dissolved in atmospheric water vapour to form sulphuric acid and nitric acid. These acids come down on earth with rain water called-
   a. Basic rain c. **Acid rain**
   b. Storm d. Flood

148. The pH of acid rain water is lesser than-
   a. **5.6** c. 7.0
   b. 8.6 d. 6.6

Answers

139. c	140. d	141. a	142. b	143. c	144. a
145. b	146. b	147. c	148. a		

149. Usually carbon dioxide is not consider as pollutant but its higher concentration forms a thick layer above the earth's surface and checks the radiation of the heat from the earth surface. Because of this temperature of earth's surface increase this is called-
a. Green- house effect
b. CO_2 effect
c. Global warming
d. **All of the above**

150. Main green-house gases which are responsible to increase the green-house effect are-
a. CO_2 (60%)
b. CFC (14%)
c. CH_4 (20%)
d. **All of above**

151. At normal temperature and pressure thickness of ozone layer-
a. **3 mm**
b. 5 mm
c. 4 mm
d. 6 mm

152. The aerosols like chloro floro carbon (CFC) release into atmosphere from the refrigerators, air conditioners and jet planes deplete or reduce the-
a. Oxygen layer
b. **Ozone layer**
c. Carbon layer
d. Chlorine layer

153. Due to ozone depleting substance the thickness of ozone is reduced this is known as-
a. Ozone spot
b. Ozone width
c. **Ozone hole**
d. Ozone state

154. Ozone hole was first discovered by a satellite-
a. **Nimbus-7**
b. Nimbus-9
c. Nimbus-8
d. Nimbus-6

155. Ozone hole was first discovered by a satellite in 1985-
a. **Antarctica region**
b. Africa region
c. Artic region
d. Pacific region

156. Maximum ozone depleting substance is-
a. CFC-18
b. CFC-15
c. CFC-12
d. **CFC-14**

Answers

149. d	150. d	151. a	152. b	153. c	154. a
155. a	156. d				

157. In this process of ozone depletion one chlorine atom of CFC convert-
a. One lakh O_3 molecules into O_2 by photo dissociation
b. One lakh O_2 molecules into O_3 by photo dissociation
c. One lakh O molecules into O_3 by photo dissociation
d. One lakh O_3 molecules into O by photo dissociation

158. Thickness of ozone layer is measured in-
a. Decibel unit (Db)
b. Hertz (Hz)
c. **Dobson unit (Du)**
d. Joule (J)

159. The amount of dissolved oxygen (D.O) needed by bacteria in decomposing the organic wastes present in water called-
a. **Biochemical oxygen demand**
b. Bio magnification
c. Chemical oxygen demand
d. Bioremediation

160. Water having D.O content below 8.0 mgL^{-1} may consider as-
a. Hard water
b. Polluted water
c. Fresh water
d. **Contaminated water**

161. Water having D.O content below 4.0 mgL^{-1} may consider as-
a. Fresh water
b. **Heavily polluted**
c. Polluted water
d. Soft water

162. B.O.D for fresh is less than-
a. **1.0 mgL^{-1}**
b. 3.0 mgL^{-1}
c. 2.0 mgL^{-1}
d. 4.0 mgL^{-1}

163. Who is the indicator of B.O.D. in a water ecosystem-
a. Traught
b. Gambusia
c. **Daphnia**
d. Lichen

164. D.O is measured by-
a. **Oximeter**
b. Lysimeter
c. Spectrophotometer
d. Doximometer

165. The amount of oxygen requirement by chemical for oxidation of total organic matter (biodegradable and non-biodegradable) in water called-
a. **Chemical oxygen demand**
b. Bioremediation
c. Bio magnification
d. Biochemical oxygen demand

Answers

157. a	158. c	159. a	160. d	161. b	162. a
163. c	164. a	165. a			

166. C.O.D value is always-
   a. Lower than B.O.D value
   b. Equal to B.O.D value
   c. **Higher than B.O.D value**
   d. Neither higher nor lower than B.O.D

167. The non-biodegradable pollutant such as DDT, Al, Hg, Fe etc. are not decomposed by micro-organism. They get accumulated in tissue in increasing concentration along the food chain is called-
   a. **Biological magnification**
   b. Sublimation
   c. Acclimation
   d. Eutrophication

168. The highest concentration of non-biodegradable pollutant occurs in-
   a. **Top consumer**
   b. Secondary consumer
   c. Primary consumer
   d. Tertiary consumer

169. High concentration of DDT disturb calcium metabolism, which cause thinning of egg shell and their premature breaking, eventually decline in-
   a. Human population
   b. Microorganism population
   c. **Avian population**
   d. Insect population

170. D.D.T (Dichloro Diphenyl Tri-chloroethane) and B.H.C (Benzene Hexa Chloride) are-
   a. **Non-degradable pollutants**
   b. Eco-friendly degradable pollutants
   c. Bio-degradable pollutants
   d. Secondary pollutants

171. The process of nutrient enrichment of water and consequent loss of species diversity (or death of aquatic animals) is referred-
   a. Eutrophication
   b. Ageing of lake
   c. Biological magnification
   d. **Both a and b**

172. The main reason of Eutrophication of lake is-
   a. **Algal Bloom**
   b. Lichen Bloom
   c. Fungi Bloom
   d. Mycorrhizal Bloom

Answers

166. c 167. a 168. a 169. c 170. a 171. d
172. a

173. A lake having high amount of organic materials and little amount of oxygen is called-

a. **Eutrophic lake**
b. Dystrophic lake
c. Oligotrophic lake
d. Polytrophic lake

174. A lake having less amount of organic materials and little more amount of oxygen is called-

a. Dystrophic lake
b. **Oligotrophic lake**
c. Eutrophic lake
d. Polytrophic lake

175. A lake having maximum amount of undecomposed organic materials and more amount of oxygen is called-

a. Eutrophic lake
b. Oligotrophic lake
c. **Dystrophic lake**
d. Polytrophic lake

176. The process in which water of Pacific Ocean get warm, in this process warm water current flows to equador and Peru in between 5 to 8 year Christmas time called-

a. **ELNino effect**
b. La -Nina effect
c. EL-Nina effect
d. La -Nino effect

177. The process in which water of Pacific Ocean get cold, in this process cold water current flows to equador and Peru in between 5 to 8 year Christmas time called-

a. La -Nino effect
b. EL-Nina effect
c. ELNino effect
d. **La -Nina effect**

178. Prairies grassland found in-

a. South America
b. **North America**
c. Europe and Asia
d. New Zealand

179. Pampas grassland found in-

a. North America
b. Europe and Asia
c. **South America**
d. New Zealand

180. Steppes grassland found in-

a. **Europe and Asia**
b. Japan
c. South America
d. North America

181. Tussocks grassland found in-

a. North America
b. **New Zealand**
c. Africa
d. New York

Answers

173. a	174. b	175. c	176. a	177. d	178. b
179. c	180. a	181. b			

182. Veldts grassland found in-
   a. Brazil
   b. Pakistan
   c. **Africa**
   d. India

183. Down grassland found in-
   a. Japan
   b. **Australia**
   c. India
   d. America

184. Savanah grassland found in-
   a. **Africa**
   b. Brazil
   c. India
   d. Holland

185. The study of interaction and inter-relationship of organism with their environment-
   a. **Ecology**
   b. Morphology
   c. Physiology
   d. Anatomy

186. MoEF means-
   a. Ministry of Forests
   b. Ministry of Fuel and Energy
   c. **Ministry of Environment and Forests and climate change**
   d. Management of Environment and Forestry

187. "Earth provides enough to satisfy every man's need, but not for every man's greed" words was given by-
   a. Tagore
   b. Nehru
   c. **Gandhi Ji**
   d. Indira Gandhi

188. The ocean covers percentage of Earth's surface-
   a. 51%
   b. **71%**
   c. 61%
   d. 95%

189. The main energy source for the environment is-
   a. **Solar energy**
   b. Hydro energy
   c. Chemical energy
   d. Electric energy

190. Moisture in the air is known as-
   a. Water
   b. Dew
   c. Snow
   d. **Humidity**

Answers

182. c	183. b	184. a	185. a	186. c	187. c
188. b	189. a	190. d			

191. Ramsar Convention refers to the conservation of-
a. Grasslands c. **Wetlands**
b. Forests d. Deserts

192. The largest brackish water lake situated in Asia is in Orissa-
a. Dal lake c. Woolar lake
b. **Chilka lake** d. Manasbal lake

193. Plants adapted to open, sunny habitats are called-
a. **Heliophytes** c. Sciophytes
b. Mesophytes d. Epiphytes

194. Plants adapted to shady (shade loving) habitats are called-
a. Mesophytes c. Epiphytes
b. **Sciophytes** d. Heliophytes

195. The relationship between nitrogen fixing bacteria and leguminous plants is an example for-
a. Parasitism c. Predation
b. **Symbiosis** d. Commensalism

196. The word Biophilia was coined by-
a. E. Hackle c. Darwin
b. Reiter d. **E. O. Wilson**

197. Ganga Action Plan in India was launched in the year-
a. **1985** c. 1895
b. 1976 d. 1927

198. The phenomenon of accumulation of non-biodegradable pesticides in human beings-
a. Bio magnification c. **Bioaccumulation**
b. Biodegradation d. Bioremediation

199. Which state proposed a ban on all types of polythene packing for the first time in India-
a. **Himachal Pradesh** c. Uttar Pradesh
b. Karnataka d. Maharashtra

200. Who coined the slogan of 'Chipko Movement ' 'Ecology is Permanent economy'-
a. Jawaharlal Nehru c. Salim Ali
b. **Sunderlal Bahuguna** d. Rachel Carson

Answers

191. c	192. b	193. a	194. b	195. b	196. d
197. a	198. c	199. a	200. b		

CHAPTER 3

Forest Management

1. Forest management is defined as, 'the practical application of the scientific, technical and economic principles of forestry' by-

   a. Ram Prakash c. **BCFT**
   b. MoEF & CC d. Jerram

2. Management of forest broadly classified into-

   a. Control of composition and structure of the growing stock
   b. Harvesting and marketing of forest produce
   c. Administration of forest property personnel
   d. **All of the above**

3. How much percent of Indian forests are under the state ownership-

   a. 33% c. **95.8%**
   b. 98.5% d. 71%

4. India's first forest policy was enunciated in

   a. **1894** c. 1952
   b. 1988 d. 2011

5. The sole objective of forest management of public forest in 1894 policy was-

   a. Forest conservation c. **Public benefits**
   b. Forest protection d. Private benefits

6. Indian Republic formulated its first National Forest Policy in-

   a. **1952** c. 1947
   b. 1988 d. 1994

7. National Forest Policy in **1952** classified the India's forest into-

   a. Two classes c. Three classes
   b. **Four classes** d. Six classes

Answers

1. c 2. d 3. c 4. a 5. c 6. a
7. b

8. 1952 National Forest Policy classified the India's forest into-
   a. Protection forest
   b. Village forest
   c. National Forest
   d. **All of above**

9. National Forest Policy **1952** suggested to keep a minimum of country's total land area under forests-
   a. **One third**
   b. Half
   c. One fourth
   d. One eighth

10. National Forest Policy 1952 suggested to keep how much percent forest in Himalayas-
    a. 20%
    b. **60%**
    c. 40%
    d. 98%

11. National Forest Policy 1952 suggested to keep how much percent forest in plain-
    a. 40%
    b. 60%
    c. 50%
    d. **20%**

12. The parliament by the 42nd Amendment to the Indian Constitution in 1976, brought Forests and Wildlife on the Concurrent List in the-
    a. **Seventh Schedule**
    b. Ninth Schedule
    c. Eight Schedule
    d. Fifth Schedule

13. A written scheme of management aiming at continuity of policy, controlling the treatment of a forest is called-
    a. Forest act
    b. Forest Policy
    c. **Working plan**
    d. Working circle

14. Working plan is based on the principle of-
    a. Progressive yield
    b. Final yield
    c. Intermediate yield
    d. **Sustained yield**

15. Which is the unit of forest management-
    a. Coupe
    b. **Working plan**
    c. Beat
    d. Working circle

16. A forest area organised with particular objects, and under one Silviculture System and one set of Working Plan prescriptions is called-
    a. **Working Circle**
    b. Beat
    c. Working plan
    d. Coupe

Answers

8. d	9. a	10. b	11. d	12. a	13. c
14. d	15. b	16. a			

17. On the basis of geographical and climatic (ecological) forest are classified into-
    a. 4 c. **5**
    b. 6 d. 7

18. On the basis of functional classification forest are classified into-
    a. **4** c. 5
    b. 6 d. 8

19. On the basis of legal classification forest are classified into-
    a. 2 c. 3
    b. **4** d. 5

20. On the basis of territorial classification forest are classified into-
    a. 2 c. **3**
    b. 4 d. 5

21. On the basis of management (Silvicultural) classification forest are classified into-
    a. 3 c. 4
    b. 6 d. **5**

22. A main territorial division of forest having a boundary and bearing a local proper name called as-
    a. **Block** c. Compartment
    b. Sub-Compartment d. Coupe

23. The smallest permanent Working plan unit of management for administration purpose called-
    a. Block c. Sub-Compartment
    b. Coupe d. **Compartment**

24. The size of compartment depends on
    a. **Intensity of management** c. Need of management
    b. Required output d. Conservation management

25. A Sub division of compartment designated with small letters for temporary administration purpose is called-
    a. Coupe c. Compartment
    b. **Sub-Compartment** d. Block

Answers

17. c	18. a	19. b	20. c	21. d	22. a
23. d	24. a	25. b			

26. Head quarter of Inspector General of Forest is situated in-
    a. Dehra Dun
    b. Kolkata
    c. **New Delhi**
    d. Uttarakhand

27. Sub division of block is called-
    a. **Compartment**
    b. Sub-compartment
    c. Coupe
    d. Beat

28. Head of forest force (HOFF) known as-
    a. C.C.F
    b. **P.C.C.F**
    c. D.C.F
    d. A.C.F

29. Kingpin of forest department is –
    a. **D.F.O**
    b. R.F.O
    c. A.C.F
    d. Forester

30. In our Indian set-up, direct recruitment to the Gazetted rank is made at the level of a/an-
    a. Assistant Conservator of Forest (A.C.F)
    b. District Forest Officer (D.F.O)
    c. Range Forest Officer (R.F.O)
    d. Conservator of Forest (C.F)

31. A forest area forming the whole or part of a Working circle for distribute felling and regeneration to maintain or create a normal distribution of age-classes is called-
    a. Felling cycle
    b. **Felling Series**
    c. Regeneration cycle
    d. Seeding felling

32. A felling area usually one of an annual series designated with Roman numerals-
    a. **Coupe**
    b. Compartment
    c. Block
    d. Working circle

33. Sub division of felling series formed with the object of regulating cuttings in some special manner is called-
    a. Felling cycle
    b. Final felling
    c. **Cutting section**
    d. Coupe

34. The part or parts of forest set aside to be regenerated, or otherwise treated during a specified period is called-
    a. **Periodic block**
    b. Block
    c. Coupe
    d. Compartment

Answers

26. c	27. a	28. b	29. a	30. a	31. b
32. a	33. c	34. a			

35. The regeneration block when it is the only periodic block allotted at each Working Plan revision known as-
a. Protected block
b. **Floating or single block**
c. Regenerated block
d. Fixed or permanent periodic blocks.

36. When all periodic blocks are allotted and retain their territorial identity at Working Plan revision, they are called-
a. **Fixed or permanent periodic blocks**
b. Regenerated block
c. Protected block
d. Floating or single block

37. The period required to regenerate the whole of a periodic block called-
a. **Regeneration period** c. Rotation period
b. Felling period d. Growing period

38. Area of a coupe is calculate (A)=
a. Rotation period/Area of felling series
b. Area of felling series × Rotation period
c. **Area of felling series/ Rotation period**
d. Area of felling series - Rotation period

39. The time that elapses between successive main fellings on the same area-
a. Felling series c. Felling period
b. **Felling cycle** d. Felling sections

40. The material or cash return obtained from time to time from a forest not organised for continuous production called-
a. **Intermediate Yield** c. Progressive yield
b. Sustained yield d. Final yield

41. The sum (by number or volume) of all the trees growing in the forest or a specified part of forest is called-
a. Stock c. Volume
b. Clump d. **Growing stock**

Answers

35. b 36. a 37. a 38. c 39. b 40. a
41. d

42. The concept of Progressive Yield was given by-
   a. **German forester Hartig**
   b. British Forester Hartig
   c. Indian forester Hartig
   d. Dutch forester Hartig
43. The principle of Progressive Yield as against the Sustained Yield principle was discussed at the-
   a. Fourth Indian Silviculture Conference
   b. **Sixth Indian Silviculture Conference**
   c. Fifth Indian Silviculture Conference
   d. Seventh Indian Silviculture Conference
44. Third World Forestry Conference was held in 1948-
   a. New Delhi
   b. Brazil
   c. **Helsinki**
   d. Geneva
45. The period which a forest crop takes between its formation and final felling is known as-
   a. Rotation period
   b. Growing Period
   c. Production period
   d. **Both a and c**
46. Rotation which coincides with the natural lease of life of a species on a given site called-
   a. **Physical rotation**
   b. Technical rotation
   c. Silviculture rotation
   d. Financial rotation
47. Rotation which is applicable in case of protection and amenity forest, park lands, roadside avenues-
   a. Silviculture rotation
   b. **Physical rotation**
   c. Technical rotation
   d. Financial rotation
48. Rotation through which a species remains satisfactory vigour of growth and reproduction on a given site called-
   a. Technical rotation
   b. Financial rotation
   c. **Silviculture rotation**
   d. Rotation of maximum volume production
49. Which rotation is useful in forests managed primarily for aesthetic and recreational purpose-
   a. Technical rotation
   b. Rotation of maximum volume production
   c. Financial rotation
   d. **Silviculture Rotation**

Answers

42. a	43. b	44. c	45. d	46. a	47. b
48. c	49. d				

50. Rotation under which a species yields the maximum material of a specified size or suitability for economic conversion or for special use called-
    a. Rotation of maximum volume production
    b. Financial rotation
    c. **Technical rotation**
    d. Physical rotation

51. Rotation which is adopted, particularly by wood based industry such as paper-wood, match-wood industry etc. for maximum raw material's production is-
    a. Rotation of maximum volume production
    b. Financial rotation
    c. **Technical rotation**
    d. Physical rotation

52. Rotation that yields the maximum annual quantity of material; i.e., the age at which Mean annual increment culminates, called-
    a. **Rotation of maximum volume production**
    b. Physical rotation
    c. Financial rotation
    d. Technical rotation

53. Rotation which is useful for maximum volume production where the total quantity of woody material is important and not the size and specification is-
    a. Physical rotation
    b. Financial rotation
    c. **Rotation of maximum volume production**
    d. Technical rotation

54. Rotation which yields the highest average annual gross or net revenue irrespective of the capital value of the forest called-
    a. Rotation of highest income
    b. Rotation of highest forest rental
    c. Rotation of highest revenue
    d. **All the above**

Answers

50. c 51. c 52. a 53. c 54. d

55. Which rotation is important from all national point of view-
 a. **Rotation of highest income**
 b. Rotation of maximum volume production
 c. Financial rotation
 d. Technical rotation

56. Rotation which yields the highest net return on the invested capital called-
 a. Financial rotation
 b. Technical rotation
 c. Economic rotation
 d. **Both a and c**

57. According to, financial or economic rotation defined as, "the rotation which is most profitable-
 a. **Hiley**
 b. L.S Khanna
 c. Ram Prakash
 d. Mohit Husain

58. There are two prominent methods to determine the financial rotation, these are-
 a. Based on Soil Expectation Value (S_e) of the land
 b. Based on the financial yield
 c. Based on Intermediate yield
 d. **Both a and b**

59. If a piece of land is expected to provide a continual net income of X rupees yearly, then that land can be valued at a sum, which at an acceptable rate of interest gives the same yearly income of Rs. X; that value is known as-
 a. Soil correction value (S_c)
 b. **Soil Expectation Value (S_e)**
 c. Soil adaption value (S_a)
 d. Soil calculated value (S_c)

60. Soil Expectation Value (S_e) is-
 a. $S_c = x/0.0k$
 b. $S_e = x/0.0d$
 c. $S_c = x/0.0h$
 d. **$S_e = x/0.0p$**

61. The length of rotation depends on which of the following parameters-
 a. Rate of growth of species
 b. Silvicultural characteristics of the species
 c. Characteristics of soil
 d. **All of above**

Answers

55. a	56. d	57. a	58. d	59. b	60. d
61. d					

62. The period during which a change from one Silvicultural system to another Silvicultural system is effected called-
    a. **Conversion period**
    b. Rotation period
    c. Payback period
    d. Growing period
63. Conversion period is usually-
    a. More than rotation
    b. Equal to rotation
    c. **less than rotation period**
    d. Always greater than rotation period
64. Conversion period may be sometimes-
    a. More than rotation
    b. Less than rotation
    c. Equal to rotation
    d. **All of the above**
65. Conversion period is usually kept ....the rotation period when mature crop is desirable to remove earlier than its rotation period-
    a. More than
    b. Equal
    c. **Less than**
    d. Any time
66. A forest which has normal increment, normal distribution of age-gradation and normal growing forest (NGS) called-
    a. **Normal Forest**
    b. Virgin Forest
    c. Protect forest
    d. Reserved Forest
67. Trinity of a normal forest is not includes-
    a. Normal Increment
    b. Normal Distribution of Age-Gradation
    c. Normal Growing Stock (NGS)
    d. **All of above**
68. The volume of a stands in a forest with normal age-classes and a normal increment called-
    a. Abnormal growing stock
    b. **Normal Growing Stock (NGS)**
    c. Growing stock
    d. Sustained yield
69. Kinds of abnormality in a forest are-
    a. They may be over/under stocked
    b. The increment may be sub-normal
    c. Normal increment volume in an abnormal forest
    d. **All of the above**

Answers

62. a	63. c	64. d	65. c	66. a	67. d
68. b	69. d				

70. The increase in girth, diameter, basal area, height, volume, quality and price of individual trees or crops during a given period called-
   a. **Increment**
   b. Maturity
   c. Growth
   d. Development

71. The increment in growth that takes place in a particular year called-
   a. Current Annual Increment (CAI)
   b. Periodic Annual Increment (PAI)
   c. Mean Annual Increment (MAI)
   d. Current Increment Percent

72. The increment in growth that takes place for any short period called-
   a. Current Annual Increment (CAI)
   b. **Periodic Annual Increment (PAI)**
   c. Mean Annual Increment (MAI)
   d. Current Increment Percent

73. The total increment up to a given age divided by that age called-
   a. Mean Annual Increment (MAI)
   b. Current Annual Increment (CAI)
   c. Periodic Annual Increment (PAI)
   d. Current Increment Percent

74. The Mean Annual Increment (MAI) at rotation age is called-
   a. Current Annual Increment (CAI)
   b. Current Increment Percent
   c. **Final Mean Annual Increment**
   d. Mean Annual Increment (MAI)

75. The average annual growth in volume over a specified period expressed as a percentage of the volume either at beginning or, more usually, half way through the period called-
   a. Increment Percent
   b. Current Annual Increment (CAI)
   c. Current Increment Percent
   d. Periodic Annual Increment (PAI)

76. In the beginning M.A.I keeps below-
   a. **C.A.I.**
   b. C.I.
   c. P. A. I.
   d. All of the above

Answers

70. a 71. a 72. b 73. a 74. c 75. a
76. a

77. The C.A.I attains the maximum value before-
a. P.A.I. c. **M.A.I**
b. C.I. d. None

78. Which is never falls up to zero in whole life of a tree-
a. P.A.I c. C.A.I
b. **M.A.I.** d. C.I.

79. Which may falls up to zero in whole life of a tree-
a. P.A.I. c. C.I.
b. M.A.I. d. **C.A.I.**

80. The relation of the increment during a given year to the volume at the beginning of the year called-
a. Current Increment Percent (C.I.P)
b. Mean Increment Percent (M.I.P)
c. Periodic Increment Percent (P.I.P)
d. Annual Increment Percent (A.I.P)

81. The relation of the increment during a given year to a basic volume which may be taken as the mean or average volume for the period, or the volume at the beginning of the period called-
a. Current Increment Percent
b. **Periodic Increment Percent**
c. Mean Increment Percent
d. Annual Increment Percent

82. The percent ratio which the M.A.I for a given age bears to the total volume at that age called-
a. Mean Annual Increment Percent
b. Current Increment Percent
c. Periodic Increment Percent
d. Mean Increment Percent

83. Pressler's formula for Increment Percent is given as-
a. Increment Percent (p) c. Increment Percent (p)
b. **Increment Percent (p)** d. Increment Percent (p)

84. The most convenient formula for finding the increment percent (p) of standing trees is that developed by-
a. **Professor Schneider in 1853** c. Professor Schneider in 1953
b. Professor Schneider in 1753 d. Professor Schneider in 1653

Answers

77. c 78. b 79. d 80. a 81. b 82. a
83. b 84. a

85. . Schneider formula is based on the determination of the-
   a. Diameter at breast height
   b. The number of the rings in the last centimetre of the radius
   c. Tree volume
   d. **Both a and b**

86. Schneider formula for increment percent is-
   a. Increment Percent (p)
   b. **Increment Percent (p)**
   c. Increment Percent (p)
   d. Increment Percent (p)

87. A fast growing species is one which yields a minimum-
   a. **10 m^3 /ha/year**
   b. 20 m^3 /ha/year
   c. 15 m^3 /ha/year
   d. 30 m^3 /ha/year

88. World forest average mean annual increment is-
   a. 4.1 m^3/ha
   b. 3.1 m^3/ha
   c. **2.1 m^3/ha**
   d. 1.2 m^3/ha

89. A fast growing species is one which height increment must not be less than-
   a. 40 cm per year
   b. **60 cm per year**
   c. 50 cm per year
   d. 70 cm per year

90. Mean annual increment of Indian forest is-
   a. 0.5-0.7 m^3/ha
   b. 0.6-0.8 m^3/ha
   c. 1.0-1.9 m^3/ha
   d. **0.7-0.9 m^3/ha**

91. The increment in the value per unit volume of a tree or a crop, independent of any increase in the price of forest produce resulting from any change in money value in general, or the supply and demand position in particular called-
   a. **Quality Increment**
   b. Increment Percent
   c. Price Increment
   d. Increment

92. The increment in price independently of quality increment, resulting from any change in the market on an account of change in money value in general, the demand and supply position in particular called-
   a. **Price Increment**
   b. Increment Percent
   c. Price Increment
   d. Quality Increment

Answers

85. d	86. b	87. a	88. c	89. b	90. d
91. a	92. a				

93. Price Increment and Quality Increment can be determined and expressed in the same way as volume increment percent by using-
    a. Von Mental method
    b. **Pressler's method**
    c. Schneider method
    d. Howard method

94. Increment from C.A.I. can be calculated by using-
    a. Area Method
    b. Increment Percent method
    c. Per Tree Method
    d. **All the above**

95. Determination of increment in **irregular crop** is done by using-
    a. **Andre's formula**
    b. Huber's formula
    c. Behre's formula
    d. Von Mental formula

96. Biolley's check method is used to determine increment and based on-
    a. Diameter
    b. **Successive inventories**
    c. Tree height
    d. Tree volume

97. According to Biolley the M.A.I can be calculated by using formula-
    a. M.A.I =
    b. M.A.I =
    c. **M.A.I =**
    d. M.A.I =

98. The intervals into which the range of age of the trees growing in a forest is divided for classification or use-
    a. Age gradation
    b. **An age-class**
    c. Gradation series
    d. Class interval

99. The unit of Yield Regulation in both regular and irregular forests is a-
    a. **Felling series**
    b. Coupe
    c. Felling cycle
    d. Periodic block

100. The total volume of trees in a fully stocked forest with normal distribution of age-classes for a given rotation called-
    a. Normal yield
    b. **Normal growing stock**
    c. Growing stock
    d. Progressive yield

101. At the beginning of the growing season the volume of the growing stock will be-
    a. **N.G.S =**
    b. N.G.S =
    c. N.G.S =
    d. N.G.S =

Answers

93. b	94. d	95. a	96. b	97. c	98. b
99. a	100. b	101. a			

102. At the middle of the growing season the volume of the growing stock will be-

a. N.G.S = c. **N.G.S =**
b. N.G.S = d. N.G.S =

103. At the **end** of the growing season the volume of the growing stock will be-

a. **N.G.S =** c. N.G.S =
b. N.G.S = d. N.G.S =

104. Flury suggest that in the mid-season formula a variable constant 'C' should be substituted for 1/2, to obtain more correct results called-

a. Flury correction c. **Flury constant**
b. Flury formula d. None

105. Value of Flury constant 'C' will be less than 1/2 for-

a. **Short rotation** c. Full rotation
b. Half rotation d. Mid rotation

106. Value of Flury constant 'C' will be more than 1/2 for-

a. Short rotation c. Mid rotation
b. Half rotation d. **Long rotation**

107. Fischer evolved a formula for a condition when part of the growing stock has been removed under regeneration felling is-

a. N.G.S = c. N.G.S =
b. **N.G.S =** d. N.G.S =

108. For the determination of N.G.S in selection system Munger evolved a formula which was based-

a. **C.A.I.** c. M.A.I.
b. P.A.I. d. C.I

109. The percent ratio of normal yield to the normal growing stock called-

a. **Utilization percent** c. Increment percent
b. Quality percent d. Yield percent

110. Yield table is constructed assuming sample plots are-

a. Simply stocked c. Partially stocked
b. **Fully stocked** d. Stocked

Answers

102. c	103. a	104. c	105. a	106. d	107. b
108. a	109. a	110. b			

111. Ratio to M.A.I of the quality to be reduced to M.A.I of the standard quality is known as-
a. Reducing factor for utilization
b. **Reducing factor for quality**
c. Reducing factor for increment
d. All of the above

112. The diameter or girth decided upon as the normal size for felling in order to fulfil the objects of management called-
a. **Exploitable size**
b. Commercial size
c. Exploitable volume
d. Commercial volume

113. How many percent of Indian forest are irregular in diameter and density distribution-
a. About 50%
b. About 95%
c. About 60%
d. **About 70%**

114. According to ....in a fully stocked selection forest, the number of stems falls off from one diameter class to the next in geometrical progression, which means that the percentage reduction in the stem number from on diameter class to the next is constant-
a. L.S Khanna
b. Hartig
c. **De Liocourt's law**
d. Ram Prakash

115. Meyer simplified De Liocourt's law in the form of an exponential function, which is-
a. $\mathbf{Y = K\,e^{-ax}}$
b. $Y = K\,e^{ax}$
c. $K = Y\,e^{-ax}$
d. $K = Y\,e^{ax}$

116. The volume or number of stems that can be removed annually or periodically, or the area over which fellings may pass annually or periodically, consistent with the attainment of objects of management called-
a. Periodic Block
b. **Yield**
c. Felling
d. Thinning

117. All the material that counts against the prescribed yield and which is derived from the main fellings in a regular forest called-
a. **Final yield**
b. Periodic yield
c. Intermediate yield
d. Sustained yield

Answers

111. b 112. a 113. d 114. c 115. a 116. b
117. a

118. All material from thinning or operations preceding the main felling in a regular forest, or its cash equivalent called-

a. Periodic yield
b. Sustained yield
c. **Intermediate yield**
d. Final yield

119. The yield from a normal forest called-

a. Sustained yield
b. Periodic yield
c. Prescribed yield
d. **Normal yield**

120. The material that a forest can yield annually in perpetuity called-

a. Progressive yield
b. **Sustained yield**
c. Periodic yield
d. Normal yield

121. The standing volume of a crop plus the total volume removed in the thinning since its establishment as a more or less even-aged stand; or the sum of the final and intermediate yields called-

a. **Total yield**
b. Periodic yield
c. Normal yield
d. Sustained yield

122. The total quantity of material per annum, of given species, that an area is capable of producing under normal conditions, so long as the factors of locality remain unchanged called-

a. Productivity
b. Growing capacity
c. Site productivity
d. **Yield capacity**

123. The calculation of the amount of material which may be removed from a forest, annually or periodically over a stated period, or of the annual or periodic area over which fellings may be made, consistent with the treatment prescribed called-

a. **Yield determination**
b. Yield capacity
c. Total Yield
d. Site productivity

124. A term generally applied to the determination of the yield and the prescribed means of releasing it, called-

a. **Yield regulation**
b. Yield determination
c. Yield capacity
d. Total Yield

125. Objects of yield regulation is/are-

a. To cut each crop or tree at maturity to obtained maximum yield of the desired produce
b. To cut, approximately, the same quantity of material annually or periodically
c. To limit the area to be felled to that which can be regenerated
d. **All of the above**

Answers

118. c	119. d	120. b	121. a	122. d	123. a
124. a	125. d				

126. The yield is usually regulated for the period-
   a. Five years
   b. Six Year
   c. Eight years
   d. **Ten year**

127. Thinning, pruning, cleaning and improvement fellings are the examples of intermediate-
   a. Final yield
   b. **Intermediate yield**
   c. Progressive yield
   d. Total yield

128. Which is the birth-place of scientific forestry-
   a. Asian continent
   b. African continent
   c. **European continent**
   d. Arabian continent

129. Explain S/α/M/0.6 abbreviation –
   a. Middle aged (M), Sal Forest (S), α quality, with a density 0.6
   b. Medium aged (M), Sal Forest (S), α quality, with a density 0.6
   c. Medium aged (M), Sal Forest (S), α class, with a density 0.6
   d. Middle aged (M), Sal Forest (S), α class, with a density 0.6

130. Two stage stratified random sampling method is used for-
   a. Teak enumeration
   b. Sal enumeration
   c. **Bamboo** enumeration
   d. Deodar enumeration

131. Period of Working Plan in India is-
   a. 5 years
   b. **10 years**
   c. 8 years
   d. 15 years

132. The forest survey of India (FSI) conducts forest cover survey once in every-
   a. One year
   b. Five years
   c. **Two** years
   d. Ten years

133. Which is based only on area for yield regulation in regular forest-
   a. Annual coupe by gross area method (A1)
   b. Annual coupe by reduced area method (A2)
   c. Annual coupe by total area method (A1)
   d. **Both a and b**

134. Which is/are based only on volume for yield regulation in regular forest-
   a. Von Mental's Formula
   b. Simons' modification
   c. Howard's modification
   d. **All of the above**

Answers

126. d	127. b	128. c	129. a	130. c	131. b
132. c	133. d	134. d			

135. Smythies modification for yield regulation in regular forest is based on-
a. Area
b. Area and volume
c. **Volume**
d. Increment

136. Burma modification for yield regulation in regular forest is based on-
a. Area
b. **Volume**
c. Area and volume
d. Increment and volume

137. Which is/are based on area and volume for yield regulation in regular forest-
a. Permanent Periodic block method
b. Single Periodic block method
c. Revocable Periodic block method
d. **All of the above**

138. Floating Periodic block method for yield regulation in regular forest is based on-
a. Volume
b. **Area and volume**
c. Area
d. Increment and volume

139. Judeich's Stand Selection Method for yield regulation in regular forest is based on-
a. Volume
b. Volume and Increment
c. Area and Increment
d. **Area and volume**

140. Which is/are based only on increment for yield regulation in regular forest-
a. Increment method
b. Biolley's Check Method
c. Swiss method
d. **All of above**

141. Which is/are based on volume and increment for yield regulation in regular forest-
a. Hufnagal's methods
b. Volume unit method
c. Formula methods
d. **All of the above**

142. Which is/are based on volume and increment for yield regulation in regular forest-
a. French Method
b. Chaturvedi's modification
c. Melard's method
d. **All of the above**

Answers

135. c	136. b	137. d	138. b	139. d	140. d
141. d	142. d				

143. Formula methods of yield regulation in regular forest includes-
a. Austrian formula
b. Hundeshagen's formula
c. Heyer's formula
d. **All of the above**

144. Hufnagal's Diameter class method for yield regulation in regular forest is based on-
a. **Volume and Increment**
b. Area and volume
c. Area and Increment
d. Diameter

145. Brandis' **Diameter class** method or The Indian method for yield regulation in regular forest is based on-
a. Area and Increment
b. Diameter
c. **Volume and Increment**
d. Area and volume

146. Smythies' Safe-guarding formula or U.P Safe-guarding formula for yield regulation in regular forest is based on-
a. Diameter
b. **Volume and Increment**
c. Area and volume
d. Area and Increment

147. Which letter is used to denote the methods which are based only on area for yield regulation in regular forest-
a. **A**
b. C.
c. B
d. D

148. Which letter is used to denote the methods which are based only on volume for yield regulation in regular forest-
a. A
b. **B**
c. V
d. E

149. Which letter is used to denote the methods which are based on area and volume for yield regulation in regular forest-
a. A
b. **C**
c. V
d. B

150. Which letter is used to denote the methods which are based only on increment for yield regulation in regular forest-
a. I
b. V
c. C
d. **D**

151. Which letter is used to denote the methods which are based on volume and increment for yield regulation in regular forest-
a. V
b. **E**
c. C
d. I

Answers

143. d	144. a	145. c	146. b	147. a	148. b
149. b	150. d	151. b			

152. E5 and E7 Which methods is/are purely of Indian origin for yield regulation in regular forest-

a. B2
c. B3
b. B4
d. **All of the above**

153. Which methods is/are purely of Indian origin for yield regulation in regular forest-

a. E5
c. E7
b. E2
d. **Both a and c**

154. Annual coupe by gross area (A1) is the oldest and simplest form of yield regulation in regular forest, It was first applied in-

a. **France in the 14th century**
c. German in the 14th century
b. France in the 19th century
d. German in the 18th century

155. In annual coupe by gross area (A1) method the area of forest or the felling series is divided into a-

a. Number of annual coupes to the number of years to regeneration
b. **Number of annual coupes to the number of years in a rotation**
c. Number of annual periodic block to the number of years in a rotation
d. Number of annual periodic block to the number of years to regeneration

156. Cotta's formula for annual yield determination is-

a. $\mathbf{Y_a = \left(\frac{V}{P}\right) + \frac{I}{2}}$
c. $Y_a = \left(\frac{V}{P}\right) - \frac{I}{2}$
b. $Y_a = \left(\frac{P}{V}\right) + \frac{I}{2}$
d. $Y_a = \left(\frac{P}{V}\right) - \frac{I}{2}$

157. In revocable periodic block method rotation is divided into........ and the fellings series into corresponding.......respectively-

a. Rotation periods, periodic blocks
b. **Regeneration periods, periodic blocks**
c. Regeneration periods, coupes
d. Rotation periods, coupes

158. Floating Periodic block method also called as-

a. **Quartier Bleu Method**
c. Quartier girth Method
b. Quartier Blanc Method
d. Quartier map Method

Answers

152. d 153. d 154. a 155. b 156. a 157. b
158. a

159. The uncoloured thinning area on map in floating periodic block (F.P.B) method called-

a. **Quartier Blanc**
b. Quartier Block
c. Quartier Space
d. Quartier coupe

160. Formula of Duchafour's Crown Cover which is used to calculate the modified or reduced area (C) is given below-

a. $C = K^2g^2n^2$
b. $C = K^2gn^2$
c. $C = Kg^2n^2$
d. $\mathbf{C = K^2g^2n}$

161. Floating Periodic block method is not suitable for-

a. Regular forests
b. Sustained yield
c. **Irregular forest**
d. Progressive yield

162. Von Mental's Formula (based on volume) for the determination of annual yield is-

a. $Y_a = \frac{N.G.S.}{r}$
b. $\mathbf{Y_a = \frac{2N.G.S.}{r}}$
c. $Y_a = \frac{N.G.S.}{2r}$
d. $Y_a = \frac{2N.G.S.}{4r}$

163. Von Mental's Formula also known as-

a. Formula of glorious infinity
b. Formula of glorious complexity
c. Formula of glorious accuracy
d. **Formula of glorious simplicity**

164. Von Mental's Formula is best applicable to-

a. Regular, even-aged or nearly even-aged normal forests
b. Regular, uneven-aged normal forests
c. Regular, even-aged or nearly even-aged abnormal forests
d. Irregular, uneven-aged normal forests

165. Mason's formula is identical with that

a. Biolley Formula
b. Howard's formula
c. Heyer's formula
d. **Von Mental's Formula**

166. Mason's formula also known-

a. Exploitation percent
b. Utilization percent
c. Masson's ratio
d. **Both a and c**

Answers

159. a	160. d	161. c	162. b	163. d	164. a
165. d	166. d				

167. Howard's formula for yield is-

a. $\mathbf{Y = \frac{8V}{r}}$

b. $Y = \frac{V}{8r}$

c. $Y = \frac{4V}{8r}$

d. $Y = \frac{V}{4r}$

168. Which modification is applicable to all fractional enumeration of G.S, down to girth/diameter, equivalent to 1/n th of the rotation-

a. Howard's modification

b. Burma's modification

c. **Simons' modification**

d. Smythies modification

169. In case of Howard's modification enumerated is carried out down to the...corresponding to half the rotation.

a. Height

b. **Diameter**

c. Volume

d. Girth

170. The formula of Simons' modification is-

a. $Y_a = \frac{n^2}{r(n^2-2)} \times V$

b. $Y_a = \frac{2n^2}{(n^2-1)} \times V$

c. $Y_a = \frac{2n^2}{r(n^2-2)} \times V$

d. $\mathbf{Y_a = \frac{2n^2}{r(n^2-1)} \times V}$

171. Smythies' modification is the modification of-

a. Burma's modification

b. **Von Mental's Formula**

c. Simons' modification

d. Howard's modification

172. The formula of Smythies modification is-

a. $\mathbf{Y_a = \frac{2V}{r-x}}$

b. $Y_a = \frac{V}{r-x}$

c. $Y_a = \frac{4V}{r-x}$

d. $Y_a = \frac{2V}{r+x}$

173. Like Howard's modification which was also evolved as a special case in which enumerations and measurement of the G.S was done down to one third rotation, instead half the rotation-

a. Simons' modification

b. Smythies modification

c. **Burma's modification**

d. Von Mental formula

Answers

167. a	168. c	169. b	170. d	171. b	172. a
173. c					

174. The formula of Burma modification is given as-

a. $Y_a = \frac{4V}{9r}$ c. $Y_a = \frac{8V}{4r}$

b. $Y_a = \frac{V}{4r}$ d. $\mathbf{Y_a = \frac{9V}{4r}}$

175. Historically which method was the first method which was based on the yield on a knowledge of increment and Growing stock-

a. **Austrian method** c. Burma method

b. Von Mental method d. Howard method

176. The formula of Austrian method is given as-

a. $Y = i + \frac{(Va + Vn)}{3r}$ c. $Y = i + \frac{(Va - Vn)}{2r}$

b. $\mathbf{Y = i + \frac{(Va - Vn)}{r}}$ d. $Y = i + \frac{(Va + Vn)}{4r}$

177. Austrian method is applicable to determine yield in-

a. Regular forests c. Irregular forests

b. Normal forests d. **All of the above**

178. Heyer's modification was based on-

a. Actual current annual increment

b. Actual periodic annual increment

c. **Actual mean annual increment**

d. Both a and b

179. The formula of Heyer's which was tried in Thanu forests of Dehra Dun by M.P Bhola is-

a. $\mathbf{Y = I_a + \frac{(Va - Vn)}{x}}$ c. $Y = I_a - \frac{(Va + Vn)}{x}$

b. $Y = I_a - \frac{(Va - Vn)}{2x}$ d. $Y = I_a + \frac{2(Va + Vn)}{x}$

Answers

174. d 175. a 176. b 177. d 178. c 179. a

180. Hundeshagan's method was based on that actual yield (increment) bears the same proportion to the actual (real) growing stock (G_r) as the normal yield does to the normal growing stock (G_n) –

a. $Y_r = G_r \times \frac{G_n}{Y_n}$

c. $Y_r = Y_n \times \frac{G_n}{G_r}$

b. $Y_r = Y_n \times \frac{G_r}{G_n}$

d. $\mathbf{Y_r = G_r \times \frac{Y_n}{G_n}}$

181. Which is quotient of utilization factor or Hundeshagan factor-

a. $\frac{Y_r}{G_n}$

c. $\frac{Y_n}{G_r}$

b. $\mathbf{\frac{Y_n}{G_n}}$

d. $\frac{G_n}{Y_n}$

182. Breymann's method is based on the same principle as Hundeshagan's method. The formula of Breymann's is given as-

a. $Y = A \times \frac{Y_n}{A_n}$

c. $\mathbf{Y = Y_n \times \frac{A}{A_n}}$

b. $Y = A_n \times \frac{A}{Y_n}$

d. $Y = Y_n \times \frac{A_n}{A}$

183. French method (1883) is sometimes called as-

a. **Quartier bleu method**
b. Formula of glorious complexity
c. Quartier blanc method
d. Formula of glorious simplicity

184. The formula of French method (1883) is given as-

a. $Y_a = \frac{V_0}{r/2} + \frac{V_0 \times t_1}{2}$

c. $Y_a = \frac{V_0}{r/2} + \frac{V_0 + t_1}{2}$

b. $\mathbf{Y_a = \frac{V_0}{r/3} + \frac{V_0 \times t_1}{2}}$

d. $Y_a = \frac{V_0}{r/2} + \frac{V_0 - t_1}{2}$

Answers

180. d 181. b 182. c 183. a 184. b

185. In French method rotation 'r' is divided into three equal parts viz. young, medium and old in the ratio-

a. 1 : 2 : 3
c. 2 : 4 : 6
b. **1 : 3: 5**
d. 1 : 1 : 1

186. If $V_o > V_{m,}$ then the crop in V_o is examined and transfer...........trees to V_m the proportion of 5:3 is adjusted-

a. Upper diameter
c. Middle diameter
b. **Lower diameter**
d. Randomly in any one

187. If $V_o < V_{m,}$ then the crop in V_m is examined and transfer..........trees to V_o the proportion of 5:3 is adjusted-

a. **Upper diameter**
c. Middle diameter
b. Lower diameter
d. Randomly in any one

188. If V_o+V_m is less than normal (5:3), in this case practically nothing should be removed from......except the dead and dying trees, and nothing will be removed from......respectively-

a. V_m and V_o
c. **V_o and V_m**
b. V_o and V_o
d. V_m and V_m

189. 1894 modification of French method called Melard's formula which is given as-

a. $\mathbf{Y = \frac{V_0}{r/3} + \frac{V_0 \times t_1}{2} + \frac{1}{n}(V_m \times t_2)}$

b. $Y = \frac{V_0}{r/2} + \frac{V_m \times t_1}{3} + \frac{1}{n}(V_0 \times t_2)$

c. $Y = \frac{V_0}{r/2} + \frac{V_0 \times t_1}{3} + \frac{1}{n}(V_0 \times t_2)$

d. $Y = \frac{2V_0}{r/3} + \frac{V_0 + t_1}{2} + \frac{1}{n}(V_m \times t_2)$

190. Chaturvedi's formula for annual yields is-

a. $Y = \frac{V_0}{R + r} + \frac{1}{2}(V_0 - i)$
c. $Y = \frac{2V_0}{R - r} + \frac{1}{2}(V_0 + i)$

a. $\mathbf{Y = \frac{V_0}{R - r} + \frac{1}{2}(V_0 \times i)}$
d. $Y = \frac{V_0}{R - r} + \frac{1}{2}(V_0 + i)$

Answers

185. b 186. b 187. a 188. c 189. a 190. b

191. Formula of Increment method, simplified form of Biolley's Check Method is-

a. $\mathbf{Y = \frac{V + a - V_n}{r}}$

c. $Y = \frac{V + a - V_n}{2r}$

b. $Y = \frac{V - a + V_n}{3r}$

d. $Y = \frac{V - a - V_n}{4r}$

192. According Swiss method only the annual increment is to be removed from the-

a. Youngest diameter classes

c. Mideast diameter classes

b. Mature diameter classes

d. **Oldest diameter classes**

193. According Swiss method annual yield is calculated using formula-

a. Annual yield = $\frac{Y - Z}{2c.c}$

c. Annual yield = $\frac{Y \times Z}{c.c}$

b. **Annual yield =** $\mathbf{\frac{Y + Z}{c.c}}$

d. Annual yield = $\frac{Y + Z}{2c.c}$

194. The principle behind the method is that the composition and distribution of the G.S can be moulded; and., by correct harvesting the timber production can be improved. This method is called-

a. Biolley's "Check method

c. Method-de-Controle

b. Smythies modification

d. **Both a and c**

195. Biolley's "Check method" or Method-de-Controle Formula to determine annual increment or yield is-

a. **Annual increment or yield =** $\mathbf{\frac{(V_2 + N) - (V_1 + P)}{n}}$

b. Annual increment or yield = $\frac{(V_2 - N) - (V_1 + P)}{n}$

c. Annual increment or yield = $\frac{(V_2 + N) - (V_1 - P)}{n}$

d. Annual increment or yield = $\frac{(V_2 + N) + (V_1 + P)}{n}$

Answers

191. a 192. d 193. b 194. d 195. a

196. Hufnagal's Diameter class method formula for calculating the annual yield is-

a. $Y_a = \frac{n_4}{a_4 - a_3} \times V_4 + \frac{n_3 + n_4}{a_4 - a_3} \times V_3 + \frac{n_2 + n_3}{a_3 - a_2} \times V_2 + \frac{n_1 - n_2}{a_2 - a_1} \times V_1$

b. $Y_a = \frac{n_4}{a_4 - a_3} \times V_4 + \frac{n_3 - n_4}{a_4 + a_3} \times V_3 + \frac{n_2 - n_3}{a_3 - a_2} \times V_2 + \frac{n_1 - n_2}{a_2 - a_1} \times V_1$

c. $Y_a = \frac{n_4}{a_4 - a_3} \times V_4 + \frac{n_3 - n_4}{a_4 - a_3} \times V_3 + \frac{n_2 - n_3}{a_3 - a_2} \times V_2 + \frac{n_1 + n_2}{a_2 - a_1} \times V_1$

d. $\mathbf{Y_a = \frac{n_4}{a_4 - a_3} \times V_4 + \frac{n_3 - n_4}{a_4 - a_3} \times V_3 + \frac{n_2 - n_3}{a_3 - a_2} \times V_2 + \frac{n_1 - n_2}{a_2 - a_1} \times V_1}$

197. Brandis' **Diameter class** method or The Indian method is suitable for extensively-

a. Temperate forests
b. **Tropical forests**
c. Deciduous forests
d. Dry deciduous forests

198. Smythies' Safe-guarding formula is also known as U.P Safe-guarding formula, It was formed mainly for sal forest in U.P. is-

a. $X = \frac{t}{f}(II + Z\%\ II)$
b. $X = \frac{t}{f}(II - Z\%\ II)$
c. $X = \frac{f}{t}(II \times Z\%\ II)$
d. $\mathbf{X = \frac{f}{t}(II - Z\%\ II)}$

199. Which scale is used for management map in India-

a. 1 : 5000
b. **1 : 50000**
c. 1 : 15000
d. 1 : 40000

200. Which scale is used for regeneration survey map in India-

a. **1 : 50000**
b. 1 : 5000
c. 1 : 15000
d. 1 : 25000

201. Which scale is used for stock map in India-

a. 1 : 50000
b. 1 : 4000
c. **1 : 15000**
d. 1 : 25000

Answers

196. d 197. b 198. d 199. b 200. a 201. c

202. Which scale is used for Forest type map & Working Plan map in India-

a. 1 : 25000
b. 1 : 20000
c. 1 : 5000
d. **1 : 50000**

203. Colour of tropical semi-evergreen forest group on the forest type map shows-

a. Violet
b. Red
c. **Purple**
d. Pink

204. Colour of alpine forest group on the forest type map shows-

a. **Sepia**
b. Green
c. Red
d. Purple

205. Colour of littoral & swamp forest group on the forest type map shows-

a. Purple
b. Red
c. Green
d. **Prussian blue**

206. Colour of regeneration map is blue then its regeneration status is-

a. Fair
b. Excellent
c. **Good**
d. Deficient

207. Colour of regeneration map is green then its regeneration status is-

a. **Excellent**
b. Fair
c. Good
d. Moderate

Answers

202. d 203. c 204. a 205. d 206. c 207. a

CHAPTER 4

Wildlife Management

1. Wildlife comprises all living organisms such as plants, animals, micro-organisms) in their natural habitats which are-
   a. Neither cultivated
   b. Not tamed
   c. Not domesticated
   d. **All of the above**
2. The science and art of changing the characteristics an interactions of habitats, wild animal population, and mean in order to achieve specific human goals by means of wildlife resources-
   a. Forest Management
   b. Rangeland Management
   c. **Wildlife Management**
   d. Watershed Management
3. Book "Wildlife Management Techniques" was written by-
   a. **Robert H. Giles**
   b. L. S. Khanna
   c. S. K. Singh
   d. Michel Clerk
4. Rewa national park in Madhya Pradesh is famous for the-
   a. **White tigers**
   b. Asiatic Lion
   c. Tigers
   d. Chinkara
5. Largest animal in the world of family Bovidae is-
   a. Elephant (*Elephus maximus*)
   b. Chinkara *(Gazella gazella)*
   c. **Gaur (*Bos gaurus*)**
   d. Sambhar *(Cervus unicolor*)
6. Resources which are regenerated through natural cycle, Such as air, water, soil, plants, animals and wind etc. are called-
   a. Non-renewable natural resources
   b. Artificial natural resources
   c. **Renewable natural resources**
   d. All of the above

Answers

1. d	2. c	3. a	4. a	5. c	6. c

7. Resources which are not replaced in the environment after their utilization. Such as metals, coal, natural gas and minerals etc. are called-
   a. Artificial natural resources
   b. **Non-renewable natural resources**
   c. Artificial natural resources
   d. Renewable natural resources

8. Book " **A Text book of** Wildlife Management" was written by-
   a. L. S. Khanna
   c. Michel Clerk
   b. Robert H. Giles
   d. **S. K. Singh**

9. First biosphere reserve of India is-
   a. Nanda Devi
   c. Manas
   b. **Nilgiri**
   d. Gulf of Mannar

10. Biosphere Reserve has how many zones-
   a. **3**
   c. 4
   b. 5
   d. 6

11. Area of Biosphere Reserve where no human activity is allowed called-
   a. Restoration zone
   c. Buffer zone
   b. Manipulated zone
   d. **Core zone**

12. Area of Biosphere Reserve where a large number of human activities would go on is called-
   a. Transition zone
   c. Manipulated zone
   b. Buffer zone
   d. **Both a and c**

13. Area of Biosphere Reserve where limited human activity is allowed is called-
   a. Manipulated zone
   c. **Buffer zone**
   b. Restoration zone
   d. Transition zone

14. First wildlife conservation movement in India was started by-
   a. **BNHS**
   c. IUCN
   b. WWF
   d. CITES

15. Predators have an important role in regulating-
   a. Mineral cycle
   c. Environment conservation
   b. **Balancing of nature**
   d. Hydrological cycle

Answers

7. b	8. d	9. b	10. a	11. d	12. d
13. c	14. a	15. b			

16. Which is not a types of predation in nature-
    a. Chance predation c. Habit predation
    b. Sanitary predation d. **All of the above**

17. The study of analysis of the changes in the numbers of animals in wildlife population at a particular time is called-
    a. Wildlife census c. **Population-dynamics**
    b. Wildlife science d. Population structure

18. Conservation of fauna and flora at their native place is called-
    a. Ex-situ conservation c. Germplasm conservation
    b. **In-situ conservation** d. Cryopreservation

19. Conservation of fauna and flora outside their native place is called-
    a. **Ex-situ conservation** c. Germplasm conservation
    b. In-situ conservation d. Cryopreservation

20. World's first national park which is situated in America-
    a. Dachigam national park c. **Yellow stone national park**
    b. Jim Corbett national park d. Kanha national park

21. India's first national park which is situated in Nainital, Uttarakhand-
    a. **Jim Corbett National Park** c. Dachigam national park
    b. Bennergutta national park d. Dudhwa national park

22. Smallest tiger reserve in India is-
    a. Tandoba national park c. Gir national park
    b. Kaziranga national park d. **Ranthambore national park**

23. Largest tiger reserve in India which is located in Andhra Pradesh-
    a. **Nagarjuna Sagar sanctuary** c. Periyar sanctuary
    b. Chilka lake Bird sanctuary d. Annamalai sanctuary

24. Book "**Silent Spring**" was written by-
    a. **Rachel Carson** c. Mohd Tariq
    b. Robert H. Giles d. S. K. Singh

25. Silent valley is situated in-
    a. Tamil Nadu c. **Kerala**
    b. West Bengal d. Karnataka

26. Hot Spot concept was given by-
    a. **Norman Mayer** c. P. S. Chauhan
    b. Alfred Mayer d. Michel Gills

Answers

16. d	17. c	18. b	19. a	20. c	21. a
22. d	23. a	24. a	25. c	26. a	

27. According to Robert Mayer, global species diversity is about-
a. 4 million
c. **7 million**
b. 8 million
d. 9 million

28. Amazon rain forest called-
a. **Lungs of the planet**
b. Heaven on earth
b. Green mufflcrs
d. Valley of flowers

29. India's first and only butterfly park was established in 1992-
a. Gujarat
c. Kerala
b. Uttarakhand
d. **Sikkim**

30. The modification done in the habitat are the requirement and benefit of the wildlife especially for food, water, shelter and area is called-
a. Shelter manipulation
c. **Habitat-manipulation**
b. Conservation
d. All of the above

31. Salt lick is a common methods of artificial-feeding to the wild animals in-
a. Favourable period
c. summer season
b. **Pinch period**
d. Winter season

32. In Betula National Park of Jharkhand during night a powerful torch with coloured papers is used giving signals like.......colour means fire is on, and ....colour means fire is adjoining jurisdiction respectively-
a. Red & Green
c. **Red and Blue**
b. Red & Yellow
d. Green and Blue

33. The animals whose body temperature does not fluctuate with the changes in the temperature of the environment and are able to regulate and maintain the body temperature at a constant level called-
a. Eurythermal
c. Homoeothermic
b. Endothermal
d. **All the above**

34. The animals whose body temperature fluctuate or varies with the changes in the temperature of the environment called-
a. Poikilothermic
c. stenothermal
b. Ectothermal
d. **All the above**

Answers

27. c	28. a	29. d	30. c	31. b	32. c
33. d	34. d				

35. According to which rule the animals living in cold region are much longer than the warmer region-
    a. Allen rule
    b. **Bergmann rule**
    c. Jordan rule
    d. Meyer rule

36. According to which rule the fishes living in low water temperature have more vertebrae than those living in warm water-
    a. Bergmann rule
    b. Meycr rulc
    c. Allen rule
    d. **Jordan rule**

37. Animals which can tolerate narrow fluctuation of the salt-concentration is called-
    a. Euryhaline
    b. **Stenohaline**
    c. Regulators
    d. Conformers

38. Animals which can tolerate wide range fluctuation of the salt-concentration is called-
    a. Regulators
    b. Conformers
    c. **Euryhaline**
    d. Stenohaline

39. The most tasteful and liked food of wild animals is called-
    a. **Preferred-food**
    b. Staple food
    c. Tasty food
    d. Habitual food

40. After preferred food, food which is important for the living of the animal being for long time called-
    a. Pinch food
    b. Urgent food
    c. **Staple food**
    d. Habitual food

41. When there is scarcity of staple food, the species depends upon certain food which is neither so tasteful nor nutritive called-
    a. Staple food
    b. **Emergency food**
    c. Stuffing food
    d. Survival food

42. The food which is totally non-nutritive and is consumed by the animals only to fulfil its stomach is called-.
    a. **Stuffing food**
    b. Habitual food
    c. Staple food
    d. Emergency food

Answers

35. b	36. d	37. b	38. c	39. a	40. c
41. b	42. a				

43. The period in which the food is not sufficiently available and causes trouble to the animal is called-
    a. **Pinch period**
    b. Survival period
    c. Dormancy period
    d. All of the above

44. The place or area which gives protection and serves other biological needs of the species is known as-
    a. Territory
    b. Cover
    c. Home Range
    d. **Shelter**

45. Which are recently extinct animals from India as well as world-
    a. Dodo (Mauritius)
    b. Stellars sea cow (Russia)
    c. Tigers subspecies (Bali, Javan, Caspian)
    d. **All of the above**

46. Antlers are entirely solid structure, can regrow, branched and found only-
    a. **Male**
    b. Both
    c. Female
    d. Neither male nor in female

47. The number of animals per unit area is called-
    a. Frequency
    b. Abundance
    c. **Density**
    d. Dominance

48. The phenomena of regular movement of a species from one place to other and back is called-
    a. **Migration**
    b. Dispersion
    c. Immigration
    d. Dispersal

49. World famous example of birds migration from Russia to Bharatpur (Rajasthan, India) is-
    a. Penguin
    b. **Siberian crane**
    c. Peacock
    d. Ostrich

50. The phenomena of movement of a species from one place to other permanently called-
    a. **Dispersal**
    b. Migration
    c. Dispersion
    d. Emigration

Answers

43. a	44. d	45. d	46. a	47. c	48. a
49. b	50. a				

51. The status and position of distribution of the individuals of a species in the particular habitat is called dispersion-

a. Emigration c. Migration
b. **Dispersion** d. Dispersal

52. Which is not a type of dispersion-

a. Random c. Uniform
b. Clumped d. **All of the above**

53. When the individuals of a species move into an area leaving its original habitat is called-

a. Emigration c. **Immigration**
b. Migration d. Adaptation

54. When the individuals of a population move out from its original habitat permanently is called-

a. **Emigration** c. Migration
b. Immigration d. Dispersal

55. The rate of birth in a population is called-

a. Mortality c. Immigration
b. **Natality** d. Emigration

56. The rate of death of individuals in a population is called-

a. **Mortality** c. Immigration
b. Natality d. Emigration

57. Fecundity is used to indicate-

a. Reproduction capacity
b. Capacity to tolerate drought
c. **Capacity to bear young ones**
d. Capacity to give birth to young ones

58. Animals living singly is called-

a. **Solitary** c. Herds
b. Flocks d. Troops

59. The species which are in danger of extinction and whose survival is unlikely if the causal factors continue to be operating called-

a. **Endangered species** c. Extinct species
b. Threatened species d. Rare species

Answers

51. b	52. d	53. c	54. a	55. b	56. a
57. c	58. a	59. a			

60. The species likely to move into the endangered categories in the near future if the causal factors continue to be operate called-
   a. Threatened species c. Rare species
   b. **Vulnerable species** d. Endangered species

61. The species with small population in the world at restricted area is called-
   a. Endangered species c. **Rare species**
   b. Threatened species d. Vulnerable species

62. A book contains a record of animals & plants which are known to be in danger called-
   a. Wildlife book c. Green data book
   b. **Red data book** d. Forestry book

63. A book containing a list of rare plants in a protected area like as Botanical garden called-
   a. **Green data book** c. Wildlife book
   b. Red data book d. Forestry book

64. Kaziranga National Park is situated in-
   a. Tamil Nadu c. Gujarat
   b. Kerala d. **Assam**

65. Kaziranga National Park is famous for-
   a. Sambhar c. **Rhinoceros**
   b. Tiger d. Lion

66. Gir National Park is situated in-
   a. West Bengal c. Madhya Pradesh
   b. **Gujarat** d. Assam

67. Gir National Park is famous for-
   a. Blackbuck c. **Asiatic Lion**
   b. Tiger d. Shanghai

68. BNHS (Bombay Naturel History Society) headquarter is located in-
   a. New Delhi c. Kolkata
   b. Dehra Dun d. **Bombay**

69. Scientific name/Zoological name of rhino is-
   a. *Panthera leo persica* c. *Felis caracal*
   b. *Bos gaurus* d. ***Rhinoceros unicornis***

Answers

60. b	61. c	62. b	63. a	64. d	65. c
66. b	67. c	68. d	69. d		

70. Corbett National Park is situated in-
    a. Madhya Pradesh   c. Assam
    b. **Uttarakhand**   d. Uttar Pradesh

71. CITES headquarter is located in which city-
    a. **Washington**   c. Ned Delhi
    b. Nairobi   d. Geneva

72. Dachigam National Park is famous for-
    a. Tiger   c. Elephant
    b. Rhinoceros   d. **Hangul**

73. Scientific name/Zoological name of Indian Tiger is-
    a. *Panthera tigris*   c. ***Panthera tigris tigris***
    b. *Panthera uncia*   d. *Panthera pardus*

74. Scientific name/Zoological name of Indian elephant is-
    a. *Hystrix indica*   c. *Panthera leo*
    b. ***Elephus maximus***   d. *Elephus indica*

75. Symbol of WWF is-
    a. Tiger   c. Lion
    b. Deer   d. **Red panda**

76. Desert National Park is located in-
    a. Jaipur   c. **Jaisalmer**
    b. Ned Delhi   d. Gujarat

77. Kanha National Park is famous for-
    a. **Tiger**   c. Chinkara
    b. Elephant   d. Mangoose

78. Kanha National Park is located in-
    a. Kerala   c. Odisha
    b. Uttar Pradesh   d. **Madhya Pradesh**

79. Scientific name/Zoological name of Red Panda is-
    a. *Gazella gazella*   c. ***Ailurus fulgens***
    b. *Cervus unicolor*   d. *Platanista gangetica*

80. Scientific name/Zoological name of Indian Porcupine is-
    a. ***Hystrix indica***   c. *Vulpes vulpes*
    b. *Selenarctos thibetanus*   d. *Capra hircus*

Answers

70. b	71. a	72. d	73. c	74. b	75. d
76. c	77. a	78. d	79. c	80. a	

81. IUCN (Now as WCU-World Conservation Union) headquarter is located in-
 a. Bombay c. **Morgis**
 b. Geneva d. Dehra Dun

82. WWF (Worldwide Fund for Nature -1990) headquarter is located in-
 a. **Gland** c. Dehra Dun
 b. Bombay d. Geneva

83. Indian Institute of wildlife is situated in-
 a. Ned Delhi c. Kolkata
 b. Bombay d. **Dehra Dun**

84. National Biodiversity Authority (NBI) headquarter is located in-
 a. **Chennai** b. Dehra Dun
 b. Bombay d. New Delhi

85. Central Zoo authority (CZA) headquarter is located in-
 a. Kolkata c. **New Delhi**
 b. Bombay d. Dehra Dun

86. National Biodiversity Board (NBB) is situated in-
 a. Haryana c. **New Delhi**
 b. Dehra Dun d. Kerala

87. National Research Centre on Camel is situated in-
 a. Jaipur c. Jodhpur
 b. Udaipur d. **Bikaner**

88. World Wildlife Day is celebrated on-
 a. 21 March c. **3 March**
 b. 21 May d. 8 October

89. World Biodiversity Day is celebrated on-
 a. 21 March c. 21 May
 b. 5 June d. **22 May**

90. International Tiger Day is celebrated on-
 a. 11 July c. 22 May
 b. **29 July** d. 5 October

91. World Animal Day is celebrated on-
 a. **4 October** c. 22 May
 b. 5 October d. 21 March

Answers

81. c	82. a	83. d	84. a	85. c	86. c
87. d	88. c	89. d	90. b	91. a	

92. World wildlife week is celebrated on-
    a. First week of May
    b. **First week of October**
    c. First week of June
    d. Second week of October

93. Wildlife Protection Act was enacted in-
    a. 1927
    b. 1976
    c. **1972**
    d. 1984

94. Wildlife Protection Act contains-
    a. 6 chapters, 7 schedules and 66 sections
    b. 7 chapters, 6 schedules and 65 sections
    c. 6 chapters, 7 schedules and 68 sections
    d. **7 chapters, 6 schedules and 66 sections**

95. Chapter which deals with preliminary definition of wildlife act is-
    a. ***Chapter i***
    b. *Chapter iii*
    c. *Chapter ii*
    d. *Chapter iv*

96. Authorities to be appointed or constituted under the chapter of WPA-
    a. *Chapter iii*
    b. ***Chapter ii***
    c. *Chapter iv*
    d. *Chapter vi*

97. Hunting of Wild Animals is deal in which chapter of WPA-
    a. *Chapter vii*
    b. *Chapter vi*
    c. *Chapter iv*
    d. ***Chapter iii***

98. Sanctuaries, National Parks and Closed Areas are created under the chapter of WPA-
    a. ***Chapter iv***
    b. *Chapter vii*
    c. *Chapter vi*
    d. *Chapter iii*

99. Prevention and Detection of Offences is deal in which chapter of WPA-
    a. *Chapter iii*
    b. ***Chapter vi***
    c. *Chapter vi*
    d. *Chapter vii*

100. Which schedule of WPA deals with Conservation of rare and endangered species-
    a. **Schedule *i***
    b. *Schedule iii*
    c. Schedule *ii*
    d. *Schedule vi*

Answers

92. b	93. c	94. d	95. a	96. b	97. d
98. a	99. b	100. a			

101. Which schedule of WPA deals with Special game animals-

a. Schedule *iii* c. **Schedule *ii***

b. ***Schedule** iv* d. Schedule *vi*

102. Which schedule of WPA deals with Big game animals-

a. Schedule *i* c. Schedule *ii*

b. **Schedule *iii*** d. Schedule *iv*

103. Project Lion was started in the year-

a. **1972** c. 1973

b. 1974 d. 1977

104. Project Tiger was started in the year-

a. 1974 c. 1976

b. **1973** d. 1972

105. Project Hangul was started in the year-

a. 1986 c. 1975

b. 1983 d. **1970**

106. Crocodile breeding Project was started in the year-

a. 1970 c. **1975**

b. 1993 d. 1991

107. Project Elephant was started in the year-

a. **1991** c. 1993

b. 1995 d. 1992

108. Gestation period of Tiger and Whale in days are respectively-

a. 535 and 209 c. 435 and 209

b. 335 and 309 d. **109 and 535**

109. Gestation period of Bear and Chimpanzee in days are respectively-

a. **220 and 240** c. 60 and 130

b. 150 and 300 d. 200 and 320

110. Gestation period of deer and Elephant in days are respectively-

a. 120 and 580 c. **201 and 617**

b. 150 and 610 d. 150 and 720

111. Gestation period of Leopard and Lion in days are respectively-

a. 120 and 210 c. 150 and 160

b. 150 and 180 d. **93 and 108**

Answers

101. c	102. b	103. a	104. b	105. d	106. c
107. a	108. d	109. a	110. c	111. d	

112. Gestation period of Rhinoceros and Kangaroo in days are respectively-
   a. **450 and 42**
   b. 450 and 120
   c. 550 and 90
   d. 650 and 92

113. The Salim Ali Bird sanctuary is located at-
   a. Pondicherry
   b. Anakkatti
   c. **Thattekad**
   d. Kalakkad

114. Who among the following is commonly called 'Bird Man of India'-
   a. **Salim Ali**
   b. Mohit Husain
   c. M.S. Swaminathan
   d. S. K. Singh

115. The inherent ability of organisms to reproduce and multiply is called-
   a. Fertilization
   b. Carrying capacity
   c. **Biotic potential**
   d. K value

116. The Indian Parliament passed the Biodiversity Bill in the year-
   a. 2000
   b. 2005
   c. **2002**
   d. 2006

117. The only known breeding colony of greater and lesser Flamingos in our country is found-
   a. **The Great Rann of Kutch**
   b. Laddakh desert
   c. The Thar desert
   d. Atacama desert

118. In which year Silent Valley was declared as National Park-
   a. 1988
   b. **1984**
   c. 1982
   d. 1981

119. The Red Data book which lists endangered species is maintained by-
   a. UNO
   b. **IUCN**
   c. BNHS
   d. WWF

120. Herpetology is a branch of Science which deals with-
   a. Aves
   b. **Reptiles**
   c. Mammals
   d. Fishes

121. Bears are usually hunted and killed for their –
   a. Teeth
   b. Gall bladder
   c. **Skin**
   d. Nails

Answers

112. a	113. c	114. a	115. c	116. c	117. a
118. b	119. b	120. b	121. c		

122. Coral reefs in India can be seen in-
   a. Himalayan region
   b. Uttar Pradesh
   c. **Andaman and Nicobar Islands**
   d. Kerala

123. The only ape found in India is-
   a. Gorilla
   b. **Hoolock gibbon**
   c. Chimpanzee
   d. Oranguttan

124. Which one is the most endangered species of Indian birds-
   a. Pigeon
   b. Owl
   c. Paradise fly catcher
   d. **The great Indian bustard**

125. Eravikulam National Park conserves which species-
   a. Lion tailed macaque
   b. Elephant
   c. Tiger
   d. **Nilgiri Tahr**

126. Name the endangered animal which is protected in Rajamalai National
   a. Park
   b. Black buck
   c. Chital
   d. **Nilgiri Tahr**

127. Many wild plant and animals are on the verge of extinction due to-
   a. **Habitat destruction**
   b. Scarcity of food
   c. Climate change
   d. Hunting

128. The most endangered ecosystem in India is-
   a. Shola forest ecosystem
   b. Pond ecosystem
   c. **Evergreen forest ecosystem**
   d. Tundra Ecosystem

129. Earth summit of Rio de Janeiro (1992) was made for-
   a. Compilation of Red list
   b. **Conservation of biodiversity**
   c. Establishment of biosphere reserves
   d. Reduced green-house gasses

130. Science which deals with study of birds called-
   a. Mammology
   b. **Ornithology**
   c. Herpetology
   d. Ichthyology

Answers

122. c	123. b	124. d	125. d	126. d	127. a
128. c	129. b	130. b			

131. Science which deals with study of amphibians called-
   a. Ichthyology   c. **Herpetology**
   b. Ornithology   d. Serpentology

132. Science which deals with study of pices called-
   a. **Ichthyology**   c. Ornithology
   b. Herpetology   d. Ophiology

133. Science which deals with study of snakes called-
   a. Ornithology   c. Herpetology
   b. Ichthyology   d. **Ophiology**

134. Science which deals with study of mammals-
   a. Ichthyology   c. **Mammology**
   b. Herpetology   d. Humanology

135. Who is the father of zoology –
   a. **Aristotle**   c. Rishabh Dev
   b. Theophrastus   d. Robert Hook

Answers

131. c   132. a   133. d   134. c   135. a

CHAPTER 5

Forest Utilization

1. Axes, saws and wedges are the conventional implements used in-
   a. Felling c. Conversion
   b. Splitting d. **All the above**

2. The axes are used for-
   a. Felling c. Splitting
   b. Grubbing d. **All the above**

3. Axe head carries a socket called-
   a. **Eye** c. Hole
   b. Back d. Neck

4. The portion of the iron head infront of the eye is known as-
   a. **Blade** c. Face
   b. Gullet d. Back

5. For hardwood blade of an axe should be-
   a. Comparatively lighter and thicker than for softwoods
   b. Comparatively heavier and thinner than for softwoods
   c. **Comparatively lighter and thinner than for softwoods**
   d. Comparatively heavier and thicker than for softwoods

6. The weight of an ordinary felling axe head varies from-
   a. 1.7 kg to 1.8 kg c. 0.7 kg to 1.2 kg
   b. **0.7 kg to 1.8 kg** d. 0.5 kg to 2.8 kg.

7. Generally the weight of the axe is-
   a. **1.5 to 2.0 kg** c. 1.0 to 2.0 kg
   b. 1.5 to 2.5 kg d. 1.0 to 2.5 kg

Answers

1. d 2. d 3. a 4. a 5. c 6. b
7. a

8. Which is used for felling bamboo and small poles and for cutting brushwood-
   a. Axe
   b. Spade
   c. Saw
   d. **Bill-hook**

9. Saws are used for-
   a. Felling
   b. Ripping
   c. cross-cutting
   d. **All the above**

10. The great advantage of saws over axes in felling and conversion is-
    a. Saw cause much less wastage of wood
    b. Saw cause less wastage of wood
    c. Saw is cheap than axe
    d. Saw is lighter than axe

11. Which one is not the popular kinds of saws used for felling and conservation-
    a. M-tooth saw
    b. Raker-tooth saw
    c. The triangular or Peg-tooth saw
    d. **All the above**

12. The frame saw is commonly used in-
    a. Tropical forest
    b. Deciduous forest
    c. **Himalaya Forests**
    d. All the above

13. Edge of the tooth which faces the cutting direction called-
    a. Space
    b. **Face**
    c. Back
    d. Gullet

14. Distance between two adjacent teeth called-
    a. Face
    b. Pitch
    c. Gullet
    d. **Space**

15. Entire opening between two adjacent teeth is called-
    a. Gauge
    b. Pitch
    c. **Gullet**
    d. Kerf

16. Angle tween the face of a toot and the line passing through points of the teeth called-
    a. **Pitch**
    b. Gauge
    c. Gullet
    d. Kerf

Answers

8. d	9. d	10. a	11. d	12. c	13. b
14. d	15. c	16. a			

17. The thickness of the saw blade called-
    a. Gullet
    b. **Gauge**
    c. Kerf
    d. Back

18. Width of the saw cut is called-
    a. Gauge
    b. **Kerf**
    c. Gullet
    d. Pitch

19. The opposite end of the teeth-
    a. **Back**
    b. Pitch
    c. Face
    d. Gauge

20. Crosscut saw also have triangular shape teeth called-
    a. Cutters
    b. Trakers
    c. **Rakers**
    d. None

21. Which cut the wood fibres along both sides of the groove during cutting the log-
    a. Rakers
    b. **Cutters**
    c. Trakers
    d. All of the above

22. Which one works like a chisel and break off the fibres during cutting the log-
    a. Trakers
    b. **Rakers**
    c. Cutters
    d. None

23. The techniques of proper levelling of the saw teeth called-
    a. **Jointing**
    b. Sharpening
    c. Balancing
    d. Buckling

24. Which is used for splitting logs or fuel billets, and for assisting in the felling of trees and longitudinal sawing of timber-
    a. Axe
    b. Saw
    c. **Wedge**
    d. Power chain saw

25. The height of the stumps should be as low as possible and not more than-
    a. 10-20 cm
    b. **15-30 cm**
    c. 20-30 cm
    d. 25-35 cm

26. The standard length of axe handle is-
    a. 50 cm
    b. 80 cm
    c. 75 cm
    d. **90 cm**

Answers

17. b	18. b	19. a	20. c	21. b	22. b
23. a	24. c	25. b	26. d		

27. Bow saw is very useful for a single man cutting logs up to diameter-
    a. **20 cm** c. 30 cm
    b. 40 cm d. 50 cm
28. A meter-long wooden stick made of seasoned timber called-
    a. Measuring Pole c. Cantilever
    b. **Measuring stick** d. None
29. The lengths of the standard blades of power saw is-
    a. **50-80 cm** c. 10-30 cm
    b. 50-100 cm d. 30-80 cm
30. The best season for felling is-
    a. Rainy season c. Summer season
    b. Spring season d. **Winter season**
31. In the plain and submontane tracts of India, the felling season is start from-
    a. January to April c. November to March
    b. **October to March** d. September to April
32. Tree should be felled as near the-
    a. As ground as possible b. At a stumpage height
    c. Fluting region d. Buttressing region
33. Tree should be felled from..........on slope or hilly ground-
    a. Downhill side c. **Uphill side**
    b. Both side d. None
34. Which is/are not a methods of felling a tree with axes and saws-
    a. With the axe alone
    b. With the saw alone
    c. With the saw and axe combined
    d. **All of the above**
35. The felling cut is made exactly opposite to the-
    a. Undercut about 10 to 15 cm above the undercut
    b. Undercut about 20 to 25 cm above the undercut
    c. Undercut about 15 to 25 cm above the undercut
    d. Undercut about 10 to 20 cm above the undercut

Answers

27. a	28. b	29. a	30. d	31. b	32. a
33. c	34. d	35. a			

36. The width of undercut should be-
 a. One-third of the tree diameter
 b. Half of the tree diameter
 c. One-fourth of the tree diameter
 d. **Two-third of the tree diameter**

37. Felling by root is always desirable in the case of very valuable trees such as-
 a. Sandal
 b. Walnut
 c. Khair
 d. **All of the above**

38. Standard sleeper size of Broad Gauge is-
 a. **275 cm × 26 cm × 13 cm**
 b. 153 cm × 18 cm × 12 cm
 c. 183 cm × 21 cm × 12 cm
 d. 153 cm × 16 cm × 11 cm

39. Standard sleeper size of Metre Gauge is-
 a. 275 cm × 26 cm × 13 cm
 b. 153 cm × 18 cm × 12 cm
 c. **183 cm** × 21 cm × 12 cm
 d. 153 cm × 16 cm × 11 cm

40. Standard sleeper size of Narrow Gauge is-
 a. 275 cm × 26 cm × 13 cm
 b. **153 cm × 18 cm × 12 cm**
 c. 183 cm × 21 cm × 12 cm
 d. 153 cm × 16 cm × 11 cm

41. Standard sleeper size of Light Narrow Gauge is-
 a. 275 cm × 26 cm × 13 cm
 b. 153 cm × 18 cm × 12 cm
 c. 183 cm × 21 cm × 12 cm
 d. **153 cm × 16 cm × 11 cm**

42. Total wastage due to all fellings is –
 a. 5-10%
 b. **9-14%**
 c. 15-20%
 d. 5-15%

43. Total wastage due to sawing is-
 a. 15-20%
 b. 9-14%
 c. **31-37 %**
 d. 5-15%

44. Total wastage due to minor transportation is-
 a. **5-6%**
 b. 9-14%
 c. 15-20%
 d. 31-37%

45. Total wastage due to major transportation is-
 a. 5-14%
 b. 9-10%
 c. 31-37%
 d. **3-5%**

Answers

36. d	37. d	38. a	39. c	40. b	41. d
42. b	43. c	44. a	45. d		

46. Total wastage due to all transportation is-
a. 5-14%
b. **8-11%**
c. 31-37%
d. 5-15%

47. Total average wastage due to all wastage-
a. **56%**
b. 20-30 %
c. 10-47%
d. 31-37%

48. The total volume of usable rough timber from the average tree is only-
a. **37-52%**
b. 9-10%
c. 10-47%
d. 13-25%

49. First of all in India logging work was started by-
a. Mohit Husain and Rohit Kumar
b. Jainil Amiki and Hershberger
c. **Dr. A.Huber and A. Koroleff**
d. Michel Clerk and Johnson

50. First Logging branch in India was established in June, 1957 in-
a. Forest Research Institute - Dehra Dun during 2nd five year Plan
b. Forest Research Institute - Dehra Dun during 3nd five year Plan
c. Forest Research Institute - Dehra Dun during 4nd five year Plan
d. Forest Research Institute - Dehra Dun during 5nd five year Plan

51. First Logging Training Centre in India was established in 1958-59 at-
a. Bangalore
b. Coimbatore
c. Bombay
d. **Batote**

52. Square or rectangular shape having 15 cm under section called-
a. Hakries
b. **Scantlings**
c. Squares
d. Poles

53. Wooden slabs having maximum thickness 5 cm is called-
a. Hakries
b. Scantlings
c. Squares
d. **Planks**

54. Minor transportation also called as-
a. On road transportation
b. Heavy transportation
c. **Off road transportation**
d. All the above

Answers

46. b	47. a	48. a	49. c	50. a	51. d
52. b	53. d	54. c			

55. Transportation capacity of a motor truck is generally-
   a. 1.5 to 2.5 tons
   b. **1.5 to 5 tons**
   c. 3 to 5 tons
   d. 2.5 to 5.5 tons

56. Which are the methods of water transportation-
   a. Floating
   b. Wet slides
   c. Rafting & booms
   d. **All of the above**

57. Depots can be classified into-
   a. Forest depots
   b. Sale depots
   c. Transit depots
   d. **All of the above**

58. Sales of forest produces may be grouped into-
   a. Lum-sum sale
   b. Hand to hand
   c. Payment on outturn
   d. **Both a and c**

59. Security deposit for auction is....... of total contract value or money-
   a. 25%
   b. 50%
   c. **10%**
   d. 20%

60. Forest labour is classified into-
   a. Local labour
   b. Forest villages or settlements
   c. Imported labour
   d. **All of the above**

61. Cutch (catechu-tannic acid) and katha (catechin) are obtained from which tree-
   a. ***Acacia catechu***
   b. *Acacia senegal*
   c. *Acacia nilotica*
   d. *Acacia japonica*

62. Cutch (catechu-tannic acid) and katha (catechin) are obtained from which part of tree-
   a. Sapwood
   b. Early wood
   c. **Heartwood**
   d. Springwood

63. Kheersal is obtained from the tree-
   a. *Shorea robusta*
   b. ***Acacia catechu***
   c. *Madhuca latifolia*
   d. *Butea monosperma*

64. Which is/are not the extraction method of cutch and katha-
   a. Country method
   b. Factory method
   c. Modified method
   d. **All of the above**

Answers

55. b	56. d	57. d	58. d	59. c	60. d
61. a	62. c	63. b	64. d		

65. Cutch yield from the modified method is-
a. **10-20%**
b. 30-40%
c. 20-30%
d. 40-50%

66. Katha yield from modified method by weight of wood is-
a. 1-2%
b. **3-4%**
c. 2-3%
d. 4-5%

67. Katha yield from improved or factory method by weight of wood is-
a. 1-2%
b. 3.0-3.5%
c. 2.0—2.5%
d. **4.0-4.5%**

68. Cutch is soluble in-
a. Cold water
b. **Both water**
c. Hot water
d. None

69. Katha is most soluble in-
a. **Cold water**
b. Both water
c. Hot water
d. None

70. Which concept is used to extract the katha from hot concentrate solution of cutch-
a. **Water solubility concept**
b. Conductivity concept
c. Density concept
d. All of the above

71. Katha is used in the preparation of-
a. Chewing gum
b. Bidi
c. **Chewing pan**
d. Tobacco

72. Paper-making in India was started in the year-
a. 1810
b. 1910
c. **1830**
d. 1930

73. The first paper mill in India was established in which state-
a. Assam
b. **West Bengal**
c. Maharashtra
d. Uttar Pradesh

74. The present (2015-16) per capita consumption of paper in India and world are respectively-
a. 4.0 kg and 50.0 kg
b. 2.0 kg and 10.0 kg
c. 5.0 kg and 10.0 kg
d. **9.0 kg and 60.0 kg**

Answers

65. a	66. b	67. d	68. b	69. a	70. a
71. c	72. c	73. b	74. d		

75. Which is used as a raw material for making paper-
a. Populus
c. Azadirache
b. Mahua
d. **Bamboo**

76. The process of paper making followed how many steps-
a. 4.
c. **5**
b. 6
d. 8

77. Which process is not a pulping process-
a. Mechanical
c. Chemical
b. Semi-chemical
d. **All the above**

78. Which is not a paper making process-
a. Pulping
c. **Waxing**
b. Bleaching
d. Stock preparation

79. Which process is not used in stock preparation during paper making-
a. Beating
c. sizing
b. Loading
d. **Filling**

80. Isolation of cellulosic fibres from raw materials is called-
a. **Pulping**
c. Cleaning
b. Bleaching
d. Sock reparation

81. In single stage bleaching which chemical is used-
a. **Calcium hypochlorite ($CaOCl_2$)**
b. Calcium chloride ($CaCl_2$)
c. Calcium oxide (CaO)
d. Chlorine (Cl_2)

82. In multi stage bleaching which chemical is used-
a. Calcium hypochlorite ($CaOCl_2$)
b. Sulphur di oxide (SO_2) and Sulphur tri oxide (SO_3)
c. **Chlorine gas (Cl_2) and Chlorine dioxide (ClO_2)**
d. Calcium chloride ($CaCl_2$) and Calcium oxide (CaO)

83. Thus physical characteristics of paper depends on-
a. Sizing
c. Loading
b. **Beating**
d. Colouring

Answers

75. d	76. c	77. d	78. c	79. d	80. a
81. a	82. c	83. b			

84. The process which is used to making paper more or less impervious to ink penetration-
   a. **Sizing**
   b. Beating
   c. Loading
   d. Colouring

85. In sizing process which is used to make paper more or less impervious to ink penetration-
   a. Resin
   b. **Rosin**
   c. Wax
   d. Gum

86. The process of addition of non-fibrous mineral matters to improve the dimensional stability as well as brightness of paper is called-
   a. Beating
   b. Bleaching
   c. Sizing
   d. **Loading**

87. The substance which are used in the process of loading to improve the dimensional stability as well as brightness of paper is called-
   a. China clay
   b. **Both a and c**
   c. Talc powder
   d. Alum

88. Mechanical pulp differs from chemical pulp having which components in original form-
   a. Lignin
   b. Cellulose
   c. Wood fibres
   d. **Both a and c**

89. Acidic process also known as-
   a. **Sulphite process**
   b. Alkaline process
   c. Sulphate process
   d. Soda process

90. Alkaline process also known as-
   a. Sulphate process
   b. Soda process
   c. Sulphite process
   d. **Both a and c**

91. Which agent is used as cooking agent in acidic process-
   a. Calcium sulphate + SO_3
   b. Calcium bisulphate + SO_2
   c. Calcium bicarbonate + SO_2
   d. Calcium carbonate + SO_3

92. Which agent is used as cooking agent in alkaline process-
   a. Sodium carbonate
   b. **Sodium hydroxide**
   c. Calcium hydroxide
   d. Calcium carbonate

Answers

84. a	85. b	86. d	87. b	88. d	89. a
90. d	91. b	92. b			

93. Which agent is used as cooking agent in soda (alkaline) process-
a. Calcium hydroxide
b. Sodium hydroxide + Na_2S
c. **Sodium hydroxide**
d. Sodium hydroxide + SO_3

94. Which agent is used as cooking agent in Sulphate (alkaline) process-
a. Calcium hydroxide
b. **Sodium hydroxide + Na_2S**
c. Sodium hydroxide
d. Sodium hydroxide + SO_3

95. Oleo-resins obtained from pines are insoluble in-
a. Benzene
b. Ether
c. Alcohol
d. **Water**

96. Turpentine oil (essential oil) and rosin (non-volatile solid) are obtained from resins by-
a. **Distillation**
b. Vaporization
c. Filtration
d. Fermentation

97. The only species regularly tapped for resin in India is-
a. *Pinus wallichiana*
b. *Pinus resiana*
c. ***Pinus roxburghii***
d. *Pinus kesiya*

98. Total annual resin production in India is about-
a. 28,000 tonnes
b. 50,000 tonnes
c. 80,000 tonnes
d. **48,000 tonnes**

99. Most commonly used resin tapping method is-
a. **Cup & lip method**
b. Improved method
c. Rill method
d. Country method

100. Which is/are not a types of resin tapping-
a. Light continuous tapping
b. **Moderate taping**
c. Heavy tapping
d. None

101. Most commonly used cup & lip method of tapping also known as-
a. **French method**
b. Factory method
c. Improved method
d. German method

102. Light continuous tapping also known as-
a. **Living method of tapping**
b. Indian method of tapping
c. Tapping to death
d. German method of tapping

Answers

93. c	94. b	95. d	96. a	97. c	98. d
99. a	100. b	101. a	102. a		

103. Heavy continuous tapping also known as-
a. German method of tapping
b. Indian method of tapping
c. **Tapping to death**
d. French method of tapping

104. In Light continuous tapping tree girth should be above-
a. **0.9 m**
b. 0.6 m
c. 0.3 m
d. 0.5

105. In heavy tapping tree girth should be above-
a. 20 cm
b. **60 cm**
c. 50 cm
d. 100 cm

106. Rill method also called as new method was developed in-
a. **FRI**
b. ICAR
c. FSI
d. IUCN

107. Rill methods gives how much percent more resin yields as compare to cup and lip method-
a. 15%
b. **22%**
c. 18%
d. 25%

108. Crude pine resin consists of two principal constituents a liquid known as.......and a solid known as........respectively-
a. Turpentine oil and resin
b. **Turpentine oil and rosin**
c. Rosin and turpentine oil
d. Rosin and turpentine oil

109. The yield of rosin and turpentine by weight is respectively-
a. 17-20% and 74%
b. **74% and 17-20%**
c. 30% and 74%
d. 20% and 50%

110. Which is/are the major commercial host plants for lac insect in India-
a. Butea monosperma (palas)
b. Schleichera oleosa (Kusum)
c. Zizyhus mauritiana (ber)
d. **All of the above**

111. There are two distinct strains of lac insects in India called-
a. Kusumi
b. Rangeeni
c. Mausami
d. **Both a and b**

112. Lac producing insect is-
a. ***Laccifer lacca***
b. *Laccifer kusumi*
c. *Laccifer mauritiana*
d. *Laccifer monosperma*

Answers

103. c	104. a	105. b	106. a	107. b	108. b
109. b	110. d	111. d	112. a		

113. Shellac is obtained from lac after how many proper washing-
a. Usually two washing
b. **Usually three washing**
c. Usually four washing
d. Usually six washing

114. All Forest Products other than *Major Forest Products* which consists of timber, small wood and fuel wood which includes grass, fruits, leaves barks, animal and mineral products found in forest and collected from them called-
a. Major forest products
b. Forest products
c. Timber products
d. **Minor forest products**

115. A long, narrow pointed end having thick walls & small cavities and provide rigidity to the plant called-
a. Cell wall
b. **Fibres**
c. Microfibrills
d. Companion cells

116. Longest cell in plant body is-
a. Xylem
b. Root
c. Phloem
d. **Fibre**

117. The process of separation or extracting of fibres from one another is called-
a. **Ratting**
b. Distillation
c. Swarming
d. Conversion

118. Stem fibre is not obtained from which species-
a. *Sterculia villosa*
b. *Hardwickia binata*
c. *Grewia tiliaefolia*
d. ***Musa textilis***

119. Leaf fibre is obtained from which species-
a. *Caryota urens*
b. *Agave sislana*
c. *Pandanus species*
d. **All of the above**

120. Stem fibre is obtained from which species-
a. *Bauhinia vahlii*
b. *Boehmeria nivea*
c. *Calotropis gigantean*
d. **All of the above**

121. Elephant harness rope is made from the fibre of which species-
a. ***Sterculia villosa***
b. *Hardwickia binata*
c. *Grewia tiliaefolia*
d. *Bauhinia vahlii*

Answers

113. b	114. d	115. b	116. d	117. a	118. d
119. d	120. d	121. a			

122. *Boehmeria nivea* fibre also known as-
   a. Rhea fibre c. Ramie fibre
   b. **Both a and c** d. Rome fibre

123. Whitish-pink colour fibre is obtained from-
   a. *Agave sislana* c. *Caryota urens*
   b. ***Sterculia villosa*** d. *Grewia tiliaefolia*

124. Flosses is/are obtained from which species-
   a. *Ceiba patendra* c. *Bombax ceiba*
   b. *Cochlospermum religiosum* d. **All of the above**

125. Which tree species known as Indian kapok-
   a. *Ceiba patendra* c. ***Bombax ceiba***
   b. *Calotropis gigantea* d. *Cochlospermum religiosum*

126. Coir fibres are obtained from which species-
   a. *Ceiba patendra* c. *Bombax ceiba*
   b. ***Cocos nucifera*** d. *Cochlospermum religiosum*

127. Coir fibres are obtained from which layer of coconut-
   a. Endoderm c. **Mesoderm**
   b. Ectoderm d. Cortex

128. Oils from plants are obtained by which methods-
   a. Distillation c. Extraction
   b. Fermentation d. **Both a and c**

129. Palmrosa oil is obtained from which species-
   a. ***Cymbopogon martini*** c. *Cymbopogon Flexosus*
   b. *Cymbopogon nardus* d. *Cymbopogon nardus*

130. Lemon oil is obtained from which species-
   a. *Cymbopogon martini* c. ***Cymbopogon Flexosus***
   b. *Cymbopogon nardus* d. *Cymbopogon java*

131. Citronella oil is obtained from which species-
   a. Cymbopogon martini c. Cymbopogon Flexosus
   b. ***Cymbopogon nardus*** d. *Cymbopogon java*

132. Khus oil is obtained from which species-
   a. Cymbopogon martini c. Cymbopogon Flexosus
   b. *Cymbopogon nardus* d. ***Vetiver zizaniodes***

Answers

122. b	123. b	124. d	125. c	126. b	127. c
128. d	129. a	130. c	131. b	132. d	

133. Ginger oil is obtained from which species-
a. ***Cymbopogon martini***
b. *Cymbopogon nardus*
c. *Cymbopogon Flexosus*
d. *Vetiver zizaniodes*

134. Geraniol is obtained from which species-
a. *Cymbopogon nardus*
b. ***Cymbopogon martini***
d. *Vetiver zizaniodes*
c. *Cymbopogon Flexosus*

135. Wood is/are obtained from which species-
a. *Santalum album*
b. *Cedrus deodara*
c. *Aquilaria agallocha*
d. **All of the above**

136. Wood oil is/are obtained from which wood of the tree-
a. Sapwood
b. **Heartwood**
c. Softwood
d. All of the above

137. Santalol is obtained from which species-
a. ***Santalum album***
b. *Cymbopogon nardus*
c. *Aquilaria agallocha*
d. *Vetiver zizaniodes*

138. Leaf oil is/are obtained from which species-
a. Pinus
b. Cedrus
c. **Eucalyptus**
d. Sandal

139. Which is called as blue gum tree-
a. *Acacia nilotica*
b. *Eucalyptus citriodora*
c. ***Eucalyptus globulus***
d. *Acacia catechu*

140. Citriodora leaf oil is obtained from which species-
a. *Acacia nilotica*
b. ***Eucalyptus citriodora***
c. *Eucalyptus globulus*
d. *Acacia catechu*

141. Camphor leaf oil is obtained from which species-
a. *Cinnamomum cardamomum*
b. *Gaultheria fragrantissima*
c. *Mentha arvensis*
d. ***Cinnamomum camphora***

142. Mint leaf oil is obtained from which species-
a. *Cinnamomum cardamomum*
b. *Gaultheria fragrantissima*
c. ***Mentha arvensis***
d. *Cinnamomum camphora*

143. Wintergreen oil is obtained from which species-
a. *Cinnamomum cardamomum*
b. ***Gaultheria fragrantissima***
c. *Mentha arvensis*
d. *Cinnamomum camphora*

Answers

133. a	134. b	135. d	136. b	137. a	138. c
139. c	140. b	141. d	142. c	143. b	

144. Root oil is/are obtained from which species-
   a. ***Sassuria lappa*** c. *Santalum album*
   b. *Gaultheria fragrantissima* d. *Cinnamomum camphora*

145. Valerian oil is obtained from the rhizome of which species-
   a. *Sassuria lappa* c. *Gaultheria fragrantissima*
   b. ***Valerian wallichii*** d. *Valerian chinensis*

146. Flower oil is/are obtained from which species-
   a. *Pandanus tectorius* c. *Accacia farnesiana*
   b. *Michelia champa* d. **All of the above**

147. Keora oil is obtained from which species-
   a. *Michelia champa* c. *Accacia farnesiana*
   b. ***Pandanus tectorius*** d. *Gaultheria fragrantissima*

148. Linaloe oil is obtained from the heartwood of-
   a. ***Bursera delpechiana*** c. *Cedrus deodara*
   b. *Shorea robusta* d. *Bursera wallichii*

149. During carbonization of wood through destructive distillation method yield of charcoal, tar, acetic acid and alcohol is respectively-
   a. 25 percent charcoal, 3% tar, 4-5% acetic acid and 3% alcohol
   b. 15 percent charcoal, 6% tar, 1-5% acetic acid and 6% alcohol
   c. 20 percent charcoal, 5% tar, 4-9% acetic acid and 2% alcohol
   d. 30 percent charcoal, 2% tar, 2-8% acetic acid and 12% alcohol

150. Tung oil is obtained from which species-
   a. ***Aleurites fordii*** c. *Azadirachta indica*
   b. *Madhuca butyracea* d. *Garcinia indica*

151. Margo oil is obtained from which species-
   a. *Aleurites fordii* c. ***Azadirachta indica***
   b. *Pongamia pinnata* d. *Schleichera oleosa*

152. Phulwara Butter is obtained from which species-
   a. *Aleurites fordii* c. *Azadirachta indica*
   b. ***Madhuca butyracea*** d. *Garcinia indica*

153. Kokam Butter is obtained from which species-
   a. *Madhuca indica* c. *Madhuca butyracea*
   b. *Schleichera oleosa* d. ***Garcinia indica***

Answers

144. a	145. b	146. d	147. b	148. a	149. a
150. a	151. c	152. b	153. d		

154. Chaulmugra is obtained from which species-
a. *Aleurites fordii* c. *Azadirachta indica*
b. ***Hydnocarpus kurzii*** d. *Garcinia indica*

155. Mahua Butter is obtained from which species-
a. ***Madhuca indica*** c. *Madhuca butyracea*
b. *Hydnocarpus kurzii* d. *Azadirachta indica*

156. Kanju oil is obtained from which species-
a. ***Pongamia pinnata*** c. *Madhuca indica*
b. *Hydnocarpus kurzii* d. *Azadirachta indica*

157. Kusum oil is obtained from which species-
a. *Vateria indica* c. *Kadamba chinensis*
b. *Pongamia pinnata* d. ***Schleichera oleosa***

158. Sal Butter is obtained from which species-
a. *Tectona grandis* c. *Pinus species*
b. ***Shorea robusta*** d. *Terminalia arjuna*

159. Piney tallow is obtained from which species-
a. ***Vateria indica*** c. *Valerian chinensis*
b. *Hydnocarpus kurzii* d. *Garcinia indica*

160. which species fruit produces Japan wax-
a. *Rhus succedonea* c. *Melia azadarach*
b. **Both a and c** d. *Hydnocarpus kurzii*

161. Hides and skins treated with tannins is called-
a. Plastics c. **Leather**
b. Raxin d. Thermoplastic

162. Indian tree do not produce-
a. **Wood tan** c. Resin
b. Gum d. Fruit

163. Myrobalans is mainly obtained from which species-
a. *Accacia nilotica* c. ***Terminalia chebula***
b. *Aleurites fordii* d. *Azadirachta indica*

164. Divi-Divi tans obtained from which species-
a. *Ceriops roxburghiana* c. *Aleurites fordii*
b. ***Caesalpinia coriaria*** d. *Cassia auriculata*

Answers

154. b	155. a	156. a	157. d	158. b	159. a
160. b	161. c	162. a	163. c	164. b	

165. Which species yield tans from the barks-
a. ***Cassia auriculata*** c. *Terminalia chebula*
b. *Anogeissus latifolia* d. *Caesalpinia coriaria*

166. Which species yield tans from the fruits-
a. ***Terminalia bellirica*** c. *Anogeissus latifolia*
b. *Carissa spinarum* d. *Ceriops roxburghiana*

167. Which species yield tans from the leaves-
a. ***Anogeissus latifolia*** c. *Terminalia chebula*
b. *Ceriops roxburghiana* c. *Aleurites fordii*

168. Tannin percent in the young leaves and dry leaves of the *Anogeissus latifolia* are respectively-
a. 5% and 16% c. 25% and 30%
b. **55% and 16%** d. 18% and 25%

169. Bark of *Terminalia arjuna* and *Ceriops roxburghiana* contain tannin percent respectively-
a. 20-37% and 20-24% c. **20-24% and 20-37%**
b. 10-20% and 20-40% d. 20-25% and 50-65%

170. Divi-Divi pods contain tannin percent-
a. **22-40** c. 40-75
b. 30-65 d. 10-40

171. Myrobalans contain tannin percent-
a. 5% c. 15%
b. 10% d. **22%**

172. Santalinus dye is obtained from which species-
a. *Caesalpinia sappan* c. ***Pterocarpus santalinus***
b. *Santalum album* d. *Atrocarpus hetrophyllus*

173. Kamela dye is obtained from which species-
a. *Bixa orellana* c. *Nyctanthes arbortristis*
b. ***Mallotus philippensis*** d. *Caesalpinia sappan*

174. Annatto dye is obtained from which species-
a. ***Bixa orellana*** c. *Nyctanthes arbortristis*
b. *Santalum album* d. *Atrocarpus hetrophyllus*

Answers

165. a	166. a	167. a	168. b	169. c	170. a
171. d	172. c	173. b	174. a		

175. Indigo dye is obtained from which species-
a. *Lawsonia inermis* c. *Nyctanthes arbortristis*
b. ***Indigofera species*** d. *Caesalpinia sappan*

176. Heena dye is obtained from which species-
a. ***Lawsonia inermis*** c. *Nyctanthes arbortristis*
b. *Mallotus philippensis* d. *Caesalpinia sappan*

177. The acacia gums largely known as-
a. Gum Chinese c. **Gum arabic**
b. Gum Americana d. Indian Gum

178. Acacia gum is obtained from which species-
a. ***Acacia catechu*** c. *Acacia senegal*
b. *Acacia nilotica* d. *Acacia leucocephala*

179. Indian gum arabic is obtained from which species-
a. *Acacia catechu* c. *Acacia senegal*
b. *Acacia leucocephala* d. ***Acacia nilotica***

180. True gum arabic is obtained from which species-
a. *Acacia leucocephala* c. *Acacia nilotica*
b. ***Acacia senegal*** d. *Acacia catechu*

181. Gum kino is obtained from which species-
a. *Lannea coromandelica* c. *Moringa pterygosperma*
b. ***Pterocarpus marsupium*** d. *Boswellia serrata*

182. Katira gum is obtained from which species-
a. *Boswellia serrata* c. ***Sterculia urens***
b. *Lannea coromandelica* d. *Acacia senegal*

183. Jhingan gum is obtained from which species-
a. ***Lannea coromandelica*** c. *Moringa pterygosperma*
b. *Anogeissus latifolia* d. *Bauhinia retusa*

184. Salai gum is obtained from which species-
a. ***Boswellia serrata*** c. *Bauhinia retusa*
b. Sterculia urens d. Acacia senegal

185. Dhaura gum is obtained from which species-
a. *Boswellia serrata* c. *Sterculia urens*
b. *Bauhinia retusa* d. ***Anogeissus latifolia***

Answers

175. b	176. a	177. c	178. a	179. d	180. b
181. b	182. c	183. a	184. a	185. d	

186. Semla gum is obtained from which species-
   a. *Anogeissus latifolia* c. ***Bauhinia retusa***
   b. Sterculia urens d. Acacia senegal

187. The gum and resins that exude from cracks or cuts and solidify with air exposure called-
   a. Ral c. Shellac
   b. **Dammar** d. Tannin

188. The true dammar is obtained from which species-
   a. *Canarium strictum* c. *Hopea odorata*
   b. *Vateria indica* d. ***Agathis loranthifolius***

189. Soft form of *Vateria indica* (vellapine) resin called as-
   a. Piney resin c. Indian copal
   b. Dhupa d. **All of the above**

190. Gum which is not soluble in water but absorb water and swell up into a mucilaginous mass called-
   a. **Gum-tetracanth** c. Dammar
   b. True gum d. Indian copal

191. Black dammar is obtained from which species-
   a. ***Canarium strictum*** c. *Hopea odorata*
   b. *Vateria indica* d. *Shorea robusta*

192. Rock dammar is obtained from which species-
   a. *Boswellia serrata* c. *Commiphora mukul*
   b. ***Hopea odorata*** d. *Vateria indica*

193. White dammar is obtained from which species-
   a. Shorea robusta c. Hopea odorata
   b. *Kingiodendron pinnatum* d. ***Vateria indica***

194. Ral dammar is obtained from which species-
   a. Hopea odorata c. Boswellia serrata
   b. *Terminalia arjuna* d. ***Shorea robusta***

195. Guggal/salai oleoresin is obtained from which species-
   a. *Dipterocarpus turbinatus* c. *Kingiodendron pinnatum*
   b. ***Boswellia serrata*** d. Sterculia urens

Answers

186. c	187. b	188. d	189. d	190. a	191. a
192. b	193. d	194. d	195. b		

196. Gurjan oil is obtained from which species-
   a. *Commiphora mukul*
   b. *Pterocarpus marsupium*
   c. ***Dipterocarpus turbinatus***
   d. *Grewia tiliaefolia*

197. Gamboge is obtained from which species-
   a. *Dipterocarpus turbinatus*
   b. ***Garcinia morella***
   c. *Kingiodendron pinnatum*
   d. *Garcinia uncia*

198. Leaves of *Butea monosperma* are used for making-
   a. Bidi
   b. Rope
   c. **Platters & Cups**
   d. Dyes

199. Leaves of *Diospyros melanoxylon* are used for making-
   a. **Bidi**
   b. Rope
   c. Platters & Cups
   d. Dyes

200. Saponin (substitute of soap) is obtained from-
   a. *Madhuca indica*
   b. *Sapindus emarginatus*
   c. *Madhuca latifolia*
   d. ***Sapindus mukorossi***

201. Bark drugs is/are obtained from which species-
   a. *Cinchona hybrida*
   b. *Soymida febrifuga*
   c. *Holarrhena antidysentrica*
   d. **All of the above**

202. Root drugs is/are obtained from which species-
   a. *Rauwolfia serpentine*
   b. *Acorous calamus*
   c. *Dioscorea deltoidea*
   d. **All of the above**

203. Root drugs is/are obtained from which species-
   a. *Saussuria lappa*
   b. *Glycyrhiza glabra*
   c. *Picrorhiza kurrooa*
   d. **All of the above**

204. Root drugs is/are obtained from which species-
   a. *Valerian wallichii*
   b. *Berberis aristata*
   c. *Aconitum species*
   d. **All of the above**

205. Laves drugs is/are obtained from which species-
   a. *Ephedra gerardiana*
   b. *Ocimum killimandscharicum*
   c. *Gaultheria fragrantissima*
   d. **All of the above**

206. Laves drugs is/are obtained from which species-
   a. *Cannabis sativa*
   b. *Datura species*
   c. *Atropa acuminata*
   d. **All of the above**

Answers

196. c	197. b	198. c	199. a	200. d	201. d
202. d	203. d	204. d	205. d	206. d	

207. Which is the example of minute structure of the wood-
a. Colour
b. Texture
c. Lustre
d. **Tracheids**

208. Which is the example of mechanical property of the wood-
a. Pith
b. **Fissibility**
c. Vessels or pores
d. Texture

209. Which is the example of physical property of the wood-
a. Flexibility
b. **Odour**
c. Fissibility
d. Elasticity

210. Which is the example of minute structure of the wood-
a. **Pith flecks**
b. Figure
c. Grain
d. Hardness

211. Which is the example of mechanical property of the wood-
a. Fibres
b. Weight
c. Tracheids
d. **Hardness**

212. Which is the example of physical property of the wood-
a. **Grain**
b. Strength
c. Tylosis
d. Fibres

213. Which is/are the example of minute structure of the wood-
a. Pith flecks
b. Ripple marks
c. Wood rays
d. **All of the above**

214. The capacity of wood to catch fire and continue to burn until it is consumed called-
a. Fissibility
b. Flexibility
c. **Combustibility**
d. Elasticity

215. Combustibility of wood signifies-
a. Elasticity
b. **Readiness**
c. Flexibility
d. Hardness

216. The quantity of heat emitted by a given weight of wood during combustion process called-
a. Combustibility
b. Calorific value
c. Heating power
d. **Both b and c**

Answers

207. d	208. b	209. b	210. a	211. d	212. a
213. d	214. c	215. b	216. d		

217. The number of grams of water to raise 1^0C temperature of 1 gram wood when completely burn is called-
a. **Calorific value** c. Specific heart
b. Combustibility d. Oxidation value

218. Calorific value (gram calorie or calories) of deodar wood and babul wood are respectively-
a. 4070 and 5094 c. **5294 and 4870**
b. 4950 and 4948 d. 5264 and 4939

219. Calorific value (gram calorie or calories) of casuarina wood and axlewood wood are respectively-
a. **4950 and 4948** c. 5245 and 4000
b. 4500 and 4548 d. 5064 and 4738

220. Calorific value (gram calorie or calories) of sal wood and imli wood are respectively-
a. 4950 and 4948 c. 5264 and 4939
b. **5264 and 4939** d. 5245 and 4000

221. Calorific value (gram calorie or calories) of oak wood is-
a. **3990** c. 4500
b. 5600 d. 3500

222. Which wood is good for fuel purpose-
a. Have less resinous substances
b. **Have more resinous substances**
c. Have more gummy substances
d. Have less gummy substances

223. Which wood is not a suitable fuel wood for cooking due to its smokiness-
a. Bamboo wood c. Popular wood
b. **Teak wood** d. Sissoo wood

224. Any of the various imperfections that can be observed in lumber and wood products called-
a. Wood Grains c. Wood lustre
b. **Wood defects** d. Wood quality

Answers

217. a	218. c	219. a	220. b	221. a	222. b
223. b	224. b				

225. Wood defects arise due to which reasons-
a. Abnormal growth c. Rupture of tissue
b. Normal growth d. **Both a and c**

226. Which defect is/are the example of abnormal growth-
a. Waviness c. Twisted fibre
b. Knots d. **All the above**

227. Which defect is/are the example of rupture of tissue-
a. Checks c. Split
b. Shakes d. **All of the above**

228. Which defect is/are the example of rupture of tissue-
a. Burr c. Interior bark growth
b. **Checks** d. Constriction due to climber

229. A part of a branch which is embedded in wood called-
a. Node c. **Knot**
b. Tylosis d. Buttress

230. A portion of branch which is living at the time of its inclusion and establishes a fibrous connection with the surrounding wood called-
a. Septum c. Clamp connection
b. **Live knots** d. Dead knots

231. Which knots do not separate when wood dries-
a. **Live knots** c. Dead knots
b. Both knots d. Fixed knots

232. Dead knots do not have a.........with the main stem and reduce the strength of wood-
a. Clamp connection c. **Fibrous connection**
b. Connectives tissues d. Septa connection

233. The knots which are detached during seasoning of wood called-
a. Clamp knots c. Live knots
b. **Dead knots** d. Intermediate Knots

234. Defect caused by fungal attack and decay in wood called as-
a. Decomposition c. Fungal knots
b. **Unsoundness** d. Cracking

Answers

225. d	226. d	227. d	228. b	229. c	230. b
231. a	232. c	233. b	234. b		

235. The most serious fungal defects of wood are-
a. Rots
c. Stains
b. Mildew
d. **Both a and c**

236. When the wood grain twisted spirally making an angle 40^0 with vertical axis called-
a. **Twisted fibre**
c. Twisted shake
b. Ripple marks
d. Flecks

237. When wood fibres shows a wavy appearance instead of true vertical straight line with main axis called-
a. Burr
c. Pith flecks
b. Checks
d. **Waviness**

238. A complex knot formed at the point where dormant bud shows abnormal growth without developed into branches resulting formation of concentrated mass around bud called-
a. **Burr**
c. Pith flecks
b. Checks
d. Split

239. Separation of fibres forming a cracks or fissure in the piece of wood not extending to other face or end of wood piece called-
a. Split
c. **Checks**
b. Burr
d. Shakes

240. Separation of fibres forming a cracks or fissure in the piece of wood and extending to other face or end of wood piece called-
a. Waviness
c. **Split**
b. Shakes
d. Checks

241. A crack starting from the pith region and extending radially outwards in the wood called-
a. **Heart or star shakes**
c. Radial shakes
b. Rings-shakes
d. All of the above

242. Cracks which start from the periphery and extend radially inward towards the centre of the wood called-
a. Rings-shakes
c. **Radial shakes**
b. Heart shakes
d. Star shakes

Answers

235. d	236. a	237. d	238. a	239. c	240. c
241. a	242. c				

243. When the cracks follows the direction of annual rings due to shrinkage of central tissues and looks like as cup called-

a. **Cup and rings-shakes** c. Cup and **radial shakes**
b. Cup and star shakes d. Cup and rings-shakes

244. Two main defects in the wood due to fungal attack are-

a. Decay c. Stain
b. Waviness d. **Both a and c**

245. In which defects fungi attacks only sapwood and responsible for sapwood discolouration-

a. Decay c. **Stain**
b. Decomposition d. Unsoundness

246. In which defects fungi attacks on sapwood as well as heartwood and responsible for wood decomposition-

a. Stain c. **Decay**
b. Rot d. Mildew

247. Main food of termites is-

a. Carbohydrate c. **Cellulose**
b. Chitin d. Protein

248. Marine borer which attack on wood surface is/are-

a. *Limnoria* c. *Chelura*
b. *Sphaeroma* d. **All the above**

249. Which genera of insects also known as shipworms-

a. *Tredo* c. *Bankia*
b. *Martesia* d. **All of the above**

250. Unlignifierd area in primary cell wall which is responsible for intercommunication between two adjacent cells known as-

a. Septum c. **Pits**
b. Plasmodesmata d. Tylosis

251. Features of wood which can be seen with naked eyes or with an ordinary lens called-

a. Gross features c. Minute features
b. Macro features d. **Both a and b**

Answers

243. a	244. d	245. c	246. c	247. c	248. d
249. d	250. c	251. d			

252. Features of wood which can be seen with compound microscope or more clear lens called-
a. **Minute features** c. Gross features
b. Macro features d. None

253. Central portion of the wood called-
a. Xylem c. Phloem
b. Cambium d. **Pith**

254. The ability of cell wall to reflect the light-
a. Reflectivity c. Emissivity
b. Infiltration d. **Lustre**

255. The ratio of the weight of material to the weight of an equal volume of water at 4°C-
a. **Specific gravity** c. Specific density
b. Specific heat d. Specific weight

256. Specific gravity (gm/cm^3) of water is-
a. 0.5 gm/cm^3 c. **1.0 gm/cm^3**
b. 1.5 gm/cm^3 d. 2.0 gm/cm^3

257. Moisture content in green timber vary from.......while in air dry timbers have...........moisture content respectively-
a. 10-100% and 2-11% c. 50-100% and 6-25%
b. 70-215% and 4-15% d. **50-200% and 8-15%**

258. Density of the very light timber is-
a. 100 kg/m^3 c. 200 kg/m^3
b. **300 kg/m^3** d. 400 kg/m^3

259. Density of the light timber is-
a. 101-250 kg/m^3 c. 201-350 kg/m^3
b. **301-450 kg/m^3** d. 401-650 kg/m^3

260. Density of the moderately heavy timber is-
a. 201-350 kg/m^3 c. **451-600 kg/m^3**
b. 301-450 kg/m^3 d. 401-650 kg/m^3

261. Density of the heavy timber is-
a. 401-850 kg/m^3 c. 101-350 kg/m^3
b. 501-750 kg/m^3 d. **601-800 kg/m^3**

Answers

252. a	253. d	254. d	255. a	256. c	257. d
258. b	259. b	260. c	261. d		

262. Density of the very heavy timber is-

a. **801-950 kg/m³** c. 501-750 kg/m³
b. 401-850 kg/m³ c. 501-950 kg/m³

263. Density of the extremely heavy timber is-

a. > 600 kg/m³ c. > 700 kg/m³
b. > 800 kg/m³ d. **> 900 kg/m³**

264. The alignment of cells or direction of fibres regards to vertical axis of the stem is called-

a. Lustre c. **Grain**
b. Pith d. Shakes

265. The arrangement of cells according their size in unit volume of wood is called-

a. Grain c. Lustre
b. **Texture** d. Figure

266. The distinctive pattern produce on longitudinal surface of timber due to alignment of tissues and direction of the grain is called-

a. Fissure c. Texture
b. Shake d. **Figure**

267. Which is/are the example of extremely hard wood species-

a. *Schleichera oleosa* c. *Mesua ferrea*
b. *Shorea robusta* d. **All of the above**

268. Which is/are the example of very elastic wood species-

a. *Grewia tiliaefolia* c. *Anogeissus latifolia*
b. *Mesua ferrea* d. **All of the above**

269. Which is/are the example of extremely soft wood species-

a. *Bombax ceiba* c. *Cryptomeria japonica*
b. **Both a and c** d. *Syzygium cuminii*

270. Which is exceptionally hard wood species-

a. *Tectona grandis* c. *Bombax ceiba*
b. *Picea smithiana* d. ***Mesua ferrea***

271. The capacity of wood to bend out of shape freely without rupture is called-

a. Elasticity c. **Flexibility**
b. Fissibility d. Fissility

Answers

262. a	263. d	264. c	265. b	266. d	267. d
268. d	269. b	270. d	271. c		

272. The capacity of wood to regain its original shape after the stress has produced deformation called-
a. Fissibility c. Fissility
b. **Elasticity** d. Flexibility

273. The capacity of wood to split along the fibre during conversion of wood with axe or a wedge is called-
a. Fissility c. Fissibility
b. Flexibility d. **both a and c**

274. The process of removal of excess moisture present in timber in its green state is called-
a. Conversion c. **Wood seasoning**
b. Wood preservation d. Wood kilning

275. Which methods is/are used in horizontal stacking during wood seasoning-
a. One in nine method c. Close rib method
b. Open rib method d. **All of the above**

276. During wood seasoning by horizontal stacking method woods are kept at the height from the ground-
a. **30-45 cm** c. 10-25 cm
b. 20-35 cm d. 50-100 cm

277. Woods which require protection against rapid seasoning otherwise it will cause defects-
a. **Highly refractory wood** c. Moderately refractory wood
b. Non refractory wood d. Refractory wood

278. Woods which require little protection against during rapid seasoning called-
a. Highly refractory wood c. **Moderately refractory wood**
b. Non refractory wood d. Refractory wood

279. Woods which are seasoned rapidly without getting any defects during seasoning called-
a. **Non refractory wood** d. Refractory wood
b. Highly refractory wood c. Moderately refractory wood

Answers

272. b	273. d	274. c	275. d	276. a	277. a
278. c	279. a				

280. Which is/are the example of refractory wood species-
   a. *Shorea robusta*   c. *Terminalia alata*
   b. *Hopea odorata*   d. **All of the above**

281. Which is/are the example of moderately refractory wood species-
   a. *Tectona grandis*   c. *Dalbergia sissoo*
   b. *Anogeissus latifolia*   d. **All of the above**

282. Which is/are the example of non-refractory wood species-
   a. *Boswellia serrata*   c. *Populus deltoides*
   b. *Trewia nudifolia*   d. **All of the above**

283. Required drying period for drying the Highly refractory wood-
   a. 6 month for 2.5 cm thickness   c. 12 month for 5.0 cm thickness
   b. 4 month 2.5 cm thickness   d. **Both a and c**

284. Required drying period for drying the moderately refractory wood-
   a. 4 month for 2.5 cm thickness   c. 9 month for 5.0 cm thickness
   b. 12 month 2.5 cm thickness   d. **Both a and c**

285. Required drying period for drying the non-refractory wood-
   a. 2 month for 2.5 cm thickness   c. 4 month for 5.0 cm thickness
   b. 4 month 2.5 cm thickness   d. **Both a and c**

286. Required drying period for drying the railway sleepers wood is-
   a. 6-12 month   c. 10-15 month
   b. **12-24 month**   d. 15-30 month

287. The process of improvement wood's natural durability by treatment with chemicals that are toxic to insects, fungi and other decaying agents called-
   a. **Wood preservation**   c. Wood seasoning
   b. Wood durability   d. Wood protection

288. Wood preservatives are classified into how many categories-
   a. Oil types   c. Water soluble types
   b. Organic solvent types   d. **All of the above**

289. The chemicals that are applied to woods to make it resistant to attack by the different decay agents called-
   a. Wood coating   c. Wood finisher
   b. **Wood preservatives**   d. Wood shiner

Answers

280. d	281. d	282. d	283. d	284. d	285. d
286. b	287. a	288. d	289. b		

290. Which is the example of oil types wood preservatives-

a. Borax
b. $ZnSO_4$
c. Copper napthalate
d. **Creosote**

291. Which is/are the example of organic solvent type wood preservatives-

a. PCP
b. Copper quinolate
c. Lindane
d. **All of the above**

292. Which is/are the example of water soluble type wood preservatives-

a. CCA
b. ACC
c. ASCU
d. **All of the above**

293. Which is/are the example of non-fixed water soluble type wood preservatives-

a. Creosote
b. **CCF**
c. Coal tar
d. PCP

294. Which is/are the example of fixed water soluble type wood preservatives-

a. Zn Meta Arsenate (ZMA)
b. Acid Cu-Chromate (ACC)
c. Ammonical Cu-Chromate (ACA)
d. **All of the above**

295. Which types of wood preservatives is not suitable for indoor works-

a. **Oil types**
b. Organic solvent types
c. Water soluble types
d. All of the above

296. Which types of wood preservatives is not suitable for outdoor works-

a. Organic solvent types
b. Oil types
c. **Water soluble types**
d. None

297. Which types of wood preservatives are most effectives against borer, marine organisms and termites-

a. Organic solvent types
b. **Oil types**
c. Water soluble types
d. All of the above

298. Which types of wood preservatives are most effectives against fungi and other microbes-

a. **Organic solvent types**
b. Oil types
c. Water soluble types
d. None

Answers

290. d	291. d	292. d	293. b	294. d	295. a
296. c	297. b	298. a			

299. The water which is present in the cavity of the wood cell called-
a. Wood water
c. Bound water
b. **Free water**
d. Capillary water

300. The water which is held within the cells of the wood by hydrogen bonding called-
a. **Bound water**
c. Wood water
b. Free water
d. Hydrogenated water

301. There are how many methods of wood seasoning-
a. **Two**
c. Three
b. Four
d. Five

302. Which is/are not a method of wood seasoning-
a. Air seasoning
c. Kiln seasoning
b. Heat seasoning
d. **Both a and c**

303. The drying of wood with the help of air and sunlight is called-
a. Heat seasoning
c. Sunlight seasoning
b. **Air seasoning**
d. Kiln seasoning

304. The process of air tight, thermal proof, enclosed chamber with provision for controlling temperature, air circulation and humidity of seasoning is called-
a. Thermal seasoning
c. Controlled seasoning
b. Air seasoning
d. **Kiln seasoning**

305. Which is/are not the constituents of water soluble preservatives-
a. As_2O_5
c. $CuSO_4$
b. Na/K Cr_2O_7
d. **All of the above**

306. Which is/are not the constituents of ASCU wood preservatives-
a. As_2O_5
c. $CuSO_4$
b. Na/K Cr_2O_7
d. **All of the above**

307. Which is/are not the constituents of ASCU-boric wood preservatives-
a. Na/K Cr_2O_7
c. $CuSO_4$
b. Boric acid
d. **As_2O_5**

308. Celcure is a type of wood preservative-
a. **Water soluble**
c. Organic solvent types
b. Oil types
d. Inorganic solvent types

Answers

299. b	300. a	301. a	302. d	303. b	304. d
305. d	306. d	307. d	308. a		

309. Vertical stacking is used only for the rapid surface drying of certain species of-
a. Highly refractory wood
b. **Non refractory wood**
c. Moderately refractory wood
d. Refractory wood

310. Wood preservation on scientific and modern basis was introduced in India by-
a. **Sir Ralph Pearson**
b. Bilal A. Lone
c. Sir George Joseph
d. Sir Albert Einstein

311. Wood preservation on scientific and modern basis was introduced in India in the year-
a. 1920
b. 1910
c. 1906
d. **1908**

312. Osmose process and hot & cold process are applied in-
a. **Soaking method**
b. Diffusion method
c. Steeping method
d. Heating and cooling method

313. Which is/are a non-pressure method of wood preservation-
a. Sap displacement method
b. Steeping
c. Soaking
d. **All of the above**

314. Which is/are a pressure method of wood preservation-
a. Full cell process
b. Lowry process
c. Empty cell process
d. **All of the above**

315. Which is/are a pressure method of wood preservation-
a. Heating and cooling method
b. **Rueping process**
c. Diffusion
d. Sap displacement method

316. Which method is only suitable for preservation of green timber-
a. Steeping
b. **Diffusion**
c. Soaking
d. Heating and cooling

317. Which method is suitable for preservation of green round timber & bamboo-
a. **Sap displacement method**
b. Soaking method
c. Steeping method
d. Heating and cooling method

Answers

309. b	310. a	311. d	312. a	313. d	314. d
315. b	316. b	317. a			

318. Which method is suitable for preservation of dry as well as green woods-

a. **Soaking method**
b. Sap displacement method
c. Heating and cooling method
d. Steeping method

319. Full cell process of wood preservation is applied at a pressure-

a. 1.50 to 10 kg/cm^2
b. **3.50 to 12 kg/cm^2**
c. 2.50 to 15 kg/cm^2
d. 5.0 to 10 kg/cm^2

320. In which pressure method of wood preservation maximum penetration is obtained with minimum absorption of preservatives-

a. Rueping process
b. Lowry process
c. **Empty cell process**
d. Full cell process

321. In which pressure method of wood preservation no need to create initial vacuum for applying preservatives-

a. **Lowry process**
b. Rueping process
c. Full cell process
d. Empty cell process

322. Rueping process of wood preservation is applied at a pressure-

a. 1.50 to 10 kg/cm^2
b. 3.50 to 12 kg/cm^2
c. 3.50 to 15 kg/cm^2
d. **5.0 to 12 kg/cm^2**

323. Which is/are the method of transportation of woods-

a. Water transportation
b. Overhead transportation
c. Land transportation
d. **All of the above**

324. Water transportation is extensively used in the-

a. Coniferous forest of S-W Himalayan
b. Coniferous forest of S-E Himalayan
c. Coniferous forest of N-E Himalayan
d. **Coniferous forest of N-W Himalayan**

325. Water transportation of woods include-

a. Floating
b. Wetslides
c. Rafting & boom
d. **All of the above**

326. Which is/are not the floating method of wood transportation-

a. Telescopic floating
b. Free floating
c. River floating
d. **All of the above**

Answers

318. a	319. b	320. c	321. a	322. d	323. d
324. d	325. d	326. d			

327. River floating method of wood transportation also called as-

a. Sweeping | c. Free floating
b. **Both** | d. None

328. In which method of transportation of wood *pathru* is used-

a. **Telescopic floating** | c. Free floating
b. River floating | d. Sweeping

329. Which methods is only used in the hilly forest of N-W Himalayan for wood transportation-

a. Wetslides | c. Rafting & boom
b. **Telescopic floating** | d. All of the above

330. The average distance covered daily in the floating operations is-

a. 1-2 kilometres | c. **2-3 kilometres**
b. 3-4 kilometres | d. 4-5 kilometres

331. Floating method of wood transportation is used to transport the woods a distance upto-

a. 10-20 kilometres | c. 50-100 kilometres
b. **100-200 kilometres** | d. 200-250 kilometres

332. A quantity of woods firmly bound together is termed as-

a. Culm | c. Clump
b. **Raft section** | d. Boom

333. The process of woods firmly bound together is termed as-

a. Floating | c. Clumping
b. **Rafting** | d. Preserving

334. The average distance covered in the rafting operations per day is-

a. 10-20 kilometres | c. 20-50 kilometres
b. **40-60 kilometres** | d. 100-200 kilometres

335. Rafting method of wood transportation is mostly used n –

a. Summer season | c. Spring season
b. Rainy season | d. **Winter season**

336. The rafts are classified into how many categories-

a. Two | c. **Three**
b. Four | d. Five

Answers

327. b	328. a	329. b	330. c	331. b	332. b
333. b	334. b	335. d	336. c		

337. Which is/are not a raft classification-
a. Log rafts c. Sawn timber rafts
b. Bamboo rafts d. **All of the above**

338. Logs rafts contains.......sections and each section contains....... logs respectively-
a. **6-10 sections and 16-20 logs** c. 10-20 sections and 16-20 logs
b. 5-15 sections and 16-20 logs d. 6-10 sections and 10-20 logs

339. The average number of logs in a log raft is-
a. 50-200 logs c. 100-200 logs
b. **96-200 logs** d. 150-200 logs

340. The average number of bamboos in a bamboo raft is-
a. 2000 c. 5000
b. 1000 d. **10000**

341. How many culms are present in a bamboo raft during transportation-
a. 10-100 culms c. 20-100 culms
b. 40-100 culms d. **50-100 culms**

342. Boom is a Dutch word which means-
a. **Floating obstruction** c. Free floating
b. Pathru floating d. Floating Channel

343. How many types of booms-
a. **Two** c. Three
b. Four d. Five

344. Single-arm boom also known as-
a. **One-way boom** c. Two-way boom
b. V-shape boom d. T- shaped boom

345. Two-way boom also known as-
a. **V-shape boom** c. I-shape boom
b. T-shape boom d. X-shape boom

346. Single-arm or one-way boom is attached permanently at its-
a. **Upper ends** c. Lower ends
b. Middle d. Corners

347. Single-arm or one-way boom guy ropes always attached at an angle to the boom-
a. 30° c. 45°
b. 60° d. **90°**

Answers

337. d	338. a	339. b	340. d	341. d	342. a
343. a	344. a	345. a	346. a	347. d	

348. The arms of two-way or V-shape boom incline at an angle to the apex on the downstream side-
a. 30° c. **45°**
b. 60° d. 90°

349. Boom proper and log line or *line dori* are used in which boom-
a. One-way boom c. Single-arm boom
b. T-shape boom d. **Both a and c**

350. The general term for built up bonded product consisting either wholly of natural woods or of wood in combination with metals, plastics etc. is-
a. **Composite wood** c. Plywood
b. Compressed wood d. Compregnated wood

351. Glued wood made up of veneers in such a manner that grain of each veneer is at right angle to that adjacent veneer in the assembly is called-
a. Laminated wood c. **Plywood**
b. Compressed wood d. Improved wood

352. The outer face of plywood is called-
a. **Face** c. Back
b. Core d. Cross bands

353. The centre face of plywood is called-
a. Cross bands c. Back
b. **Core** d. Face

354. The numbers of veneers in a plywood is always-
a. Even numbers c. **Odd numbers**
b. Both d. None

355. In a plywood panels having more than three plies, the layers between the core and the face or back are called-
a. Gullet c. Kerf
b. **Cross bands** d. Parallel bands

356. The simplest structure of plywood has ply number-
a. **3** c. 5
b. 7 d. 9

Answers

348. c	349. d	350. a	351. c	352. a	353. b
354. c	355. b	356. a			

357. The maximum length of logs used in plywood is about-

a. 155-200 cm
c. 185-250 cm
b. 200-280 cm
d. **255-280 cm**

358. The pressing of glued coated veneers in plywood is done at a pressure-

a. 5-10 kg/cm²
c. **7-18 kg/cm²**
b. 10-15 kg/cm²
d. 15-25 kg/cm²

359. Moisture content in plywood is bring upto-

a. 10%
c. **12%**
b. 15%
d. 25%

360. The grains direction in laminated wood is-

a. **Parallel to adjacent veneer**
b. Antiparallel to adjacent veneer
c. Perpendicular to adjacent veneer
d. Diagonally to adjacent veneer

361. The grains direction in core boards is-

a. Parallel to adjacent veneer
c. Diagonally to adjacent veneer
b. Antiparallel to adjacent veneer
d. **Perpendicular to adjacent veneer**

362. The glue laminated wood is called as-

a. Gallium
c. **Glulam**
b. Lamina
d. Lamin

363. A core board has width 7.5 cm called as-

a. **Batten board**
c. Block-board
b. Lamin board
d. Hardboard

364. A core board has width not more than 2.5 cm called as-

a. Lamin board
c. Hardboard
b. Batten board
d. **Block-board**

365. A core board has width less than 7.0 mm called as-

a. Block-board
c. **Lamin board**
b. Hardboard
d. Batten board

366. The thin faces of sandwich board called as-

a. Face
c. Back
b. Core
d. **Skins**

Answers

357. d	358. c	359. c	360. a	361. d	362. c
363. a	364. d	365. c	366. d		

367. Fibre boards are classified into how many categories-
  a. Three c. **Four**
  b. Five d. Six

368. Which is not a types of fibre board-
  a. Hard board c. Soft hardboard
  b. Medium hardboard d. **Batten board**

369. Which is/are not the examples of particles boards-
  a. Chip-boards c. Flake boards
  b. Shaving boards d. **Lamin board**

370. Improved wood also known as-
  a. **Modified wood** c. Composite wood
  b. Laminated wood d. Plywood

371. Improved wood may be classified into how many categories-
  a. Three c. Four
  b. Five d. **Six**

372. Which is/are not a type of improved wood-
  a. Impregnated wood c. Heat stabilized wood
  b. Compressed wood d. **Laminated wood**

373. Which is/are not a type of improved wood-
  a. **Composite wood**
  b. Heat stabilized compressed wood
  c. Compregnated wood
  d. Chemically modified wood

374. Veneers treated with a thermosetting fibre penetrating resin called as-
  a. **Impreg** c. Compreg
  b. Lamin d. All of the above

375. Compressed wood can be made either from solid wood or laminate wood by applying pressure at-
  a. 10-20 kg/cm^2 c. 50-100 kg/cm^2
  b. **90-140 kg/cm^2** d. 100-150 kg/cm^2

376. Compregnated wood is highly resistance to-
  a. **Solvents** c. Termites
  b. Insects d. All of the above

Answers

367. c	368. d	369. d	370. a	371. d	372. d
373. a	374. a	375. b	376. a		

377. Compregnated wood has how many times more hardness as compare to normal wood-
a. 5-10 times
b. **10-20 times**
c. 10-15 times
d. 20-25 times

378. Compregnated wood is an excellent materials for-
a. Railway sleepers
b. Particle board
c. **Aeroplane propeller blades**
d. Core boards

Answers

377. b 378. c

CHAPTER 6

Plant and Wood Anatomy

1. The branch of botany which deals with the study of internal structures and organization of plants or plant organs called-
   a. **Plant anatomy** b. Plant physiology
   c. Plant taxonomy d. Histology
2. Who is the father of plant anatomy-
   a. Dharmendra Kumar b. **N. Grew**
   c. J. C. Bose d. Habellandt
3. Father of Indian plant anatomy is-
   a. A. K. Chaudhary b. Mohit Husain
   c. **K. A. Chaudhary** d. Naseer A. Mir
4. A group of cells which is similar or dissimilar in shape, having a common origin and usually perform a common function is called-
   a. Organs b. Cells
   c. **Tissue** d. Systems
5. Term tissue was coined by-
   a. Karl Nageli b. **N. Grew**
   c. K. C. Mehta d. Amerjeet Singh
6. The meristems which occurs at the tips of roots and shoots and produce primary tissues are called-
   a. **Apical meristems** b. Lateral meristem
   c. Intercalary meristem d. None
7. Which meristematic tissues are responsible for increase in the length of plants organs-
   a. Intercalary meristem b. Lateral meristem
   c. **Apical meristems** d. None

Answers

1. a 2. b 3. c 4. c 5. b 6. a
7. c

8. The meristems which are occurs between mature tissue is called-
   a. Apical meristems
   b. Lateral meristem
   c. **Intercalary meristem**
   d. Cambium

9. Intra fascicular or fascicular cambium occurs inside the vascular bundles-
   a. Apical meristem
   b. Cortex
   c. **Vascular bundles**
   d. All of the above

10. Apical cell theory was given by-
    a. Hanstein
    b. Schmidt
    c. Newman
    d. **Karl Nageli & Hofmeister**

11. The cells of permanent tissues are composed of-
    a. Live cells
    b. Dead cells
    c. Immature cells
    d. **Both a and b**

12. Histogen theory was given by-
    a. **Hanstein**
    b. Schmidt
    c. Newman
    d. Popham & Chan

13. Phellem, phellogen and phelloderm are collectively called as-aperiderm-
    a. Cork cambium
    c. Cortex
    c. **Periderm**
    d. Bark

14. Tunica corpus theory was given by-
    a. Schuepp
    b. **Schmidt**
    c. Newman
    d. Popham & Chan

15. Simple tissues are classified into-
    a. **Three**
    b. Four
    c. Five
    d. Six

16. Which is/are an example of complex tissue-
    a. Parenchyma
    b. Collenchyma
    c. Sclerenchyma
    d. **Phloem**

17. Mantle core theory was given by-
    a. **Popham & Chan**
    b. Schmidt
    c. Newman
    d. Karl Nageli & Hofmeister

Answers

8. c	9. c	10. d	11. d	12. a	13. c
14. b	15. a	16. d	17. a		

18. Which is/are an example of simple tissue-
 a. Parenchyma b. Collenchyma
 c. Sclerenchyma d. **All of the above**

19. Korper-Kappe theory was given by-
 a. **Schuepp** b. Schmidt
 c. Newman d. Popham & Chan

20. Which tissue provides buoyancy to hydrophytes-
 a. Cholerenchyma b. **Aerenchyma**
 c. Prosenchyma d. Mucilage parenchyma

21. Which is universal tissue in plants-
 a. Cholerenchyma b. Aerenchyma
 c. Prosenchyma d. **Parenchyma**

22. The main function of storage of food is-
 a. Sclerenchyma b. Collenchyma
 c. **Parenchyma** d. Phloem

23. Which tissue provide mechanical strength to the plant-
 a. Sclerenchyma b. Collenchyma
 c. Parenchyma d. **Both a and b**

24. Which tissue provide main mechanical strength to the plant-
 a. **Sclerenchyma** b. Collenchyma
 c. Parenchyma d. Xylem

25. Term parenchyma was coined by-
 a. Mettenius b. Schleiden
 c. Robert Hook d. **N. Grew**

26. Term collenchyma was coined by-
 a. Linnaeus b. **Schleiden**
 c. Robert Hook d. Aristotle

27. Term sclerenchyma was coined by-
 a. **Mettenius** b. Schleiden
 c. Hugo de Varies d. N. Grew

28. Xylary fibres or wood fibres are hard fibres and obtained from-
 a. Phloem b. **Xylem**
 c. Cortex d. Pericycle

Answers

18. d	19. a	20. b	21. d	22. c	23. d
24. a	25. d	26. b	27. a	28. b	

29. Bast fibres or extra xylary fibres are also known as-
   a. Phloem fibre b. Commercial fibre
   c. Pericycle fibre d. **All of the above**
30. Cotton fibres are composed of-
   a. Protein c. Lipid
   c. **Cellulose** d. Lignin
31. Xylem and phloem collectively called as-
   a. Connective tissue b. Ground tissue
   c. **Vascular tissue** d. Mechanical tissue
32. Which is/are the example of complex tissues-
   a. Parenchyma b. Collenchyma
   c. **Phloem** d. Sclerenchyma
33. Term xylem was coined by-
   a. Haberlandt b. Schleiden
   c. Hugo de Varies d. **Nageli**
34. Term xylem was coined by-
   a. Robert Hook b. Schleiden
   c. **Nageli** d. J. C. Bose
35. Which is/are not an element of xylem-
   a. Tracheids b. Vessels
   c. Xylem parenchyma d. **Companion cells**
36. Which is/are not an element of phloem-
   a. Sieve cells b. Phloem fibres
   c. **Vessels** d. Sieve tubes
37. Which is the only living element of xylem-
   a. **Xylem parenchyma** b. Vessels
   c. Tracheids d. Xylem fibres
38. Which is the only dead element of phloem-
   a. Sieve cells b. **Phloem fibres**
   c. Companion cells d. Phloem parenchyma
39. The main function of xylem is-
   a. Conduction of food materials from roots to shoots
   b. Conduction of food materials from shoots to roots
   c. **Conduction of water & minerals salts from roots to shoots**
   d. Conduction of water & minerals salts from shoots to roots

Answers

29. d	30. c	31. c	32. c	33. d	34. c
35. d	36. c	37. a	38. b	39. c	

40. The main function of phloem is-
    a. Conduction of food materials from roots to shoots
    b. **Conduction of food materials from shoots to roots**
    c. Conduction of water & minerals salts from roots to shoots
    d. Conduction of water & minerals salts from shoots to roots

41. Which are the primitive conducting elements of xylem in plants-
    a. **Tracheids** b. Vessels
    c. Sieve tube d. Companion cells

42. The maximum bordered piths are found in the tracheids of-
    a. Angiosperms b. Bryophytes
    c. **Gymnosperms** d. Pteridophytes

43. Which is work as a pipe line during conduction of water in plants-
    a. **Vessels** b. Sieve tube
    c. Tracheids d. Xylem fibres

44. Tracheids and vessels are collectively known as-
    a. **Hadrom** c. Laptom
    c. Trichome d. Holard

45. Term hadrom was given by-
    a. Robert Hook b. Schleiden
    c. Nageli d. **Haberlandt**

46. Term totipotency was given by-
    a. Nageli b. **Haberlandt**
    c. Schleiden d. J. C. Bose

47. Formation of whole individual from a single cell called-
    a. Regeneration b. Vegetation
    c. **Totipotency** d. All of the above

48. Sieve elements was discovered by-
    a. Haberlandt b. Hartwig
    c. Nageli d. **Hartig**

49. Companion cells are found only in which plants-
    a. Bryophytes b. Pteridophytes
    c. Gymnosperms d. **Angiosperms**

Answers

40. b	41. a	42. c	43. a	44. a	45. d
46. b	47. c	48. d	49. d		

50. The conducting elements of phloem is termed as-
    a. Hadrom b. **Laptom**
    c. Trichome d. Holard
51. Term laptom was given by-
    a. Nageli b. **Haberlandt**
    c. Schleiden d. J. C. Bose
52. When the xylem and phloem are present separately on different radii in alternate manner called-
    a. Concentric vascular bundles b. Conjoint vascular bundles
    c. **Radial vascular bundles** d. Bicollateral vascular bundles
53. When the xylem and phloem are present on same radius called-
    a. Concentric vascular bundles b. **Conjoint vascular bundles**
    c. Radial vascular bundles d. Bicollateral vascular bundles
54. When the xylem and phloem are present separately on different radii in alternate manner called-
    a. **Exarch** b. Endarch
    c. Mesarch d. None
55. When the xylem and phloem are present on same radius called-
    a. Exarch b. **Endarch**
    c. Mesarch d. None
56. Exarch vascular bundles are found in-
    a. Stem b. **Roots**
    c. Both d. None
57. Endarch vascular bundles are found in-
    a. **Stem** b. Roots
    c. Both d. None
58. When cambium is absent between the xylem and phloem it is called as-
    a. Exarch vascular bundles b. Endarch vascular bundles
    c. Open vascular bundles d. **Closed vascular bundles**
59. When cambium is present between the xylem and phloem it is termed as-
    a. Endarch vascular bundles b. Exarch vascular bundles
    c. **Open vascular bundles** d. Closed vascular bundles

Answers

50. b	51. b	52. c	53. b	54. a	55. b
56. b	57. a	58. d	59. c		

60. Closed vascular bundles are found in-
    a. **Monocotyledon** b. Dicotyledons
    c. Bryophytes d. Pteridophytes

61. Open vascular bundles are found in-
    a. Monocotyledon b. Dicotyledons
    c. Gymnosperm d. **Both b and c**

62. Conjoint Bicollateral and open vascular bundles are found in stems of-
    a. Malvaceae c. Compositae
    c. **Cucurbitaceae** d. Lamiaceae

63. The whole central mass of vascular tissue with or without pith surrounded by endodermis is called as-
    a. Periderm b. Cortex
    c. Pericycle d. **Stele**

64. Stele concept was given by-
    a. Van Tieghem and Duoliot b. Nageli and Haberlandt
    c. Robert hook and Michel Clerk d. Von Erich and Duoliot

65. Which is the most primitive types of stele in plants-
    a. Plecto stele b. **Protostele**
    c. Eustele d. Dictyostele

66. Which is the most advance types of stele in plants-
    a. Siphonostele b. **Atactostele**
    c. Eustele d. Dictyostele

67. In dicot stem cortex is divided into-
    a. Hypodermis b. General cortex
    c. Endodermis d. **All of the above**

68. Pith is well developed in-
    a. Monocot roots and dicot stems b. Monocot stems and dicot roots
    c. Dicot roots and dicot stems d. Dicot roots and monocot roots

69. Casparian strips are found in the endodermis of-
    a. Monocot roots b. **Dicot roots**
    c. Monocot stems d. Dicot stems

Answers

60. a	61. d	62. c	63. d	64. a	65. b
66. b	67. d	68. a	69. b		

70. Pith is less developed or absent in-
    a. Monocot roots and dicot stems
    b. Dicot roots and dicot stems
    c. **Monocot stems and dicot roots**
    d. Dicot roots and monocot roots
71. The central region of secondary xylem which is dark in colour called-
    a. **Heartwood**
    b. Sapwood
    c. Alburnum
    d. Pith
72. The peripheral region of secondary xylem which is light in colour called-
    a. Heartwood
    b. Duramen
    c. **Alburnum**
    d. Pith
73. Alburnum and duramen respectively also known as-
    a. Heartwood and Sapwood
    b. **Sapwood and heartwood**
    c. Soft wood and hardwood
    d. Hardwood and softwood
74. The function of heart wood is-
    a. Provide mechanical strength to the stem
    b. Conduction of water and minerals
    c. Both a and b
    d. Provide path to ray parenchyma
75. The function of sap wood is-
    a. Provide mechanical strength to the stem
    b. **Conduction of water and minerals**
    c. Both a and b
    d. Provide insect and pest resistance
76. Manoxylic or soft woods found in-
    a. Bryophytes
    b. Pteridophytes
    c. **Gymnosperms**
    d. Angiosperms
77. A wood which has more amount of living parenchyma is called-
    a. Pycnoxylic wood
    b. **Manoxylic wood**
    c. Apotracheal wood
    d. Paratracheal wood
78. A wood which has less amount of living parenchyma is called-
    a. **Pycnoxylic wood**
    b. Manoxylic wood
    c. Apotracheal wood
    d. Paratracheal wood

Answers

70. c	71. a	72. c	73. b	74. a	75. b
76. c	77. b	78. a			

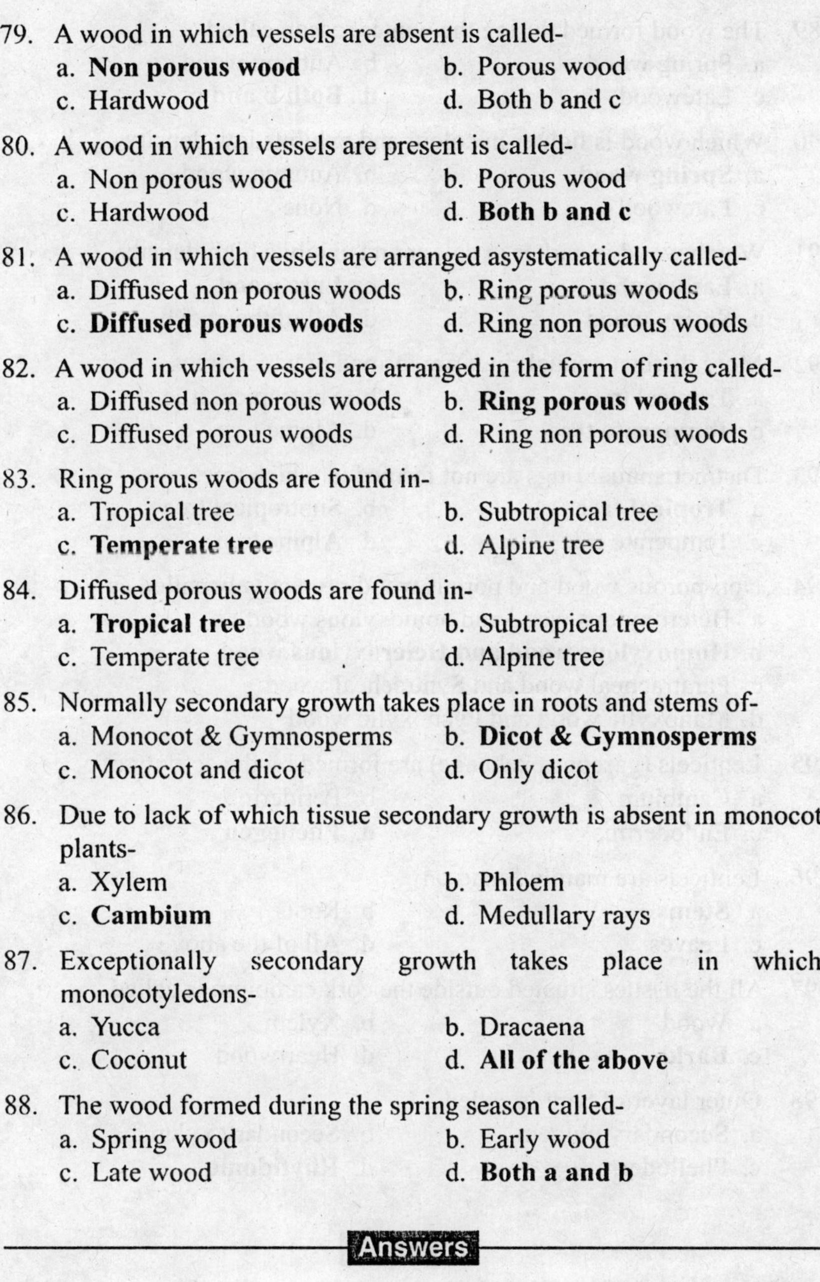

79. A wood in which vessels are absent is called-
    a. **Non porous wood** b. Porous wood
    c. Hardwood d. Both b and c

80. A wood in which vessels are present is called-
    a. Non porous wood b. Porous wood
    c. Hardwood d. **Both b and c**

81. A wood in which vessels are arranged asystematically called-
    a. Diffused non porous woods b. Ring porous woods
    c. **Diffused porous woods** d. Ring non porous woods

82. A wood in which vessels are arranged in the form of ring called-
    a. Diffused non porous woods b. **Ring porous woods**
    c. Diffused porous woods d. Ring non porous woods

83. Ring porous woods are found in-
    a. Tropical tree b. Subtropical tree
    c. **Temperate tree** d. Alpine tree

84. Diffused porous woods are found in-
    a. **Tropical tree** b. Subtropical tree
    c. Temperate tree d. Alpine tree

85. Normally secondary growth takes place in roots and stems of-
    a. Monocot & Gymnosperms b. **Dicot & Gymnosperms**
    c. Monocot and dicot d. Only dicot

86. Due to lack of which tissue secondary growth is absent in monocot plants-
    a. Xylem b. Phloem
    c. **Cambium** d. Medullary rays

87. Exceptionally secondary growth takes place in which monocotyledons-
    a. Yucca b. Dracaena
    c. Coconut d. **All of the above**

88. The wood formed during the spring season called-
    a. Spring wood b. Early wood
    c. Late wood d. **Both a and b**

Answers

79. a	80. d	81. c	82. b	83. c	84. a
85. b	86. c	87. d	88. d		

89. The wood formed during the winter season called-
   a. Spring wood b. Autumn wood
   c. Latewood d. **Both b and c**

90. Which wood is lighter in colour and exhibits low density-
   a. **Spring wood** b. Autumn wood
   c. Latewood d. None

91. Which wood is darker in colour and exhibits high density-
   a. Early wood b. **Late wood**
   c. Spring wood d. All of the above

92. More distinct annual rings are formed in which trees-
   a. Tropical tree b. Subtropical tree
   c. **Temperate tree** d. Alpine tree

93. Distinct annual rings are not formed in which trees-
   a. **Tropical tree** b. Subtropical tree
   c. Temperate tree d. Alpine tree

94. Non-porous wood and porous wood respectively called as-
   a. Heteroxylous wood and homoxylous wood
   b. **Homoxylous wood and Heteroxylous wood**
   c. Paratracheal wood and Syntracheal wood
   d. Manoxylic wood and Pycnoxylic wood

95. Lenticels (gaseous exchange) are formed by the activity of-
   a. Cambium b. Periderm
   c. Endodermis d. **Phellogen**

96. Lenticels are mainly found on-
   a. **Stems** b. Roots
   c. Leaves d. All of the above

97. All the tissues situated outside the cork cambium is called-
   a. Wood b. Xylem
   c. **Bark** d. Heartwood

98. Outer layer of bark is called-
   a. Secondary phloem b. Secondary xylem
   c. Phelloderm d. **Rhytidome**

Answers

89. d	90. a	91. b	92. c	93. a	94. b
95. d	96. a	97. c	98. d		

99. Inner layer of bark is called-
    a. **Secondary phloem** b. Secondary xylem
    c. Cork cambium d. Rhytidome

100. Continuous bark of equal thickening is called-
    a. Scaly bark b. **Ring bark**
    c. Star bark d. Rough bark

101. Discontinuous bark of unequal thickening is called-
    a. **Scaly bark** b. Ring bark
    c. Concentric bark d. Hollow bark

102. Ring bark is found in-
    a. Bhojpatra b. Eucalyptus
    c. **Both a and b** d. Mango

103. A ring of autumn wood and a ring of spring wood are collectively known as annual-
    a. Alternate ring b. **Annual ring**
    c. Common ring d. Seasonal ring

104. The number of annual rings formed in a tree gives the idea of the tree-
    a. Regeneration period c. Growth
    c. **Age** d. Rotation period

105. The study of determination of age of the plant by counting annual rings is called-
    a. Wood anatomy b. Dendrology
    c. Dendrochromonology d. **Dendrochronology**

106. Study of the wood called-
    a. Wood anatomy b. **Xylotomy**
    c. Dendrochromonology d. Histology

107. The wood is actually as-
    a. Primary phloem b. Primary xylem
    c. Secondary phloem d. **Secondary xylem**

Answers

99. a	100. b	101. a	102. c	103. b	104. c
105. d	106. b	107. d			

CHAPTER 7

Forest Mensuration

1. Mensuration is derived from-
   a. French word
   b. Greek word
   c. **Latin word**
   d. Arabic word
2. The term *Pramana* meaning is-
   a. **Measure**
   b. Management
   c. Mensuration
   d. Protection
3. The term *Pramana* was classified into-
   a. *Mana*
   b. *Tula mana*
   c. *Avamana*
   d. **All of the above**
4. The term *mana* means measure of-
   a. Weight
   b. **Volume**
   c. Linear
   d. Time
5. The term *tula mana* means measure of-
   a. **Weight**
   b. Volume
   c. Linear
   d. Time
6. The term *avamana* means measure of-
   a. Weight
   b. Volume
   c. **Linear**
   d. Time
7. The term *kala pramana* means measure of-
   a. Weight
   b. Volume
   c. Linear
   d. **Time**
8. Kautilya has derived which unit as the area unit of forest-
   a. Canal
   b. **Gorut**
   c. Hogat
   d. Nautical

Answers

1. c 2. a 3. d 4. b 5. a 6. c
7. d 8. b

9. Which is the first Indian satellite-
   a. Agni I b. SAT I
   c. **Aryabhatta I** d. Prathvi I

10. The branch of the forestry which deals with the determination of dimensions (diameter, height, volume etc.) form, age and increment of single trees, stands or whole woods, either standing or after felling called-
   a. Forest management b. Forest protection
   c. **Forest mensuration** d. Biometry

11. The most important measurement from the point of view of forest management is-
   a. Area b. **Volume**
   c. Diameter d. Height

12. Which is/are not object of forest mensuration-
   a. Basis of management b. Measurement for research
   c. Measurement of planning d. **All of the above**

13. The Standards of Weights and Measure Act was formed in-
   a. 1926 b. **1956**
   c. 1976 d. 1972

14. Metric system or C.G.S system was introduced in India in-
   a. 1975 b. 1993
   c. **1962** d. 1986

15. Which is commonly used system of measurement in India-
   a. **Metric system** b. British system
   c. F.P.S system d. All of the above

16. Tree height, crown height, crown width and crown length is measured in-
   a. Centimetre b. Nearest centimetre
   c. **Metre** d. Metre and centimetre

17. Stump height or seedling height is measured in-
   a. Centimetre b. **Nearest centimetre**
   c. Metre d. Metre and centimetre

Answers

9. c	10. c	11. b	12. d	13. b	14. c
15. a	16. c	17. b			

18. Diameter of tree is measured in-
a. Metre
b. Nearest centimetre
c. **Centimetre**
d. Metre and centimetre

19. Girth of tree is measured in-
a. Centimetre
b. Nearest centimetre
c. Metre
d. **Metre and centimetre**

20. Full for of LASER is-
a. Light Amplification by Stimulated Emission of Radio wave
b. Light Amplification by Standard Emission of Radiation
c. Light Amplification by Stimulated energy of Radiation
d. **Light Amplification by Stimulated Emission of Radiation**

21. Full for of RADAR is-
a. Radio Detection And Ranging
b. Radio Detecting And Ranging
c. Radio Detection And Radiation
d. Radio Detecting And Radiation

22. The circumference or perimeter of a circle is-
a. $2\pi r$
b. πd
c. πr
d. **Both a and b**

23. Almost universal adopted standard height for measuring girth, diameter and basal area of standing trees is called-
a. Tree height
b. **Breast-Height**
c. Tree girth
d. Breast-girth

24. In India Breast-Height is taken as above the ground level -
a. 1.30 m
b. 1.37 cm
c. 1.30 cm
d. **1.37 m**

25. Europe and United Kingdom Breast-Height is taken as above the ground level -
a. **1.30 m**
b. 1.37 cm
c. 1.30 cm
d. 1.37 m

26. Breast-Height point should be marked by intersecting vertical and horizontal lines and lines length should be long-
a. 10 cm
b. **12 cm**
c. 15 cm
d. 20 cm

Answers

18. c	19. d	20. d	21. a	22. d	23. b
24. d	25. a	26. b			

27. A mark which is marked at Breast-Height is called-
    a. Breast-Height mark
    b. Plus mark
    c. **Cross mark**
    d. Diagonal mark
28. Cross mark is painted with which paint-
    a. Blue
    b. Black
    c. Red
    d. **White**
29. On slopping ground diameter at Breast-Height should be measured on the-
    a. Low hill side
    b. **Uphill side**
    c. Aspect side
    d. Topography side
30. Diameter of leaned tree at flat ground should be measured along-
    a. **Tree leaned side**
    b. Opposite leaned side
    c. Vertically
    d. Diagonally
31. When the tree is forked above the Breast-Height, it is counted as-
    a. **One tree**
    b. More than one tree
    c. Separate tree
    d. Multiple tree
32. When the tree is forked below the Breast-Height it is counted as-
    a. Each fork should not be counted as separate tree
    b. **Each fork should be counted as separate tree**
    c. All fork should be counted as one tree
    d. All of the above
33. Which instruments is/are used for the measurement of tree diameter or girth-
    a. Wooden scale
    b. Calliper
    c. Tape
    d. **All of the above**
34. Diameter & girth of standing tree can be measured by-
    a. Wooden scale
    b. Calliper
    c. Tape
    d. **Both a and c**
35. Diameter & girth of felled tree can be measured by-
    a. Wooden scale
    b. Calliper
    c. **Tape**
    d. Spiegel Relaskop
36. Negative & positive errors are show in-
    a. Wooden scale
    b. **Calliper**
    c. Tape
    d. Spiegel Relaskop

Answers

27. c	28. d	29. b	30. a	31. a	32. b
33. d	34. d	35. c	36. b		

37. Positive errors are show in-
    a. Wooden scale b. Calliper
    c. **Tape** d. Spiegel Relaskop

38. D.O.B can be converted into D.U.B by-
    a. Adding twice bark thickness
    b. Multiplying twice bark thickness
    c. Dividing twice bark thickness
    d. **Deducting twice bark thickness**

39. G.O.B (g') can be converted into G.U.B (g) by the formula-
    a. $g' = g - 2\pi t$ b. $g = g' - \pi t$
    c. $g = g\pi/2\pi t$ d. $\mathbf{g = g' - 2\pi t}$

40. The thickness of bark can be measured by using-
    a. Spiegel Relaskop b. British bark gauge
    c. **Swedish bark gauge** d. Penta prism

41. Seth and Lohani have prepared bark thickness table for which tree-
    a. *Cedrus deodara* b. *Shorea robusta*
    c. *Tectona grandis* d. ***Picea smithiana***

42. Which is/are not an usual diameter class which is used in India-
    a. 2 cm b. **3 cm**
    c. 5 cm d. 10 cm

43. A tree falls in diameter class 2 cm then it will attain diameter at maturity-
    a. **30 cm d.b.h** b. 50 cm d.b.h
    c. 60 cm d.b.h d. 80 cm d.b.h

44. A tree falls in diameter class 5 cm then it will attain diameter at maturity-
    a. 30 cm d.b.h b. **30-50 cm d.b.h**
    c. 50 cm d.b.h d. 80 cm d.b.h

45. A tree falls in diameter class 10 cm then it will attain diameter at maturity-
    a. 30 cm d.b.h b. 30-50 cm d.b.h
    c. 40 cm d.b.h d. **≥50 cm d.b.h**

Answers

37. c	38. d	39. d	40. c	41. d	42. b
43. a	44. b	45. d			

46. The usual girth classes used in India are-
a. 5 cm b. 15 cm
c. 30 cm d. **All of the above**

47. Girth class 15 cm have tree size-
a. Small-sized b. **Medium-sized**
c. Large-sized d. Any size

48. Girth class 5 cm have tree size-
a. **Small-sized** b. Medium-sized
c. Large-sized d. Any size

49. Girth class 30 cm have tree size-
a. Small-sized b. Medium-sized
c. **Large-sized** d. Any size

50. Diameter class 0-20 is denoted by which colour-
a. **Blank** b. Green
c. Red d. Red

51. Diameter class 20-30 and 30-40 is denoted by which colour respectively-
a. Red & Green b. **Green & Red**
c. Red & Yellow d. Yellow & Black

52. Diameter class 40-50 and 50-60 is denoted by which colour respectively-
a. **Yellow & Black** b. Red & Yellow
c. White & Blue d. Black & White

53. Diameter class 60-70 and 70-80 or above is denoted by which colour respectively-
a. Red & Green b. **White & Blue**
c. Red & Yellow d. Yellow & Black

54. Which instrument is used to measure the diameter of standing tree at greater height-
a. Clinometer b. Abney's level
c. Spiegel Relaskop d. **Dendrometer**

Answers

46. d	47. b	48. a	49. c	50. a	51. b
52. a	53. b	54. d			

55. Which bands are used in diameter measurement using Spiegel Relaskop-
a. Band 1 and 2 narrow bands
b. **Band 1 and 4 narrow bands**
c. Band 2 and 4 narrow bands
d. Band 2 and 3 narrow bands

56. Which is used to measure stem diameter accurately at any point on the tree stem from any convenient distance-
a. Clinometer
b. Spiegel Relaskop
c. **Wheeler Penta Prism**
d. Digital calliper

57. The vertical straight line distance from the tip of the leading shoot of the tree to the ground level is called-
a. Tree height
b. Bole height
c. Crown height
d. **Total height**

58. The distance between ground level and crown point of a tree called-
a. Tree height
b. Bole height
c. Crown height
d. **Total height**

59. The height of bole that is usually fit for utilization as timber is called-
a. **Commercial bole height**
b. Bole height
c. Crown height
d. Standard timber

60. The height of the bole from the ground level upto the point where average diameter over bark is 20 cm-
a. Height of Standard wood bole
b. **Height of Standard timber bole**
c. Height of timber bole
d. Height of commercial timber bole

61. The vertical measurement of the tree crown from the tip to the point half way between the lowest green branches forming green crown all round and the lowest green branch on the bole-
a. **Crown length**
b. Crown height
c. Crown width
d. Bole height

62. The height of the crown as measured vertically from the ground level to the point half way between the lowest green branch and green branches forming green crown all round is called-
a. Crown length
b. **Crown height**
c. Crown width
d. Bole height

Answers

55. b	56. c	57. d	58. d	59. a	60. b
61. a	62. b				

63. The maximum spread of the crown along its widest diameter is called-

a. Crown length b. Bole height
c. **Crown width** d. Crown height

64. Which is/are the methods of height measurements-

a. Ocular method b. Non-instrumental method
c. Instrumental method d. **All of the above**

65. Shadow method is the method of height mesurement-

a. Ocular method b. **Non-instrumental method**
c. Instrumental method d. None

66. Single pole method is the method of high measurement-

a. Ocular method b. **Non-instrumental method**
c. Instrumental method d. None

67. Height measuring instrument which are based on principle of similar triangle are-

a. Christen's Hypsometer b. Smythies Hypsometer
c. Improved callipers d. **All**

68. Height measuring instrument which are based on trigonometric principle are-

a. Relaskop b. Tele Relaskop
c. Hega Altimeter d. **All of the above**

69. Height measuring instrument which are based on trigonometric principle are-

a. Brandis Hypsometer b. Abney's level
c. Blume-Leiss Hypsometer d. **All of the above**

70. Height measuring instrument which are based on trigonometric principle are-

a. Topographical Abney's level
b. Bar & Stroud Dendrometer
c. Brandis Hypsometer
d. **All of the above**

71. In India which is/are height classes are generally used-

a. 1 m b. 3 m
c. 5 m d. **All of the above**

Answers

63. c	64. d	65. b	66. b	67. d	68. d
69. d	70. d	71. d			

72. A tree falls in height class 1 m then it will attain height at maturity-
a. Less than 5 m
b. Less than 10 m
c. **Less than 15 m**
d. More than 10 m

73. A tree falls in height class 3 m then it will attain height at maturity-
a. 10 m
b. 20 m
c. 25 m
d. **30 m**

74. A tree falls in height class 5 m then it will attain height at maturity-
a. 8 m
b. 12 m
c. 10 m
d. **20 m**

75. The rate of taper of a log or stem during the growth is called-
a. **Form**
b. Taper
c. Bending
d. Frustum

76. The decrease in the diameter of a stem of a tree of a log form base upwards is called-
a. Form
b. **Taper**
c. Bending
d. Frustum

77. Basal portion of a tree Similar to-
a. Frustum of a paraboloid
b. **Frustum of a neiloid**
c. Like cone
d. Like cube

78. Middle portion of a tree Similar to-
a. **Frustum of a paraboloid**
b. Frustum of a neiloid
c. Like cube
d. Like cone

79. Top portion of a tree Similar to-
a. **Like cone**
b. Like cube
c. Frustum of a neiloid
d. Frustum of a paraboloid

80. According to which theory the tree stem should be considered as a cantilever beam of uniform size against the bending force of the wind-
a. Gause Theory
b. **Metzger theory**
c. Hartig theory
d. Newton theory

81. Metzger theory also called as-
a. Glider theory
b. Gillard theory
c. Gauss Theory
d. **Girder theory**

Answers

72. c	73. d	74. d	75. a	76. b	77. b
78. a	79. a	80. b	81. d		

82. Which is the correct equation of Metzger theory-

a. $d^2 = \frac{32 \times W \times F \times L}{2\pi \times S}$ b. $\boldsymbol{d^3 = \frac{32 \times W \times F \times L}{\pi \times S}}$

c. $d^2 = \frac{32 \times W \times F}{\pi \times S \times L}$ d. $d^3 = \frac{32 \times W \times F \times L}{2\pi \times S}$

83. According to Metzger' theory the tree stem must have the shape of a-

a. **Cubic paraboloid** b. Cubic rhomboid

c. Paraboloid d. Cone shape

84. The ratio between the volume of a tree to the product of basal area and height is called-

a. Form height b. Form quotient

c. Form value d. **Form factor**

85. Breast-height form factor also known as-

a. **Artificial form factor** b. Absolute form factor

c. Normal form factor d. True form factor

86. A form factor in which basal area is measured at the BH and volume refers to the whole tree both above and below the measurement point called as-

a. **Artificial form factor** b. Absolute form factor

c. Normal form factor d. True form factor

87. A form factor in which basal area is measured at any convenient height and volume is measured above the measurement point is called-

a. Absolute form factor b. Normal form factor

c. True form factor d. Artificial form factor

88. A form factor in which basal area is measured at a constant proportion of the total height of the tree and volume is measured whole the tree above the ground level is called-

a. Artificial form factor b. Absolute form factor

c. Breast-height d. **Normal or true form factor**

89. Which form factor is most commonly used form factor in India-

a. True form factor b. Normal form factor

c. **Artificial form factor** d. Absolute form factor

Answers

82. b	83. a	84. d	85. a	86. a	87. a
88. d	89. c				

90. The formula of form factor is-

a. $F = \frac{V \times h}{S}$ b. $F = \frac{h}{V \times S}$

c. $F = \frac{S}{V \times h}$ d. $\boldsymbol{F = \frac{V}{S \times h}}$

91. **Form factor is used to calculate or study-**

a. Growth law and to estimate the volume of standing tree

b. To estimate the volume of standing trec

c. Height and volume of standing tree

d. Growth law and to estimate the volume of felled tree

92. The product of form factor and total height of the tree is called-

a. Form factor b. Form quotient

c. **Form height** d. Form coefficient

93. The formula of form height is-

a. $\boldsymbol{Fh = \frac{V}{S}}$ b. $Fh = \frac{V}{h}$

c. $Fh = \frac{S}{V}$ d. $Fh = \frac{h}{V}$

94. The concept of form quotient was given by-

a. German forester A. Schiffel b. **Austrian forester A. Schiffel**

c. Indian forester Mohit Husain d. German forester Hartig

95. The ratio between the mid-diameter and d.b.h. is known as-

a. Form factor b. **Form quotient**

c. Form height d. Form coefficient

96. The formula of form quotient is-

a. Form quotient = $\frac{\text{mid} - \text{diameter}}{\text{volume}}$

b. **Form quotient** = $\boldsymbol{\frac{\textbf{mid} - \textbf{diameter}}{d.b.h}}$

c. Form quotient = $\frac{d.b.h}{\text{mid} - \text{diameter}}$

d. Form quotient = $\frac{\text{volume}}{d.b.h}$

Answers

90. d	91. a	92. c	93. a	94. b	95. b
96. b					

97. Who modified the form quotient-
a. Swedish forester Tor Jonson b. German forester A. Schiffel
c. Indian forester Mohit Husain d. German forester Tor Jonson

98. The ratio of diameter or girth of a stem at one half its height above the breast-height to the diameter or girth at breast-height is called-
a. Normal Form factor b. Absolute form factor
c. **Absolute form quotient** d. True form factor

99. One of the intervals in which the range of form quotients of trees is divided for classification and use is called-
a. Form coefficient b. Form factor
c. Form point d. **Form class**

100. The point in the crown at which wind pressure is estimated to be centred is called-
a. Form coefficient b. Form factor
c. **Form point** d. Form class

101. The relationship of the height of the form point above ground level to the total height of the tree is called-
a. Form coefficient b. Form factor ratio
c. **Form point ratio** d. Form class ratio

102. Hojer and Behre formula are used to determine-
a. Form coefficient b. Volume quotient
c. Form quotient d. **Diameter quotient**

103. The ratio of the diameter (d) of a stem at any given height to its breast-height diameter (d.b.h) is called-
a. **Diameter quotient** b. Volume quotient
c. Height quotient d. Form quotient

104. The area of the bole surface of a tree and log can be calculated by the formula-
a. $S = \frac{g_1 + g_2}{2} \times l$ b. $S = g \times 1$
c. $S = g \times 1$ or $\frac{g_1 + g_2}{2} \times l$ d. **All of the above**

Answers

97. a 98. c 99. d 100. c 101. c 102. d
103. a 104. d

105. The horizontal projection on the ground of the tree crown is called-

a. Crown width b. Crown spreadness
c. Crown height d. **Crown cover**

106. Which formulae is/are used to calculate the volume of logs-

a. Smalain's formula b. Huber' formula
c. Newton's formula d. **All of the above**

107. Smalain's formula for calculation volume of the logs is-

a. $\frac{s_1 - s_2}{2} \times l$ b. $S_m \times l$

c. $\frac{(s_1 + 4Sm + s_2)}{6} \times l$ d. $\mathbf{\frac{s_1 + s_2}{2} \times l}$

108. Huber' formula for calculation volume of the logs is-

a. $\frac{s_1 - s_2}{2} \times l$ b. $\mathbf{S_m \times l}$

c. $\frac{(s_1 + 4Sm + s_2)}{6} \times l$ d. $\frac{s_1 + s_2}{2} \times l$

109. Newton's formula for calculation volume of the logs is-

a. $\frac{s_1 - s_2}{2} \times l$ b. $S_m \times l$

c. $\mathbf{\frac{(s_1 + 4Sm + s_2)}{6} \times l}$ d. $\frac{s_1 + s_2}{2} \times l$

110. Newton's formula is used to calculate the error in the volume calculated by-

a. Huber' formula b. Smalain's formula
c. Lowry's formula d. **Both a and b**

111. Smalain's formula requires the which cross-sections of the log for volume calculation-

a. Side cross section b. Mid cross-sections
c. **End cross-sections** d. Corner sections

Answers

105. d	106. d	107. d	108. b	109. c	110. d
111. c					

112. Huber's formula requires the which cross-sections of the log for volume calculation-

a. **Mid cross-sections** b. Base cross-sections
c. End cross-sections d. Corner sections

113. The difference between volumes obtained by Smalain's and Newton's formula is-

a. $\Delta = \frac{l}{2} \times (s_1 - s_2 + 2Sm)\ \Delta =$ b. $\Delta = \frac{l}{3} \times (s_1 + s_2 - 2Sm)$
c. $\Delta = \frac{l}{2} \times (s_1 \times s_2 - 2Sm)$ d. $\Delta = \frac{l}{4} \times (s_1 - s_2 - 2Sm)$

114. Volumes obtained by Smalain's and Newton's formula in case of cone and neiloid the error was-

a. Negative b. **Positive**
c. Zero d. None

115. Volumes obtained by Huber's and Newton's formula in case of cone and neiloid the error was-

a. **Negative** b. Positive
c. Zero d. None

116. In India volumes of logs are calculated by using-

a. Half girth formula b. **Quarter girth formula**
c. Newton formula d. Smalian formula

117. Quarter girth formula in Britain also known as-

a. Allen's rule b. Hartig' rule
c. Hojer's rule d. **Hoppus's rule**

118. Quarter girth formula for calculation volumes of logs is-

a. $V = (g/4)^3 \times l$ b. $\mathbf{V = (g/4)^2 \times l}$
c. $V = (g/4) \times l$ d. $V = (g/2)^2 \times l$

119. Quarter girth formula gives how much percent of the true volume-

a. 75% b. **75.8%**
c. 78.5% d. 85%

120. The space occupied by a stack as distinct from the cubic contents of the wood itself i. e. solid volume called-

a. Stock volume b. Total volume
c. Actual volume d. **Stacked volume**

Answers

112. a	113. b	114. b	115. a	116. b	117. d
118. b	119. b	120. d			

121. Volume of solid firewood can be determined by using which methods-
a. Xylometric method b. Specific gravity method
c. **Both a and b** d. None

122. Which is/are the methods to determine volume of standing tree-
a. Ocular estimation method
b. Partly ocular and partly by measurement
c. Direct/Indirect measurement
d. **All of the above**

123. Which instruments is/are used to determine the volume of standing tree in indirect measurement method-
a. Spiegel Relaskop b. Tele Relaskop
c. Wheeler Penta Prism d. **All of the above**

124. Which instruments is/are not used to determine the volume of standing tree in indirect measurement method-
a. Tele Relaskop b. **Wooden scale**
c. Stroud Dendrometer d. Callipers

125. A table showing for a species the average contents of trees, logs or sawn timber for one or more given dimensions is called-
a. Yield table b. Money table
c. **Volume table** d. Stand table

126. The volume of a tree depends mainly upon how many variables-
a. Two b. **Three**
c. Four d. Five

127. The volume of a tree does not depends upon which variable-
a. Diameter b. Height
c. Form d. **Area**

128. The most important variables for volume table among the three variables is/are-
a. **Diameter (BH)** b. Height
c. Form d. Area

129. General volume table is/are based on which variables-
a. One variables b. **Two variables**
c. Three variables d. Four variables

Answers

121. c	122. d	123. d	124. b	125. c	126. b
127. d	128. a	129. b			

130. Local volume table is/are based on which variables-
   a. Area b. Height
   c. **Diameter** d. Form

131. Volume table which is based on one variable-
   a. General volume table b. Regional volume table
   c. **Local volume table** d. Standard volume table

132. Volume table which is based on two variables-
   a. **General volume table** b. Regional volume table
   c. Local volume table d. Standard volume table

133. General volume table is/are based on which variables-
   a. Diameter and volume b. Diameter and form
   c. **Diameter and height** d. Diameter

134. For standard round timber estimation which volume table is used-
   a. Commercial volume table b. Assortment table
   c. Local volume table d. **Standard volume table**

135. Volume table based on.........variables have not been prepared in India-
   a. Two b. **Three**
   c. Four d. Five

136. General volume tables are used for the deriving which table-
   a. Regional volume table b. **Local volume table**
   c. Commercial volume table d. Assortment volume table

137. Most commonly used table in India is-
   a. Standard volume table b. Regional volume table
   c. Local volume table d. **General volume table**

138. Standard volume table and Commercial volume table are the special cases of-
   a. Assortment table b. Regional volume table
   c. Sawn out turn table d. Local volume table

139. Which is/are the methods of volume table preparation -
   a. Graphical method
   b. Regression equations method or method of least squares fit
   c. Alignment chart method
   d. All of the above

Answers

130. c	131. c	132. a	133. c	134. d	135. b
136. b	137. d	138. a	139. d		

140. The portion of the tree stem or log which is unmerchantable is called-
   a. Stump b. Stack
   c. Billets d. **Cull**

141. Volume table is used to estimate the volume of-
   a. **Standing tree** b. Felled tree
   c. Leaned tree d. Forked tree

142. The ratio of density of the substance to density of water is called as-
   a. Relative density b. Specific gravity
   c. **Botha a and b** d. Specific heat

143. The moisture content when free water has been evaporated leaving only absorbed or bound water is called-
   a. Free water b. Saturated water point
   c. Water holding capacity d. **Fibre saturation point**

144. Determination of the age of single standing tree can be estimated by using which methods-
   a. From existing record or general appearance
   b. By the number of annual shoots or whorls of branches
   c. By means of Pressler's increment borer
   d. **All of the above**

145. Age of tree can be determined by counting annual ring with the help of-
   a. Wheeler Penta Prism b. Callipers
   c. **Pressler's increment borer** d. Spiegel Relaskop

146. To determine the age of tree by Pressler's increment borer boring is made……in case of smaller tree and…… for bigger tree respectively-
   a. 20 cm above ground and at breast-height
   b. **30 cm above ground and at breast-height**
   c. 40 cm above ground and at 50 cm from ground
   d. 50 cm above ground and 100 cm from the ground

147. The age of standing tree without counting annual ring can be calculated by-
   a. Taking two periodic mean b. **Taking three periodic mean**
   c. Taking four periodic mean d. Taking five periodic mean

Answers

140. d	141. a	142. c	143. d	144. d	145. c
146. b	147. b				

148. The age of tree can be found by the following formula-

a. **Age** $= \dfrac{1}{p_1 \times s}$

b. Age $= \dfrac{p_1 \times s}{2}$

c. Age $= \dfrac{1}{p+s}$

d. Age $= \dfrac{1}{p_1 - s}$

149. The C.A.I and M.A.I do not coincide with each other throughout the life of the tree or stand except twice viz-

a. One, at the end of first year
b. Second, in the year of the culmination of C.A.I
c. Second, in the year of the culmination of M.A.I
d. **Both a and c**

150. The value of C.A.I may be-

a. Positive
b. Negative
c. Zero
d. **All of the above**

151. The value of M.A.I may not be-

a. Positive
b. Negative
c. **Zero**
d. All of the above

152. Compound interest formula is used to determine-

a. Diameter increment percent
b. Volume increment percent
c. Height increment percent
d. **Both a and b**

153. Pressler's formula is used to determine-

a. Diameter increment percent
b. Volume increment percent
c. Height increment percent
d. **Both a and b**

154. Schneider's formula for increment percent is-

a. $p = \dfrac{200}{D}$

b. $p = \dfrac{200}{nD}$

c. $p = \dfrac{n \times 400}{D}$

d. $\mathbf{p = \dfrac{400}{nD}}$

155. Pressler's formula for volume increment percent is-

a. $p = \dfrac{100(D-d)}{n} \times \dfrac{(V+v)}{n(V-v)}$

b. $p = \dfrac{200(D+d)}{n} \times \dfrac{(V-v)}{n(V+v)}$

c. $\mathbf{p = 200 \times \dfrac{(V-v)}{n(V+v)}}$

d. $p = \dfrac{400(D-d)}{n} \times \dfrac{(V+v)}{n(V-v)}$

Answers

148. a	149. d	150. d	151. c	152. d	153. d
154. d	155. c				

156. Pressler's formula for diameter increment percent is-

a. $p = \frac{200(D-d)}{n} \times \frac{(V-v)}{n(V+v)}$ b. $p = \frac{200(D-d)}{(D+d)}$

c. $p = \frac{400(D+d)}{n(D-d)}$ **d. $p = \frac{200(D-d)}{n(D+d)}$**

157. Compound interest formula for volume increment percent is-

a. $P = 100\left[\left(\frac{V}{v}\right)^{n} + 1\right]$ b. $P = 200\left[\left(\frac{V}{v}\right)^{1/n} - 1\right]$

c. $P = 300\left[\left(\frac{V}{v}\right)^{1/n} + 1\right]$ **d. $P = 100\left[\left(\frac{V}{v}\right)^{1/n} - 1\right]$**

158. Compound interest formula for diameter increment percent is-

a. $P = 100\left[\left(\frac{D}{d}\right)^{1/n} + 1\right]$ b. $P = 200\left[\left(\frac{D}{d}\right)^{1/n} - 1\right]$

c. $P = 100\left[\left(\frac{D}{d}\right)^{1/n} - 1\right]$ d. $P = 200\left[\left(\frac{D}{d}\right)^{1/n} - 1\right]$

159. The analysis of a complete stem by measuring annual rings on a number of cross-section at different heights in order to determine its past rates growth is called-

a. Stump analysis **b. Stem analysis**
c. Stand analysis d. Growth analysis

160. The boring of a tree stem with Pressler's increment or any other borer to determine increment of trees with annual rings is called-

a. Increment boring b. Increment percent
c. Volume increment d. Height increment

161. Diameter corresponding to the mean basal area of a uniform, generally pure crop called-

a. Top diameter **b. Crop diameter**
c. Mean diameter d. Average diameter

Answers

156. d 157. d 158. c 159. b 160. a 161. b

162. Diameter corresponding to the mean basal area of a group of trees or a stand is called-
a. Crop diameter b. Top diameter
c. **Mean diameter** d. Average diameter

163. Diameter corresponding to the mean basal area of the 250 biggest trees per hectare in a uniform, generally pure crop is called-
a. Crop diameter b. **Top diameter**
c. Mean diameter d. Average diameter

164. Mean basal area is of a crop is calculated by using the formula-

a. Mean basal area = $\frac{n_1 s_1 + n_2 s_2 + n_3 s_3 + ...}{S_1 + S_2 + S_3 + ...}$

b. Mean basal area = $\frac{n_1 s_1 - n_2 s_2 - n_3 s_3 - ...}{S_1 + S_2 + S_3 + ...}$

c. **Mean basal area** = $\frac{n_1 s_1 + n_2 s_2 + n_3 s_3 + ...}{n_1 + n_2 + n_3 + ...}$

d. Mean basal area = $\frac{n_1 s_1 + n_2 s_2 + n_3 s_3 + ...}{n_1 - n_2 - n_3 - ...}$

165. Mean diameter of a crop is calculated by using the formula-

a. Mean diameter D = $\frac{n_1 s_1 + n_2 s_2 + n_3 s_3 + ...}{n_1 + n_2 + n_3 + ...}$

b. **Mean diameter D** = $\frac{n_1 d_1 + n_2 d_2 + n_3 d_3 + ...}{n_1 + n_2 + n_3 + ...}$

c. Mean diameter D = $\frac{n_1 d_1 + n_2 d_2 + n_3 d_3 + ...}{d_1 + d_2 + d_3 + ...}$

d. Mean diameter D = $\frac{n_1 d_1 - n_2 d_2 - n_3 d_3 + ...}{d_1 + d_2 + d_3 + ...}$

166. The average height of a regular crop is called-
a. **Crop height** b. Top height
c. Mean height d. Average height

Answers

162. c 163. b 164. c 165. b 166. a

167. Height corresponding to the mean diameter of a group of trees or the crop diameter of a stand is called-

a. Crop height b. Top height
c. **Mean height** d. Average height

168. Height corresponding to the mean diameter of the 250 biggest trees per hectare in a uniform, generally pure crop is called-

a. Crop height b. **Top height**
c. Mean height d. Average height

169. Crop height is calculated by using formula-

a. Smalian formula b. **Lorey's formula**
c. Pressler's formula d. Schneider's formula

170. Crop height is calculated by using the formula-

a. Crop height = $\frac{h_1s_1 + h_2s_2 + h_3s_3 + ...}{h_1 + h_2 + h_3 + ...}$

b. Crop height = $\frac{n_1s_1 + n_2s_2 + n_3s_3 + ...}{n_1 + n_2 + n_3 + ...}$

c. **Crop height** = $\frac{h_1s_1 + h_2s_2 + h_3s_3 + ...}{s_1 + s_2 + s_3 + ...}$

d. Crop height = $\frac{n_1d_1 + n_2d_2 + n_3d_3 + ...}{d_1 + d_2 + d_3 + ...}$

171. The formula used for determined crop age for even-aged crop is-

a. **Crop age** = $\frac{a_1s_1 + a_2s_2 + a_3s_3 + ...}{s_1 + s_2 + s_3 + ...}$

b. Crop age = $\frac{h_1s_1 + h_2s_2 + h_3s_3 + ...}{h_1 + h_2 + h_3 + ...}$

c. Crop age = $\frac{a_1s_1 + a_2s_2 + a_3s_3 + ...}{a_1 + a_2 + a_3 + ...}$

d. Crop age = $\frac{a_1s_1 + a_2s_2 + a_3s_3 + ...}{s_1 - s_2 - s_3 - ...}$

Answers

167. c 168. b 169. b 170. c 171. a

172. Who has given the formulae for calculation of mean age of uneven-aged crop-
a. Smalian & Heyer
b. Andre
c. Gumpel
d. **All of the above**

173. How many plots are considered enough for preparation of yield tables for a species-
a. 200 plots
b. 400 plots
c. **600 plots**
d. 800 plots

174. In India mostly plots are laid having plot size-
a. Rectangular plots having size 0.1 to 0.2 hectare
b. Square plots having size 0.5 to 1.0 hectare
c. Triangular plots having size 0.1 to 0.2 hectare
d. Square plots having size 0.5 to 1.0 hectare

175. An area maintained round an experimental or sample plot to ensure that the latter is not being affected by the treatment applied to the area outside the plot is called-
a. Surround
b. Isolation
c. Buffer strip
d. **All of the above**

176. All the trees in the surround are generally marked by-
a. Blue paint ring
b. **White paint oval**
c. Black paint ring
d. Green paint sphere

177. In India sample plots are generally thinned to-
a. A or D grade ordinary thinning
b. B or D grade ordinary thinning
c. A or C grade ordinary thinning
d. **C or D grade ordinary thinning**

178. Who has given the formulae/methods to determine volume of sample plot-
a. Urich
b. Hartig
c. Huber
d. **All of the above**

179. Who has given the formulae for determine volume of sample plot-
a. Hossfeld
b. Draut
c. Block
d. **All of the above**

Answers

172. d	173. c	174. a	175. d	176. b	177. d
178. d	179. d				

180. Which method is used in FRI, Dehra Dun to determine volume of sample plot-
   a. Urich b. Hartig
   c. Huber d. **Block**

181. Form factor method or Prussian institute method is used to determine-
   a. Crop volume b. Mean basal area
   c. Crop diameter d. **Sample plot volume**

182. Which method is used in temporary sample plots for volume determination-
   a. Urich b. **Hartig**
   c. Huber d. Block

183. Volume of sample plot can be calculated by using a computer programme named as-
   a. FORTUNE b. TRANSFOR
   c. **FORTRAN** d. FORTON

184. The tabulated, reliable and satisfactory tree information, related to the required unit, respectively units, of assessment in hierarchic order is called-
   a. Yield table b. Forest survey
   c. **Forest inventory** d. Forest Biometry

185. Forest inventory is synonymous which term in India-
   a. Tabulation b. **Enumeration**
   c. Planning d. Formulation

186. Which is/are the kinds of enumeration-
   a. Total enumeration b. Complete enumeration
   c. Partial or sample enumeration d. **All**

187. Enumeration is classified into how many kinds-
   a. **Two** b. Three
   c. Four d. Five

188. Total enumeration also known as-
   a. Complete enumeration b. Partial enumeration
   c. Sample enumeration d. Simple enumeration

Answers

180. d	181. d	182. b	183. c	184. c	185. b
186. d	187. a	188. a			

189. Partial enumeration also known as-
   a. Sample enumeration b. Simple enumeration
   c. Total enumeration d. Complete enumeration

190. The ratio of the sample to the whole population is called-
   a. Sampling fraction b. Sampling intensity
   c. Sampling unit d. **Both a and b**

191. Main kinds of sampling used in forest inventories are categorised into-
   a. **Two** b. Three
   c. Four d. Five

192. Which is/are not the main kinds of sampling used in forest inventories-
   a. Random sampling c. Non- random sampling
   c. **Selective sampling** d. None

193. Which is/are the kinds of random sampling-
   a. Simple random sampling b. Multi-stage sampling
   c. Stratified random sampling d. **All of the above**

194. Which is/are the kinds of non-random sampling-
   a. Selective sampling b. Systematic sampling
   c. Sequential sampling d. **All of the above**

195. Which is/are the kinds of random sampling-
   a. Multiphase sampling
   b. Sampling with varying probability
   c. List sampling
   d. **All of the above**

196. A sampling in which sampling units are selected in such a manner that all possible units of the same size have equal chance of being chosen is called-
   a. **Random sampling** b. Point sampling
   c. Non- random sampling d. Partial sampling

197. The method of sampling in which samples are selected according to the subjective judgement of the observer on the basis of certain guidelines is called-
   a. Random sampling b. Point sampling
   c. **Non- random sampling** d. Partial sampling

Answers

189. a	190. d	191. a	192. c	193. d	194. d
195. d	196. a	197. c			

198. A sampling unit whose boundaries are predominantly topographical or natural features such as *nalas*, streams and ridges is called-
   a. Point sampling units
   b. **Topographical sampling units**
   c. Geological sampling units
   d. Physiographic sampling units

199. Topographical units are mostly used as sampling units in which forest-
   a. Plain forest b. Ravine forest
   c. **Hill Forest** d. Rain forest

200. A sampling unit or group of small units is called-
   a. Sample b. Population
   c. **Cluster** d. Clump

201. The percentage of the area of the population to be included in the sample is called-
   a. Sample b. Sampling unit
   c. Sampling design d. **Sampling intensity**

202. In order to determine optimum intensity in various forests the sampling error should be-
   a. Less than 5% b. More than 10%
   c. **Less than 10%** d. More than 10%

203. Sampling percent in tropical wet evergreen forest is-
   a. **10** b. 2.5
   c. 5.0 d. 15

204. Sampling percent in sub-tropical pine forest is-
   a. 1.0 b. 2.5
   c. **5.0** d. 10

205. Sampling percent in tropical moist deciduous forest is-
   a. 1.0 b. **2.5**
   c. 5.0 d. 10

206. National Inventory Design concept was given by-
   a. FRI b. BSI
   c. **FSI** d. MoEF & CC

Answers

198. b	199. c	200. c	201. d	202. c	203. a
204. c	205. b	206. c			

207. The difference between the population mean (M) and sample mean (Y) is called-
a. Sampling mean
b. Sampling intensity
c. **Sampling error**
d. Sampling unit

208. Which method is used in plain for sapling-
a. Strip sampling
b. Line plot survey
c. Topographical units
d. **Both a and b**

209. A sampling in which the population is first divided into sub-population of different strata and then units are selected from each of them in proportion to their size-
a. Stratified random sampling
b. Selective sampling
c. Systematic sampling
d. Sequential sampling

210. A sampling in which sampling units are not taken out at one stage but are taken out in two or three stages called-
a. Stratified random sampling
b. Systematic sampling
c. Sequential sampling
d. **Multi-stage sampling**

211. A sampling in which some of the same sampling units are used at different phases of sampling to collect different information or same information by different methods is called-
a. Simple random sampling
b. Multi-stage sampling
c. Stratified random sampling
d. **Multiphase sampling**

212. Two phase sampling or double sampling is commonly used in-
a. Bamboo enumeration
b. Aerial photograph interpretation
c. Forest inventories
d. **All of the above**

213. The Bitterlich or variable plot or point sampling or probability proportional to size (PPS) and PPP or sampling with probability proportional to prediction are based on-
a. Multi-stage sampling
b. Stratified random sampling
c. Multiphase sampling
d. **Sampling with varying probability**

Answers

207. c	208. d	209. a	210. d	211. d	212. d
213. d					

214. A sampling is in which samples are select according to the subjective judgement of the observer is called-
a. Stratified random sampling
b. **Selective sampling**
c. Systematic sampling
d. Sequential sampling

215. 4 PEA sampling is one of the forms of sampling-
a. Multi-stage sampling
b. **Selective sampling**
c. Multiphase sampling
d. Sampling with varying probability

216. A sampling in which sampling units are selected acceding to a predetermined pattern is-
a. Systematic sampling
b. Stratified random sampling
c. Selective sampling
d. Sequential sampling

217. Which sampling method also known as systematic sampling with a random start-
a. Multi-stage sampling
b. Stratified random sampling
c. Multiphase sampling
d. **Systematic sampling**

218. A method of sampling in which the number of observations in the sample is not predetermined but sampling unis are taken successively from a population is called-
a. Systematic sampling
b. Stratified random sampling
c. Selective sampling
d. **Sequential sampling**

219. The usual shapes of the sampling units in India are-
a. Plots
b. Strips
c. Topographical
d. **All of the above**

220. Total enumeration required how many enumerator and mazdoors respectively-
a. 1 enumerator and 7 mazdoors
b. 1 enumerator and 13 mazdoors
c. 1 enumerator and 14 mazdoors
d. 1 enumerator and 9 mazdoors

Answers

214. b 215. b 216. a 217. d 218. d 219. d
220. a

221. Liner strip enumeration required how many enumerator and mazdoors respectively-
   a. 1 enumerator and 7 mazdoors
   b. **1 enumerator and 13-14 mazdoors**
   c. 1 enumerator and 15 mazdoors
   d. 1 enumerator and 9 mazdoors

222. Basal area is expressed in-
   a. **Square meter**
   b. Square centimetre
   c. Hectare
   d. Miles

223. A method of counting from a random point the number of trees whose breast height cross-section exceeds a certain critical angle which is multiplied by a constant factor to give an unbiased estimate of basal area per hectare ic called-
   a. Random sampling
   b. **Point sampling**
   c. Stratified random sampling
   d. Simple random sampling

224. Which is/are not a kind of point sampling-
   a. Horizontal point sampling
   b. Vertical point sampling
   c. **Straight line point sampling**
   d. All of the above

225. Most commonly used type of point sampling in India is-
   a. Horizontal point sampling
   b. Vertical point sampling
   c. Straight line point sampling
   d. Simple random sampling

226. The process of sampling in which a series of sampling points are selected randomly or systematically distributed over the whole area, when trees around these points are viewed through any angle-gauge at breast height then the tree having an angle greater than critical angle of angle-gauge are counted called-
   a. Systematic sampling
   b. Selective sampling
   c. **Horizontal point sampling**
   d. Vertical point sampling

227. The number of trees counted by horizontal point sampling and multiplied by a constant factor gives-
   a. Basal area per hectare (BA/ha)
   b. Number of trees per hectare
   c. Diameter per hectare
   d. Tree height per hectare

Answers

221. b 222. a 223. b 224. c 225. a 226. c
227. a

228. The ratio between the tree basal area & area of circular plot is-
   a. $5k^2/2$ b. $2k^2/7$
   c. $3k^2/4$ d. **$k^2/4$**

229. The multiplying factor associated with any point sampling instrument called-
   a. **Basal area factor** b. Basal area coefficient
   c. Calibration factor d. Plot radius factor

230. The factor which indicates the distance at which a given angle gauge will exactly cover a one metre wide rectangular target called-
   a. Basal area factor b. Basal area coefficient
   c. **Calibration factor** d. Plot radius factor

231. Calibration factor is more times of plot radius factor-
   a. 10 b. 25
   c. 50 d. **100**

232. Horizontal point sampling is used for determination of the following-
   a. Basal area per hectare (BA)
   b. Number of stem per hectare (N)
   c. Volume per hectare (V)
   d. **All of the above**

233. Which instruments are used in horizontal point sampling-
   a. Simple angle gauge b. Wedge prism
   c. Spiegel Relaskop d. **All of the above**

234. Wedge prism works on the principle of-
   a. Bending of light rays passing through the trees
   b. Turning of light rays passing through the prism
   c. **Bending of light rays passing through the prism**
   d. Bending of light rays passing through the eyes of person

235. The tree is counted if the image-
   a. Overlaps the directly viewed tree
   b. Do not overlaps the directly viewed tree
   c. There is a gap
   d. None

Answers

228. d	229. a	230. c	231. d	232. d	233. d
234. c	235. a				

236. A tree whose image overlap in horizontal point sampling is called-
a. **Fully tally** b. Non tally tree
c. Half tally tree d. Partially tally tree

237. A tree whose image just touch the tree stem in horizontal point sampling is called-
a. Fully tally b. Non tally tree
c. **Half tally tree** d. Partially tally tree

238. A tree whose image do not touch the tree stem in horizontal point sampling is called-
a. Fully tally b. **Non tally tree**
c. Half tally tree d. Partially tally tree

239. The Basal Area Factor of wedge-prism is calculated on the basis of the-
a. Acute angle b. Obtuse angle
c. **Critical angle of prism** d. Finite angle

240. The formula of basal area factor of wedge-prism is-
a. $BAF = \frac{10,00}{1+3\left(\frac{L}{D}\right)^2}$ b. $BAF = \frac{10,000}{1+2\left(\frac{L}{D}\right)^2}$
c. $BAF = \frac{10,00}{1+4\left(\frac{L}{D}\right)^2}$ d. $\mathbf{BAF = \frac{10,000}{1+4\left(\frac{L}{D}\right)^2}}$

241. No slope correction is required in wedge prism upto an angle-
a. **18°** b. 38°
c. 45° d. 90°

242. A slope angle of 25° cause an error about in counting of tally trees by wedge prism-
a. **10%** b. 20%
c. 30% d. 50%

243. It is necessary that the prism is held in a.......position in **level terrain** and right angle to the line of sight on.............ground respectively-
a. Horizontal and sloping ground
b. Vertical and hilly ground
c. **Vertical and sloping ground**
d. Horizontal and plain ground

Answers

236. a	237. c	238. b	239. c	240. d	241. a
242. a	243. c				

244. Fully tally trees are counted as…while two half tally trees makes…. tally tree respectively-
   a. Two tally tree and one tally tree
   b. One tally tree and two tally tree
   c. **One tally tree and one tally tree**
   d. Two tally tree and two tally tree

245. Spiegel Relaskop is widely used in point sampling has following scales-
   a. 20 m height band
   b. 25 m height band
   c. 30 m height band
   d. **All of the above**

246. Which bands is/are not used in spiegel relaskop-
   a. Band 1
   b. Band 2
   c. **Band 3**
   d. Band 4

247. Range finder bands is used in which instrument-
   a. Ravi multimeter
   b. Abney level
   c. Hega altimeter
   d. **Spiegel Relaskop**

248. Four narrow band (2 white + 2 black) are used in which instrument-
   a. Ravi multimeter
   b. Abney level
   c. Hega altimeter
   d. **Spiegel Relaskop**

249. Band 1 plus 4 narrow bands are collectively called-
   a. Band 1
   b. Band 2
   c. Band 3
   d. **Band 4**

250. Tele Relaskop is used to calculate accurate height & diameter of-
   a. Felled tree
   b. Leaned tree
   c. **Standing tree**
   d. All of the above

251. Which is/are several methods to determine the volume of standing trees-
   a. Hohenadl's formula
   b. The 'odd ten percents' method
   c. The 'Any visible diameters' method
   d. **All of the above**

Answers

244. c	245. d	246. c	247. d	248. d	249. d
250. c	251. d				

252. A process of sampling in which all trees appearing more than a critical angle are counted with in a full 360^0 sweep around sample point called-

a. Horizontal point sampling b. **Vertical point sampling**
c. Straight line point sampling d. Simple random sampling

253. The value of critical angle is fixed for tally tree in vertical point sampling-

a. **45°** b. 30°
c. 60° d. 90°

254. Conimeter is based on the principle of-

a. Horizontal point sampling
b. **Vertical point sampling**
c. Non- random sampling
d. Simple random sampling

255. 1 diopter is equal to an angle-

a. 0.27° b. 0.37°
c. **0.57°** d. 0.87°

256. 1 diopter is equal to-

a. 14.56 minutes b. 24.46 minutes
c. 44.56 minutes d. **34.36 minutes**

257. The error in height measurement will be less if the angle (θ) is equal to-

a. **45°** b. 30°
c. 60° d. 90°

Answers

252. b 253. a 254. b 255. c 256. d 257. a

CHAPTER 8

Remote Sensing

1. Term remote sensing was coined by-
   a. **Ms. Evelyn Pruitt** b. Pisharoth Rama Pisharoty
   c. C.V. Raman d. ISRO
2. Father of Indian remote sensing is-
   a. J. C. Bose b. K. S Sharma
   c. Maniknandan V. C d. **Pisharoth Rama Pisharoty**
3. First aerial photograph was taken in the year-
   a. 1820 b. **1858**
   c. 1892 d. 1976
4. First aerial photograph was taken in which country-
   a. India b. America
   c. Japan d. **France**
5. First satellite launched by India was-
   a. **ARYABHATA** b. AGNI
   c. PRATHVI d. BHASKAR
6. First satellite launched by India in the year-
   a. 1896 b. 1956
   c. **1975** d. 1993
7. Full form of ISRO is-
   a. International Space Research Organization
   b. Indian Statistics Research Organization
   c. **Indian Space Research Organization**
   d. Indian Space Research Orbit

Answers

1. a 2. d 3. b 4. d 5. a 6. c
7. c

8. The Indian Space Research Organization (ISRO) headquarter situated in-
   a. New Delhi b. Mumbai
   c. Hyderabad d. **Bengaluru**

9. The Indian Space Research Organization (ISRO) was formed in the year-
   a. 1858 b. 1920
   c. 1958 d. **1969**

10. The first country to launch an artificial satellite is-
    a. France b. U.S.A
    c. Japan d. **Russia**

11. Russia launched world first artificial satellite in the year-
    a. 1876 b. 1920
    c. 1950 d. **1957**

12. Father of Indian space programme is-
    a. **Vikram Sarabhai** b. Sanjeev Kumar
    c. P. K. Mishra d. Pisharoth Rama Pisharoty

13. The art and science of obtaining the information about an object, area or phenomena through the analysis of data acquired by a device that is not in direct contact with the object, area or phenomena under investigation is called-
    a. **Remote sensing** b. Aerial photography
    c. Forest management d. Biometry

14. The most common and important sources of energy used in remote sensing is-
    a. Solar Battery b. Photovoltaic cell
    c. Diode d. **Sun**

15. The electromagnetic radiation occurs as a continuum of wavelength (λ) and frequency from short wavelength and high frequency cosmic waves to long wavelength and low frequency radio waves is called-
    a. Electromagnetic radiation path (ERP)
    b. Energy spectrum (ES)
    c. **Electromagnetic spectrum (EMS)**
    d. Wavelength distribution chart (WDC)

Answers

8. d 9. d 10. d 11. d 12. a 13. a
14. d 15. c

16. The bands of spectrum in which there is little or no absorption found are called-

a. Spatial field b. **Atmospheric window**

c. Black hole d. Absorption zone

17. Remote sensing is classified into how many categories on the basis of distance of platform from the earth -

a. Aerial remote sensing (ARS) b. Aerial Photography

c. Space remote sensing (SRS) d. **All of the above**

18. The science of obtaining the photographs from the air using various platforms like as balloons, aircrafts etc. for getting information about the surface of the earth is called-

a. Aerial photography b. Aerial remote sensing

c. Space remote sensing d. **Both a and b**

19. In aerial remote sensing the platform height is varies from-

a. 1000 m to 5000 m b. 5000 m to 10000 m

c. 1000 m to 15000 m d. **5000 m to 15000 m**

20. In space remote sensing the platform height is varies from-

a. 50 km to 100 km b. 100 km to 300 km

c. 300 km to 800 km d. **300 km to 1000 km**

21. Aerial photographs show-

a. Perspective projection of earth surface

b. Orthographical projection of earth surface

c. Symmetrical projection of earth surface

d. Angular projection of earth surface

22. Actual image of the object & the ground features is seen-

a. **Aerial photograph** b. Map

c. Plane table survey d. Both a and b

23. Maps shows-

a. Perspective projection of earth surface

b. **Orthographical projection of earth surface**

c. Symmetrical projection of earth surface

d. Angular projection of earth surface

Answers

16. b	17. d	18. d	19. d	20. d	21. a
22. a	23. b				

24. Objects and ground features show by conventional symbol in-
    a. Aerial photograph   b. **Map**
    c. Plane table survey   d. Both a and b

25. 3-D view of earth is seen by-
    a. **Aerial photograph**   b. Map
    c. Plane table survey   d. Both a and b

26. Maps shows which type of view of earth surface-
    a. 1-D view   b. **2-D view**
    c. 3-D view   d. All of the above

27. A yellow filter absorbs which energy incident on it-
    a. Red   b. **Blue**
    c. White   d. Green

28. A red filter absorbs which energy incident on it-
    a. Blue   b. Violet
    c. Green   d. **All of the above**

29. A method of remote sensing in which cameras, sensors or other devices are attached to the satellite orbiting round the earth, take the digital photographs of earth & resources and after converted these digital photographs into photographs by computer, required information is obtained called-
    a. Aerial photography   b. Aerial remote sensing
    c. **Space remote sensing**   d. Both a and b

30. A yellow filter transmits which energy-
    a. Blue & green   b. Black & white
    c. **Green & red**   d. Yellow & violet

31. A red filter transmits which energy-
    a. Lower wavelength of visible spectrum
    b. **Higher wavelength of visible spectrum**
    c. Higher wavelength of infra-red spectrum
    d. Lower wavelength of infra-red spectrum

32. Most commonly used filter in forestry for remote sensing operation is-
    a. Red filter   b. Green filter
    c. Violet filter   d. **Yellow filter**

Answers

24. b	25. a	26. b	27. b	28. d	29. c
30. c	31. b	32. d			

33. Which filter is used for water bodies-
    a. Blue filter
    b. White filter
    c. **Red filter**
    d. Yellow filter
34. There is about.........percent overlap in between the photographs taken from two consecutive flight lines.
    a. 10%
    b. 20%
    c. **30%**
    d. 50%
35. The ideal time for take aerial photograph is-
    a. 6 to 9 A.M and 12 to 3 P.M.
    b. **9 to 11 A.M and 1 to 2 P.M.**
    c. 9 to 12 A.M and 3 to 4 P.M.
    d. 8 to 10 A.M and 1 to 3 P.M.
36. In India following agencies are responsible for aerial photography-
    a. Indian Air Force (IAF)
    b. Messers Air Survey Company of India (Pvt. Ltd.)
    c. National Remote Sensing Agency (NRSA)
    d. **All of the above**
37. National Remote Sensing Agency is situated in-
    a. Bengaluru
    b. **Hyderabad**
    c. New Delhi
    d. Bombay
38. The Surveyor General's office for aerial photography is situated-
    a. Bengaluru
    b. Hyderabad
    c. New Delhi
    d. **Dehra Dun**
39. Optimum season for aerial photography of coniferous forests of Himalayan is-
    a. Dec – February
    b. Sept – December
    c. **Oct – November**
    d. March – April
40. Optimum season for aerial photography of mixed deciduous forests is-
    a. **Dec – February**
    b. Sept – December
    c. Oct – November
    d. March – April
41. Optimum season for aerial photography of sal forests is-
    a. Sept – December
    b. **Mid-March – Mid April**
    c. Oct – November
    d. Jan - March

Answers

33. c	34. c	35. b	36. d	37. b	38. d
39. c	40. a	41. b			

42. Optimum season for aerial photography of teak forests is-
    a. **Sept – December** b. Mid-March – Mid April
    c. Oct – November d. Jan – March

43. Optimum season for aerial photography of tropical evergreen & moist deciduous forests is-
    a. Sept – December b. Mid-March – Mid April
    c. Oct – November d. **Jan – Feb**

44. Vertical photograph covered area in which form-
    a. Square b. Triangular
    c. Rhomboidal d. **Rectangular**

45. Oblique photograph covered area in which form-
    a. Square b. Triangular
    c. **Trapezoidal** d. Rectangular

46. Which type of photograph is used as map-
    a. Vertical photograph b. Horizontal photograph
    c. Linear photograph d. Oblique photograph

47. Which photographs are used to identification of tree species & disease-
    a. Panchromatic Black & White photograph
    b. Infra-red (IR) black & white photograph
    c. **Colour photograph**
    d. Infra-red colour photograph

48. Which photographs are used to identification of broad leaved & coniferous tree-
    a. Panchromatic Black & White photograph
    b. **Infra-red (IR) black & white photograph**
    c. Colour photograph
    d. Infra-red colour photograph

49. Infra-red colour photograph also called as-
    a. True colour photographs
    b. **False colour photographs**
    c. Colour photographs
    d. All of the above

Answers

42. a	43. d	44. d	45. c	46. a	47. c
48. b	49. b				

50. Which type of film is used in false colour photograph-
    a. Panchromatic film
    b. Infra-red (IR) film with proper filter
    c. Colour films
    d. **Kodak Aerochrome Infra-red**

51. Scale which is used in very large scale photograph-
    a. 1: 5,000 to 1: 10,000 b. 1: 10,000 to 1: 20,000
    c. 1: 20,000 to 1: 40,000 d. 1: 40,000 to 1: 70,0000

52. Scale which is used in large scale photograph-
    a. 1: 5,000 to 1: 10,000 b. **1: 10,000 to 1: 20,000**
    c. 1: 20,000 to 1: 40,000 d. 1: 40,000 to 1: 70,0000

53. Scale which is used in medium scale photograph-
    a. 1: 10,000 to 1: 25,000 b. 1: 20,000 to 1: 50,000
    c. **1: 20,000 to 1: 40,000** d. 1: 10,000 to 1: 20,000

54. Scale which is used in small scale photograph-
    a. 1: 5,000 to 1: 10,000 b. 1: 10,000 to 1: 20,000
    c. 1: 20,000 to 1: 40,000 d. **1: 40,000 to 1: 70,0000**

55. Which scale is used in broad land use classification and forest type identification-
    a. **<1: 70,0000** b. >1: 70,0000
    c. >1: 50000 d. <1: 20000

56. Deviation allow in vertical photograph upto an angle-
    a. 4° from perpendicular position
    b. 4° from parallel position
    c. 10° from perpendicular position
    d. 4° from parallel position

57. Oblique photographs are taken from an angle-
    a. 90° b. 60°
    c. 45° d. **30°**

Answers

50. d	51. a	52. b	53. c	54. d	55. a
56. a	57. d				

CHAPTER 9

Rangeland Management

1. The total livestock population according 19th Livestock Census of India is-
   a. 312.56 million
   b. 402.12 million
   c. **512.05 million**
   d. 645.12 million

2. The total livestock population has decreased by about over the previous census-
   a. 1.30%
   b. 2.98%
   c. **3.33%**
   d. 4.55%

3. Livestock Census-2012 was-
   a. 10th Livestock Census of India
   b. 13th Livestock Census of India
   c. 15th Livestock Census of India
   d. **19th Livestock Census of India**

4. India support about the total livestock of the world-
   a. 05%
   b. 10%
   c. **15%**
   d. 20%

5. In India the area under permanent pasture is-
   a. 10.29 million ha
   b. **12.47 million ha**
   c. 15.78 million ha
   d. 20.54 million ha

6. In India area under permanent pasture is about of the total geographical are of the country-
   a. **4.10%**
   b. 5.90%
   c. 6.67%
   d. 8.45%

Answers

1. c 2. c 3. d 4. c 5. b 6. a

7. Sehima/Dichanthium grasslands are best grown in-
   a. Paddy tracts & high rainfall belts
   b. **Black soil**
   c. Low hills
   d. Sandy loams
8. Dichanthium/Cenchrus grasslands are best grown in-
   a. Low hills b. High mountains
   c. **Sandy loams** d. Temperate Alpine climate
9. Phragmites/Saccharum grasslands are best grown in-
   a. Low hills b. **Marshy area**
   c. Sandy loams d. Temperate Alpine climate
10. Cymbopogon grasslands are best grown in-
    a. Mixed temperate climate b. Sandy loams
    c. Marshy area d. **Low hills**
11. Dabadghao and Shankar Narayan have recognised how many types of grass cover in India-
    a. Four b. **Five**
    c. Six d. Eight
12. The number of individual animals that can survive the greatest period of stress each year on a land area is called-
    a. Productivity b. **Carrying capacity**
    c. Grazing capacity d. Rangeland
13. The number of animal grazing a unit of area at a particular time is called-
    a. Carrying capacity b. Grazing capacity
    c. **Stocking rate** d. Livestock unit
14. The feed requirement used as the basis of comparison for different classes and species of stock is called-
    a. Stocking rate b. **Livestock unit**
    c. Grazing capacity d. None
15. 1 livestock unit requires approximately good quality pasture dry matter per year-
    a. 450 kg b. **520 kg**
    c. 650 kg d. 700 kg

Answers

7. b	8. c	9. b	10. d	11. b	12. b
13. c	14. b	15. b			

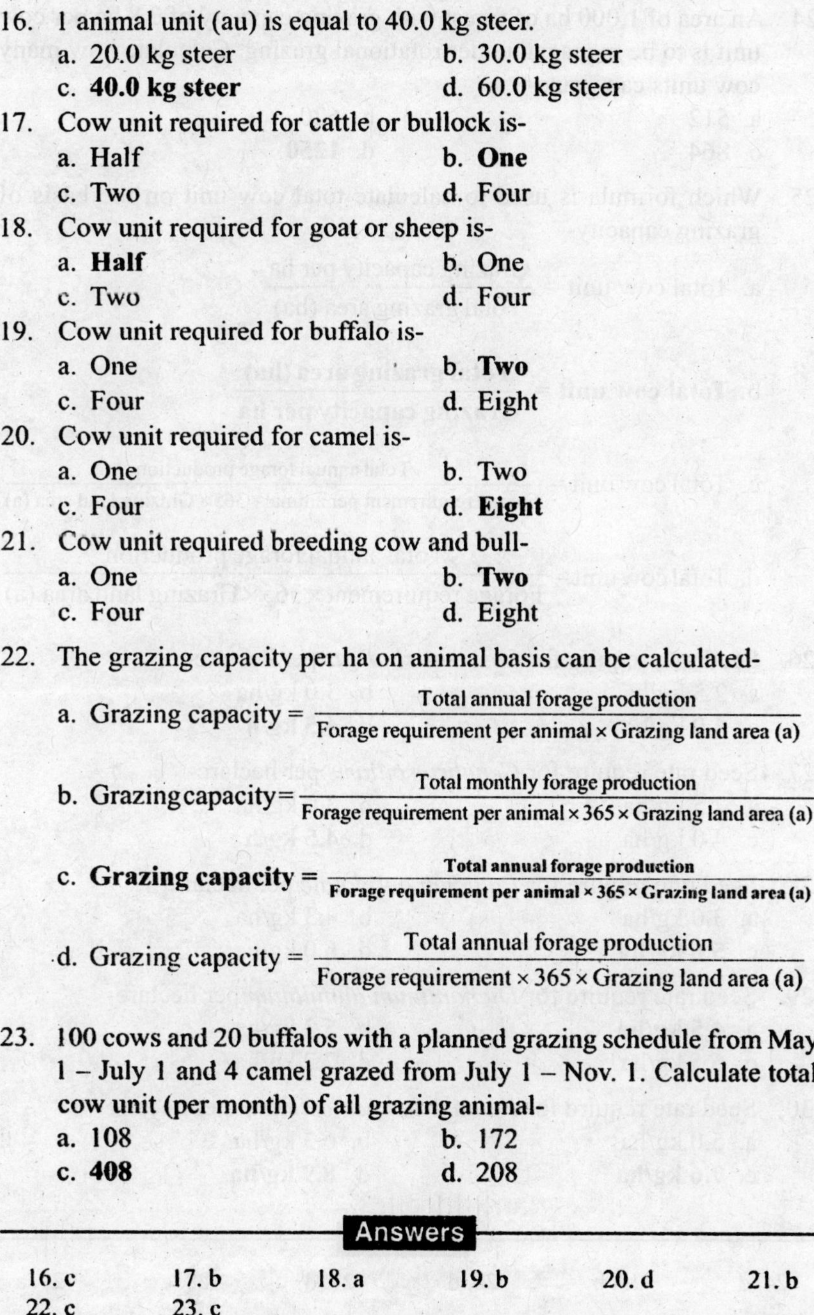

16. 1 animal unit (au) is equal to 40.0 kg steer.
    a. 20.0 kg steer b. 30.0 kg steer
    c. **40.0 kg steer** d. 60.0 kg steer
17. Cow unit required for cattle or bullock is-
    a. Half b. **One**
    c. Two d. Four
18. Cow unit required for goat or sheep is-
    a. **Half** b. One
    c. Two d. Four
19. Cow unit required for buffalo is-
    a. One b. **Two**
    c. Four d. Eight
20. Cow unit required for camel is-
    a. One b. Two
    c. Four d. **Eight**
21. Cow unit required breeding cow and bull-
    a. One b. **Two**
    c. Four d. Eight
22. The grazing capacity per ha on animal basis can be calculated-

    a. $$\text{Grazing capacity} = \frac{\text{Total annual forage production}}{\text{Forage requirement per animal} \times \text{Grazing land area (a)}}$$

    b. $$\text{Grazing capacity} = \frac{\text{Total monthly forage production}}{\text{Forage requirement per animal} \times 365 \times \text{Grazing land area (a)}}$$

    c. $$\textbf{Grazing capacity} = \frac{\textbf{Total annual forage production}}{\textbf{Forage requirement per animal} \times 365 \times \textbf{Grazing land area (a)}}$$

    d. $$\text{Grazing capacity} = \frac{\text{Total annual forage production}}{\text{Forage requirement} \times 365 \times \text{Grazing land area (a)}}$$

23. 100 cows and 20 buffalos with a planned grazing schedule from May 1 – July 1 and 4 camel grazed from July 1 – Nov. 1. Calculate total cow unit (per month) of all grazing animal-
    a. 108 b. 172
    c. **408** d. 208

Answers

16. c	17. b	18. a	19. b	20. d	21. b
22. c	23. c				

24. An area of 1,000 ha of forest with grazing capacity of 0.8 ha per cow unit is to be managed under rotational grazing. Calculate how many cow units can graze-
   a. 512 b. 620
   c. 864 d. **1250**

25. Which formula is used to calculate total cow unit on the basis of grazing capacity-
   a. $\text{Total cow unit} = \dfrac{\text{Grazing capacity per ha}}{\text{Total grazing area (ha)}}$

   b. $\textbf{Total cow unit} = \dfrac{\textbf{Total grazing area (ha)}}{\textbf{Grazing capacity per ha}}$

   c. $\text{Total cow unit} = \dfrac{\text{Total annual forage production}}{\text{Forage requirement per animal} \times 365 \times \text{Grazing land area (a)}}$

   d. $\text{Total cow unit} = \dfrac{\text{Total annual forage production}}{\text{Forage requirement} \times 365 \times \text{Grazing land area (a)}}$

26. Seed rate require for *Lasiurus sindicus* per hectare-
   a. 2.5 kg/ha b. 3.0 kg/ha
   c. 4.0 kg/ha d. **4.5 kg/h**

27. Seed rate require for *Cenchrus ciliaris* per hectare-
   a. 2.5 kg/ha b. 3.0 kg/ha
   c. 4.0 kg/ha d. **4.5 kg/h**

28. Seed rate require for *Panicum antidotale* per hectare-
   a. 3.0 kg/ha b. 4.3 kg/ha
   c. **5.6 kg/ha** d. 6.0 kg/ha

29. Seed rate require for *Dichanthium annulatum* per hectare-
   a. **4.5 kg/ha** b. 5.3 kg/ha
   c. 6.5 kg/ha d. 7.2 kg/ha

30. Seed rate require for *Pennisetum pedicellatum* per hectare-
   a. 5.0 kg/ha b. 6.3 kg/ha
   c. **7.6 kg/ha** d. **8.9 kg/ha**

Answers

24. d 25. b 26. d 27. d 28. c 29. a
30. c

31. Seed rate require for *Sehima nervosum* per hectare-
a. 4.5 kg/ha   b. 5.0 kg/ha
c. **7.8 kg/ha**   d. 8.5 kg/ha

32. The object of scientific management of grassland is to maintain the grassland in-
a. The highest state of production
b. Soil and moisture conservation
c. Sustainable production
d. Master seed production

33. When the number of animals that are allowed to graze per unit area of rangeland is fixed accordance with rangeland carrying capacity and grazing is not allowed during plants are passing through the critical stage of growth called-
a. Controlled grazing   b. Continuous controlled grazing
c. Deferred grazing   d. Rotational grazing

34. Buffalo grass or paragrass is botanically known as-
a. ***Brachiaria mutica***   b. *Cenchrus ciliaris*
c. *Sehima nervosum*   d. *Panicum maximum*

35. When the animals are kept permanently on a given area of a rangeland and allowed to move freely over this area due to this the crop composition is changed is called-
a. Deferred rotational grazing
b. **Continuous controlled grazing**
c. Deferred grazing
d. Rotational grazing

36. *Cenchrus ciliaris* commonly known as-
a. Barware grass   b. Sain grass
c. **Anjan grass**   d. Elephant grass

37. When the grassland is divided into three compartments and the grazing being allowed in two compartments while third one is given rest called-
a. Controlled grazing   b. Continuous controlled grazing
c. **Deferred grazing**   d. Rotational grazing

Answers

31. c	32. a	33. a	34. a	35. b	36. c
37. c					

38. Bermuda grass or doob grass is botanically known as-
a. Pennisetum pedicellatum b. ***Cynodon dactylon***
c. *Dichanthium annulatum* d. *Pennisetum purpureum*

39. The conservation of green forage into dry form without affecting the quality of the original material called-
a. Silage b. Fodder
c. **Hay** d. Fermentation

40. When the grassland is divided into two or more **blocks** and the grazing is allowed in one of the blocks while other block gets rest and next year cattle are moved to other block to give rest to the previous block called-
a. Continuous controlled grazing b. Deferred grazing
c. **Rotational grazing** d. Deferred rotational grazing

41. Sain grass botanically known as-
a. Thysanolaena maxima b. *Pennisetum purpureum*
c. *Sorghum sudanense* d. ***Sehima nervosum***

42. The most common practice of grazing in India is-
a. Controlled grazing
b. **Continuous controlled grazing**
c. Deferred grazing
d. Rotational grazing

43. Which system of grazing management has been found more suitable for perennial grasses-
a. Continuous controlled grazing
b. **Deferred grazing**
c. Rotational grazing
d. Deferred rotational grazing

44. Deferred rotational grazing is a combination of which system
a. Continuous controlled grazing and Rotational grazing
b. **Deferred grazing and Rotational grazing**
c. Rotational grazing and Deferred rotational grazing
d. Deferred rotational grazing and Controlled grazing

45. Deferred rotational grazing has increased the grazing capacity upto-
a. 23% b. 29%
c. 35% d. **37%**

Answers

38. b	39. c	40. c	41. d	42. b	43. b
44. b	45. d				

46. The moisture content in hay is upto-
    a. 10% b. 12%
    c. 13% d. **15%**

47. Livestock population census in India is conducted in each-
    a. 2 year b. 4 year
    c. **5 year** d. 10 year

48. *Panicum maximum* locally known as-
    a. Napier grass b. Elephant grass
    c. Tiger Grass d. **Guinea grass**

49. Population census in India is conducted in each-
    a. 1 year b. 5 year
    c. **10 year** d. 15 year

50. Tiger Grass botanically known as-
    a. Dichanthium annulatum b. *Panicum antidotale*
    c. *Sorghum sudanense* d. ***Thysanolaena maxima***

51. Wildlife population census India is conducted in each-
    a. 2 year b. **5 year**
    c. 7 year d. 10 year

52. Pennisetum purpureum is locally known as-
    a. Napier grass b. Elephant grass
    c. **Both a and b** d. Tiger Grass

53. Forest survey report is released every-
    a. **2 year** b. 5 year
    c. 7 year d. 10 year

54. Jarga grass botanically known as-
    a. Dichanthium annulatum b. Panicum antidotale
    c. Sorghum sudanense d. Thysanolaena maxima

55. The product obtained by packing fresh fodder in a suitable container and allowing it to ferment under anaerobic condition without undergoing much loss of nutrients called-
    a. Hay b. **Silage**
    c. Alcohol d. Ensilage

Answers

46. d	47. c	48. d	49. c	50. d	51. b
52. c	53. a	54. a	55. b		

56. The process of preserving green fodder in a suitable container and allowing it to ferment under anaerobic condition called-
    a. Silage b. Hay
    c. **Ensilage** d. All of the above

57. Silage is prepared in how many month-
    a. 1-2 months b. **2-3 months**
    c. 3-4 months d. 2-4 months

58. In which system of grazing area of grazing is divided into two block A and B-
    a. Continuous controlled grazing
    b. Deferred grazing
    c. Rotational grazing
    d. **Deferred rotational grazing**

59. Average forage yield of tiger grass per ha is-
    a. 2 tons/ha b. **4 tons/ha**
    c. 6 tons/ha d. 8 tons/ha

60. Average forage yield of Cyanodon grass per ha is-
    a. 200-350 Qt./ha b. 300-450 Qt./ha
    c. 200-300 Qt./ha d. **300-350 Qt./ha**

61. Average forage yield of napier grass or elephant grass is-
    a. 10 ton/ha b. 30 ton/ha
    c. **40 ton/ha** d. 50 ton/ha

Answers

56. c	57. b	58. d	59. b	60. d	61. c

CHAPTER 10

Forest Protection

1. A branch of forestry which deals with the activities directed towards the prevention and control of damage to forest by man, animals, fungi, insects, injurious plants and adverse climatic factors is called-
   a. Forest pathology
   b. Forest management
   c. **Forest protection**
   d. Forest Biometry
2. According Swaminathan about million tonnes of top soil in India are washed down or blown away every year-
   a. 1k million tonnes
   b. 3.5k million tonnes
   c. 4.5k million tonnes
   d. **6k million** tonnes
3. Loss of cultivable land due to erosion every year is-
   a. 4.5 million acre
   b. **6.0 million acre**
   c. 6.3 million acre
   d. 7.2 million acre
4. According to Wiackowski the leaf surface is times greater than the earth's surface occupied by plants-
   a. **10 to 20**
   b. 10 to 15
   c. 20 to 25
   d. 20 to 30
5. Dense plantation of which species is very beneficial in reducing burning coal pollution-
   a. *Tectona grandis*
   b. *Madhuca indica*
   c. *Cedrus deodara*
   d. ***Pitheolobium dulce***
6. Forest protection measures can be broadly classified into how many categories-
   a. **Two**
   b. Three
   c. Four
   d. Six

Answers

1. c	2. d	3. b	4. a	5. d	6. a

7. Which is a method of forest protection measures-
   a. Preventive measures b. Remedial measures
   c. **Both a and b** d. None

8. Those measures which prevent occurrence of damage to forest from damage causal agency-
   a. Preventive measures b. Remedial measures
   c. Preservative measure d. Conservation measures

9. Those measure operations which are carried out to minimize the damage after damage has occurred is called-
   a. Preventive measures b. **Remedial measures**
   c. Preservative measure d. Conservation measures

10. Illicit felling of trees from a piece of land without the intension of reforestation is called-
    a. Plantation b. Afforestation
    c. Reforestation d. **Deforestation**

11. Who has submitted his report on "Improvement of Indian Agriculture" in 1893-
    a. Dr. Swaminathan b. Dr. B.P Pal
    c. **Dr. Voelcker** d. Dr. N. E. Barlog

12. Act of seizing prossession of some forest land by the peoples around the forest is called-
    a. Tribal Act b. Fundamental rights Act
    c. **Encroachment** d. Land Act

13. A place where the timber or other forest produce required by villagers for their bonafide domestic and agricultural use are obtained called-
    a. **Nistar** b. Forest depots
    c. Timber market d. Halters

14. Forest fire percent caused by man's deliberate & intentional actions is-
    a. 5% b. 25%
    c. 50% d. **5%**

15. On the basis of causative factors forest fire is classified into how many types-
    a. Two b. **Three**
    c. Four d. Five

Answers

7. c	8. a	9. b	10. d	11. c	12. c
13. a	14. d	15. b			

16. Which is/are not a types of forest fire on the basis of causative factors-
a. Natural fires b. Accidental fires
c. Deliberate or intentional fires d. **Traditional forest fire**

17. On the basis of the place of their action forest fire is classified into how many types-
a. Two b. Three
c. **Four** d. Five

18. Which is/arc not a types of forest fire on the basis of the place of their action-
a. Creeping fire b. **Accidental fires**
c. Ground fire d. Surface fire

19. Which is/are types of forest fire on the basis of causative factors-
a. **Natural fires** b. Crown fire
c. Surface fire d. Ground fire

20. Which is/are types of forest fire on the basis of the place of their action-
a. Accidental fires b. Deliberate fire
c. **Ground fire** d. Intentional fires

21. A forest fire spreading slowly over the ground with low flame in the absence of strong wind-
a. Crown fire b. Surface fire
c. Ground fire d. **Creeping fire**

22. A forest fire that burns ground cover (herbaceous plants and shrubs) only is called-
a. **Ground fire** b. Intentional fires
c. Surface fire d. Deliberate fire

23. A forest fire which burns not only the ground cover but also undergrowth is called-
a. Crown fire b. **Surface fire**
c. Ground fire d. Creeping fire

24. A forest fire which spreads through the crowns of trees and consumes all or part of the upper branches and foliage is called-
a. **Crown fire** b. Surface fire
c. Ground fire d. Creeping fire

Answers

16. d	17. c	18. b	19. a	20. c	21. d
22. a	23. b	24. a			

25. Which forest fire most commonly occurs in plains-
    a. Ground fire   b. **Surface fire**
    c. Crown fire   d. Creeping fire

26. Which forest fire most commonly occurs in coniferous forests-
    a. **Crown fire**   b. Surface fire
    c. Ground fire   d. Creeping fire

27. Which tree is resistance to fire-
    a. Teak   b. Sissoo
    c. Siris   d. **Chirpine**

28. Which is/are the factors responsible for make up the fire environment
    a. Weather   b. Inflammable materials
    c. topography   d. **All of the above**

29. The unusable residue after logging and conversion operations called-
    a. Stump   b. Stock
    c. **Slash**   d. Cull

30. Speed of forest fire varies from 0.3 meter per minute to 107 meter per minute-
    a. 0.3 meter per minute to 107 meter per minute
    b. 0.6 meter per minute to 110 meter per minute
    c. 1.0 meter per minute to 120 meter per minute
    d. 1.3 meter per minute to 125 meter per minute

31. Fire speed under forest canopy does not exceed-
    a. 4 m/min   b. **6 m/min**
    c. 8 m/min   d. 10 m/min

32. In plains (Northern & Central India) fire season start from the month-
    a. Feb to June   b. May to June
    c. **March to June**   d. March to July

33. In sub mountain chirpine forest & deodar forests fire season start from the month-
    a. March to May   b. **May to June**
    c. March to June   d. March to July

34. In southern India forests fire season start from the month-
    a. Mid-January to April ending   b. Mid-February to May ending
    c. Mid-January to June ending   d. Mid-March to June ending

Answers

25. b	26. a	27. d	28. d	29. c	30. a
31. b	32. c	33. b	34. a		

35. Burning of leaves litter or other undergrowth before completely drying to prevent later fire damage called- early burning or controlled burning

a. Early burning b. Controlled burning
c. Deliberate burning d. **Both a and b**

36. The treatment or handling of slash for reducing hazards from fire, fungi or insects and for providing the seeds with access to soil I called-

a. Controlled burning b. **Slash disposal**
c. Slash burning d. Jhum cultivation

37. Which species act as a natural fire break in evergreen forests due to its juicy stem-

a. **Strobilanthus** b. Seabuckthorn
c. Chirpine d. Bamboo

38. A cleared permanent fire break intended to prevent fires from crossing from one area to another is called-

a. Coupe b. **Fire lines**
c. Trenches d. Fire trace

39. How many kinds of fire lines found in the forest-

a. **Two** b. Three
c. Four d. Eight

40. Which is not a kinds of firelines in the forest-

a. Internal firelines b. External firelines
c. Intermediate firelines d. **Both a and b**

41. Internal firelines (width 6m to 30 m) are set-inside the forest to prevent the fire

a. Inside the forest to provide the path to fire
b. Inside the forest to provide water supply to the forest
c. Outside the forest to prevent the fire spread from other area
d. **Inside the forest to prevent the fire**

42. Width of internal firelines are kept-

a. 2-20 m b. **6-30 m**
c. 8-30 m d. 6-20 m

Answers

35. d	36. b	37. a	38. b	39. a	40. d
41. d	42. b				

43. A cleared lines used as a base from which to counterfire is called-
a. Fire lines
b. Trenches
c. **Fire trace**
d. Watch tower

44. Map prepared by using 1:50,000 R.F scale called-
a. Regeneration map
b. Forest type map
c. Soil map
d. **Fire map**

45. *Xyleutes ceramicus* (Beehole) is more abundantly found in-
a. **Teak plantation**
b. Teak natural forest
c. Sal forest
d. Deodar forest

46. Which is teak defoliator-
a. *Hyblaea tectonae*
b. ***Hyblaea puera***
c. *Plecoptera reflexa*
d. *Tonica niviferana*

47. Which is deodar defoliator-
a. *Hyblaea cedrae*
b. *Hyblaea puera*
c. *Plecoptera reflexa*
d. ***Ectropis deodarae***

48. Which is shisham/sissoo defoliator-
a. *Plecoptera dalbergeae*
b. *Hyblaea puera*
c. ***Plecoptera reflexa***
d. *Ectropis deodarae*

49. Which is toon shoot borer-
a. *Tonica niviferana*
b. ***Hypsipyla robusta***
c. *Hypsipyla ciliae*
d. *Hyblaea puera*

50. Which is sal heartwood borer-
a. Hypsipyla robusta
b. *Hyblaea puera*
c. *Hapalia machaeralis*
d. ***Hoplocerambyx spinicornis***

51. Which is teak skeletonizer-
a. ***Hapalia machaeralis***
b. *Hypsipyla robusta*
c. *Plecoptera reflexa*
d. None

52. Which is sal skeletonizer-
a. *Hoplocerambyx spinicornis*
b. *Hypsipyla robusta*
c. *Tonica niviferana*
d. **None**

53. Which is the bark borer of *Pinus longifolia*-
a. *Hypsipyla longifolia*
b. ***Ips longifolia***
c. *Plecoptera machaeralis*
d. None

Answers

43. c	44. d	45. a	46. b	47. d	48. c
49. b	50. d	51. a	52. d	53. b	

54. Which is/are the major pest of forest nursery-
a. Cockchafers b. Cutworms
c. Crickets d. **All of the above**

55. In forest nursery cockchafers feed on-
a. **Seedlings roots** b. Seedlings leaves
c. Cut seedlings d. All of the above

56. Which is/are the example of obnoxious weeds-
a. *Lantana* b. *Parthenium*
c. *Eupartium* d. **Both a and c**

57. Which cause pink disease in eucalyptus tree-
a. Cronartium himalaynese b. ***Corticium salmonicolour***
c. *Corticium unicolour* d. *Poria monticola*

58. The National Plant Quarantine Station situated in-
a. Mumbai b. **New Delhi**
c. Odisha d. Kolkata

59. A technique for ensuring disease- and pest-free plants, whereby a plant is isolated while tests are performed to detect the presence of a problem-
a. Tissue culture b. Plant pathology
c. **Plant quarantine** d. Elisa test

60. *Cronartium himalaynese* commonly known as-
a. Chirpine Swertia felt rust b. Deodar Swertia felt rust
c. Teak Swertia felt rust d. Sal Swertia felt rust

61. Which is cause decay in heartwood of blue pine-
a. *Fusarium* b. *Trametes pini*
c. *Fomes pini* d. **Both b and c**

62. Which is responsible for developed red colour of wood in spruce and fir instead of white colour of wood-
a. ***Fusarium*** b. *Trametes pini*
c. *Fomes pini* d. Both b and c

63. Root rot disease in sissoo is caused due to-
a. ***Ganoderma lucidum*** b. *Fusarium solani*
c. *Fomes annosus* d. None

Answers

54. d	55. a	56. d	57. b	58. b	59. c
60. a	61. d	62. a	63. a		

64. Wilt disease in sissoo is caused due to-
a. *Ganoderma lucidum* b. ***Fusarium solani***
c. *Fomes annosus* d. *Fomes badius*

65. Root rot disease in khair (*Accacia catechu)* is caused due to-
a. *Armillarea mellea* b. *Fomes annosus*
c. *Fusarium solani* d. ***Ganoderma lucidum***

66. Heart rot disease in khair (*Accacia catechu)* is caused due to-
a. ***Fomes badius*** b. *Fusarium solani*
c. *Fomes annosus* d. *Fomes pini*

67. Root rot disease in sal tree is caused due to-
a. *Ganoderma lucidum* b. *Polyporus zonalis*
c. *Polyporus annosus* d. ***Polyporus shoreae***

68. Heart rot disease in sal tree is caused due to-
a. Hymenochaete rubiginosa b. Hymenochaete shoreae
c. Hymenochaete solanecearum d. Hymenochaete vesiculosum

69. Wilt disease in teak is caused due to-
a. Hymenochaete rubiginosa b. *Pseudomonas vesiculosum*
c. ***Pseudomonas solanecearum*** d. *Hymenochaete solanecearum*

70. Root rot disease in teak tree is caused due to-
a. *Ganoderma lucidum* b. ***Polyporus zonalis***
c. *Polyporus shoreae* d. *Polyporus tectonae*

71. Wilt disease in *Casuarina equisetifolia* is caused due to-
a. Trichosporium solanecearum b. ***Trichosporium vesiculosum***
c. *Hymenochaete solanecearum* d. *Ganoderma lucidum*

72. Root rot disease in silver fir and spruce tree is caused due to-
a. ***Armillarea mellea*** b. *Armillarea annosus*
c. *Armillarea zonalis* d. All of the above

73. Root rot disease in deodar tree is caused due to-
a. *Ganoderma lucidum* b. *Polyporus zonalis*
c. ***Fomes annosus*** d. *Polyporus cedrae*

74. Heart rot disease in silver fir and spruce tree is caused due to-
a. *Ganoderma lucidum* b. ***Fomes spp.***
c. *Armillarea zonalis* d. *Polyporus cedrae*

Answers

64. b	65. d	66. a	67. d	68. a	69. c
70. b	71. b	72. a	73. c	74. b	

CHAPTER 11

Silviculture

1. The word forest is derived from Latin word "*foris*" meaning-
   a. Outside the village boundary
   b. Away from inhabited land
   c. **Both a and b**
   d. None
2. The art and science of cultivating forest crops is called-
   a. Forestry b. Silvics
   c. **Silviculture** d. Forest biology
3. The study of life history and generally characteristics of forest trees crops with particular references to environmental factors as the basis for the practice of silviculture is called-
   a. Forestry b. **Silvics**
   c. Silviculture d. Plantation forestry
4. An area set aside for the production of timber and other forest produce or maintained under woody vegetation for certain indirect benefits which it provides-
   a. Forest farm b. **Forest**
   c. Forestry d. Social forestry
5. On the basis of regeneration forest is classified into-
   a. High forest b. Coppice forest
   c. Pollard forest d. **Both a and b**
6. On the basis of age forest is classified into-
   a. Even- aged b. Uneven-aged
   c. Mixed forest d. **Both a and b**

Answers

1. c 2. c 3. b 4. b 5. d 6. d

7. A forest which is regenerated from seed is called-
   a. Low forest b. **High forest**
   c. Pure forest d. Mixed forest

8. A forest which is regenerated from vegetative parts of tree either naturally or artificially is called as-
   a. Production forest b. **Coppice forest**
   c. Pollard forest d. High forest

9. On the basis of management forest is classified into-
   a. Protection forest b. Production forest
   c. Social forest d. **All of the above**

10. Pure forests are composed almost entirely.........species while mixed forests composed..... species respectively-
    a. One species & more than two species
    b. More than two species & one species
    c. Both are more than two species
    d. Two species & more than three species

11. On the basis of composition forest is classified into-
    a. Pure forest b. Mixed forest
    c. Normal forest d. **Both a and b**

12. An area with complete protection, constituted according to chapter II of Indian forest Act-1927 is called as-
    a. **Reserved forest** b. Protected forest
    c. Village forest d. Protection forest

13. On the basis of growing stock forest is classified into-
    a. Pure forest b. Abnormal forest
    c. Normal forest d. **Both b and c**

14. An area with limited degree of protection, constituted according to chapter IV of Indian forest Act-1927 is called-
    a. Reserved forest b. **Protected forest**
    c. Village forest d. Protection forest

15. On the basis of ownership forest is classified into-
    a. Public forest b. Public forest
    c. Production forest d. **Both a and b**

Answers

7. b	8. b	9. d	10. a	11. d	12. a
13. d	14. b	15. d			

16. A forest which is stable forest assigned to a village community under the provisions of chapter III of Indian forest Act-1927 is called-
    a. Reserved forest b. Protected forest
    c. **Village forest** d. Protection forest
17. The theory and practice of all that constitutes the creation, conservation and scientific management of forests and the utilization of their resources is called-
    a. Social forestry b. Silviculture
    c. **Forestry** d. Forest management
18. The foundation of scientific forestry was laid during the year-
    a. 1825 b. 1847
    c. 1856 d. **1864**
19. Who is the first Inspector General (IGF) of Indian forest-
    a. L. S. Khanna b. **Dr. Dietrich Brandis**
    c. Dr. Dietrich Hokip d. Ram Prakash
20. The Central Board of Forestry (CBF) was constituted in the year-
    a. 1864 b. 1920
    c. **1948** d. 1950
21. Van Mahotsav is a-
    a. Festival of tree planting
    b. Festival of tribal living in forest
    c. Festival of wildlife
    d. Festival of tree census in a year
22. Van Mahotsav was started by-
    a. Sunder Lal Bahuguna b. Munshi Prem Chand
    c. **K. M. Munshi** d. Mohit Husain
23. Van Mahotsav was started in the year-
    a. 1864 b. 1927
    c. **1950** d. 1976
24. About per cent of the forests of the country are owned by the Government-
    a. **95.8** b. 33.0
    c. 24.75 d. 75.87

Answers

16. c	17. c	18. d	19. b	20. c	21. a
22. c	23. c	24. a			

25. Most of the forests in India belongs to which climate type-
a. Temperate b. **Tropical**
c. Deciduous d. Alpine

26. Complex of physical and biological factors of an area that determine what vegetation it may carry is called-
a. Site productivity b. Site index
c. **Site** d. Site quality

27. The relative productive capacity of a site for particular species is called-
a. Site productivity b. Site index
c. Site value d. **Site quality**

28. The portion of the solar energy to which human eye is sensitive is known as-
a. **Light** b. Energy
c. Albido d. Spectrum

29. The sky appears blue because of-
a. Reflection b. Refraction
c. **Scattering** d. Emission

30. The ability of a surface to reflect the solar radiation is called-
a. **Albedo** b. Reflection
c. Refraction d. Transmission

31. The species which require abundant light for their optimum growth and development are called-
a. Fast growing b. Slow growing
c. **Light demander** d. Hardy species

32. A species which require shade atleast in early stages for their optimum growth and development called-
a. Shade loving b. Fast growing
c. Slow growing d. **Shade demander**

33. A species which are capable of persisting and developing under shade called-
a. **Shade bearer** b. Shade demander
c. Hardy species d. Slow growing

Answers

25. b	26. c	27. d	28. a	29. c	30. a
31. c	32. d	33. a			

34. Chilling of air below the its freezing point called-
a. Snow
b. Due
c. **Frost**
d. Fog

35. The rate at which rain falls is called-
a. Rainfall
b. **Rainfall intensity**
c. Precipitation
d. Rain coefficient

36. A day in which 2.5 mm or more rain falls called-
a. **Rainy day**
b. Dry day
c. Wet day
d. All of the above

37. Which is/are light demander species-
a. Dipterocarpus
b. Calophylum
c. Shorea
d. **All of the above**

38. Radiation frost also known as-
a. Pool frost
b. **Ground frost**
c. Advective frost
d. Convection frost

39. Which is/are shade bearer species-
a. Artocarpus
b. Pterocarpus
c. Gmelina
d. **All of the above**

40. Which is/are shade bearer species-
a. Mesua
b. Xylia
c. Schleichera
d. **All of the above**

41. A productivity which is really obtainable of at given conditions is called-
a. Real productivity
b. **Potential productivity**
c. Site productivity
d. Productivity

42. Full abbreviation of CVP index is-
a. Climate, Vegetation and Precipitation
b. Climate, Vegetation and Productivity site
c. **Climate, Vegetation and Productivity index**
d. Climate, Vegetation and Potential index

43. CVP index was given by-Paterson-
a. Dietrich Brandis
b. Backer
c. **Paterson**
d. Bruce Zobel

Answers

34. c	35. b	36. a	37. d	38. b	39. d
40. d	41. b	42. c	43. c		

44. Which is/are shade bearer species-
a. Quercus b. Cedrus
c. Cupressus d. **All of the above**

45. Which is/are light demander species-
a. Bombax b. Pinus
c. Sissoo d. **All of the above**

46. Which is/are shade demander species-
a. Syzygium b. Taxus
c. Abies d. **All of the above**

47. The height of a place from mean sea level is called-
a. Latitude b. **Altitude**
c. Aspect d. Slope

48. The direction towards which a slope faces or it is the direction of slope of land is called-
a. Latitude b. Altitude
c. **Aspect** d. Slope

49. The angle formed by the surface of the land with horizontal plane is called-
a. Slope b. Gradient
c. Altitude d. **Both a and b**

50. Most disease in plants are caused by Thallophytes (fungi, bacteria and lichens)-
a. Bryophytes b. **Thallophytes**
c. Mycota d. Protozoans

51. Thallophytes includes-
a. Fungi, Bacteria and Virus b. Fungi, Bacteria and Algae
c. Fungi, Algae and Earthworm d. **Fungi, Bacteria and Lichens**

52. Which is/are light demander species-
a. Adina b. Populus
c. Teak d. **All of the above**

53. Which is/are shade bearer species-
a. Picea b. Toona
c. *Dalbergia latifolia* d. **All of the above**

Answers

44. d	45. d	46. d	47. b	48. c	49. d
50. c	51. d	52. d	53. d		

54. Which is/are shade demander species-
a. Mallotus
b. Careya
c. Taxus
d. **All of the above**

55. Plants which grow on the trunk and branches of other plants for the purpose of supports is called-
a. Parasite
b. **Epiphytes**
c. Hygrophytes
d. Partial Parasite

56. Which is/are he examples of important epiphytes-
a. Banyan
b. Pipal
c. Orchids
d. **All of the above**

57. Accumulation of heavy cold air into natural depression is called-
a. Radiation frost
b. **Pool frost**
c. Advective frost
d. Ground frost

58. Cold air brought from outside during winters is called-
a. Radiation frost
b. Pool frost
c. **Advective frost**
d. Ground frost

59. Frost which is produced due to loss of heat by radiation is called-
a. **Ground frost**
b. Pool frost
c. Advective frost
d. Convection frost

60. Pool frost/convection frost is most common occurs in-
a. New Delhi
b. Jammu & Kashmir
c. **Dehra Dun valley**
d. Himanchal Pradesh

61. Which frost is more injurious among all frost-
a. Ground frost
b. **Pool frost**
c. Advective frost
d. Radiation frost

62. Tree with single apical meristem are termed as-
a. Abaxial tree
b. **Monoaxial tree**
c. Buttress tree
d. Wolf tree

63. Forking is most common in which species-
a. Soft wood
b. Hard wood
c. **Broad-leaved**
d. Conifers

Answers

54. d	55. b	56. d	57. b	58. c	59. a
60. c	61. b	62. b	63. c		

64. Buttressing is most common in-
 a. Tropical forest
 b. Temperate forests
 c. **Wet and moist tropical forests**
 d. Wet and moist evergreen tropical forests

65. When a stem shows irregular involution and swellings called-
 a. Forking b. Buttressing
 c. **Fluting** d. Canker

66. The symbiotic association between certain fungi and roots of higher plants which enhances the uptake of water and nutrients called-
 a. Lichen b. **Mychorrhiza**
 c. Serula d. Helotism

67. The branch of botany which deals with the study of internal structures and organization of plants or plant organs called-
 a. Plant physiology b. Plant Taxonomy
 c. Plant morphology d. **Plant anatomy**

68. From germination to the stage that the young trees has a few leaves called-
 a. Seedling b. Sapling
 c. Pole d. **Recruit**

69. From recruit stage upto a height of 1 m of the young plant called-
 a. **Seedling** b. Sapling
 c. Pole d. Recruit

70. From the time the young tree reaches 1 m height till the lower branches begin to fall and having dead bark on the stem called-
 a. Seedling b. **Sapling**
 c. Pole d. Recruit

71. From the fall of the lower branches to the time when the rate of increase in height begins to fall and crown expansion becomes marked called-
 a. Seedling b. Sapling
 c. **Pole** d. Recruit

Answers

64. c	65. c	66. b	67. d	68. d	69. a
70. b	71. c				

72. The seasonal changes in the development of foliage, flowering and fruiting is called-
a. **Phenology** b. Morphology
c. Growth c. Development

73. An area where plants are raised for eventual planting out has ordinarily both seedlings and transplants is called-
a. Forest area b. **Forest nursery**
c. Poly house d. Green house

74. Seedling Nursery is a nursery which has only seedling beds for raising-
a. **Seedling** b. Sapling
c. Pole d. Recruit

75. For coniferous species which soil is reported to be good for their growth-
a. Acidic soils having pH 3 4.0
b. Alkaline soils having pH 6 – 6.5
c. **Acidic soils having pH 5 – 6.5**
d. Alkaline soils having pH 7– 8.5

76. The size of nursery varies from percent of total plantation area-
a. 0.1 to 1.5 percent b. 0.4 to 2.0 percent
c. 0.4 to 2.5 percent d. **0.5 to 2.5 percent**

77. Distance between bed to bed in a forest nursery is-
a. 25 cm b. 25 cm
c. 35 cm d. **45 cm**

78. Normal bed size is in a forest nursery is-
a. **10 m × 1 m** b. 10 m × 12 m
c. 12 m × 10 m d. 12.5 m × 1.2 m

79. Standard bed size is in a forest nursery is-
a. 10 m × 1 m b. 10 m × 12 m
c. 12 m × 10 m d. **12.5 m × 1.2 m**

Answers

72. a	73. b	74. a	75. c	76. d	77. d
78. a	79. d				

80. Area of nursery (A) is calculated by using formula-

a. $A = \dfrac{\mathbf{1.8 \times 1.2 \times plantation\ area\ (X) \times required\ plant/ha\ (Y)}}{\mathbf{Number\ of\ plant\ in\ a\ bed\ (z)}}$

b. $A = \dfrac{1.8 \times \text{plantation area (X)} \times \text{required plant/ha (Y)}}{\text{Number of plant in a bed (z)}}$

c. $A = \dfrac{1.2 \times \text{plantation area (X)} \times \text{required plant/ha (Y)}}{\text{Number of plant in a bed (z)}}$

d. $A = \dfrac{1.8 \times \text{plantation area (X)} \times \text{required plant/ha (Y)}}{2 \times \text{Number of plant in a bed (z)} \times 1.2}$

81. The operations which are carried out in the forest crops from seedlings to mature stages to provide a healthy environment for their growth and development called-tending operations
a. Cultural operation b. **Tending operations**
c. Thinning operation d. None

82. Tending operations includes-
a. Weeding, cleaning and thinning
b. Improvement fellings, pruning and climber cutting
c. Weeding, cleaning and earthing up
d. **Both a and b**

83. Removal of all unwanted plants particularly in seedling stage at nursery or in forest crops to reduce the competitions for moisture, nutrients and to provide sufficient growing space for the desired species is called-
a. Thinning b. Pruning
c. **Weeding** d. Cleaning

84. Lantana camera is a notorious weed controlled by-
a. ***Orthezia insignis*** b. *Sporobolus tecti*
c. *Pasturella conyza* d. *Bacillus lantanae*

85. Cleaning is carried out in the nursery at-sampling stage
a. Seedling b. **Sapling**
c. Pole d. Recruit

Answers

80. a	81. b	82. d	83. c	84. a	85. b

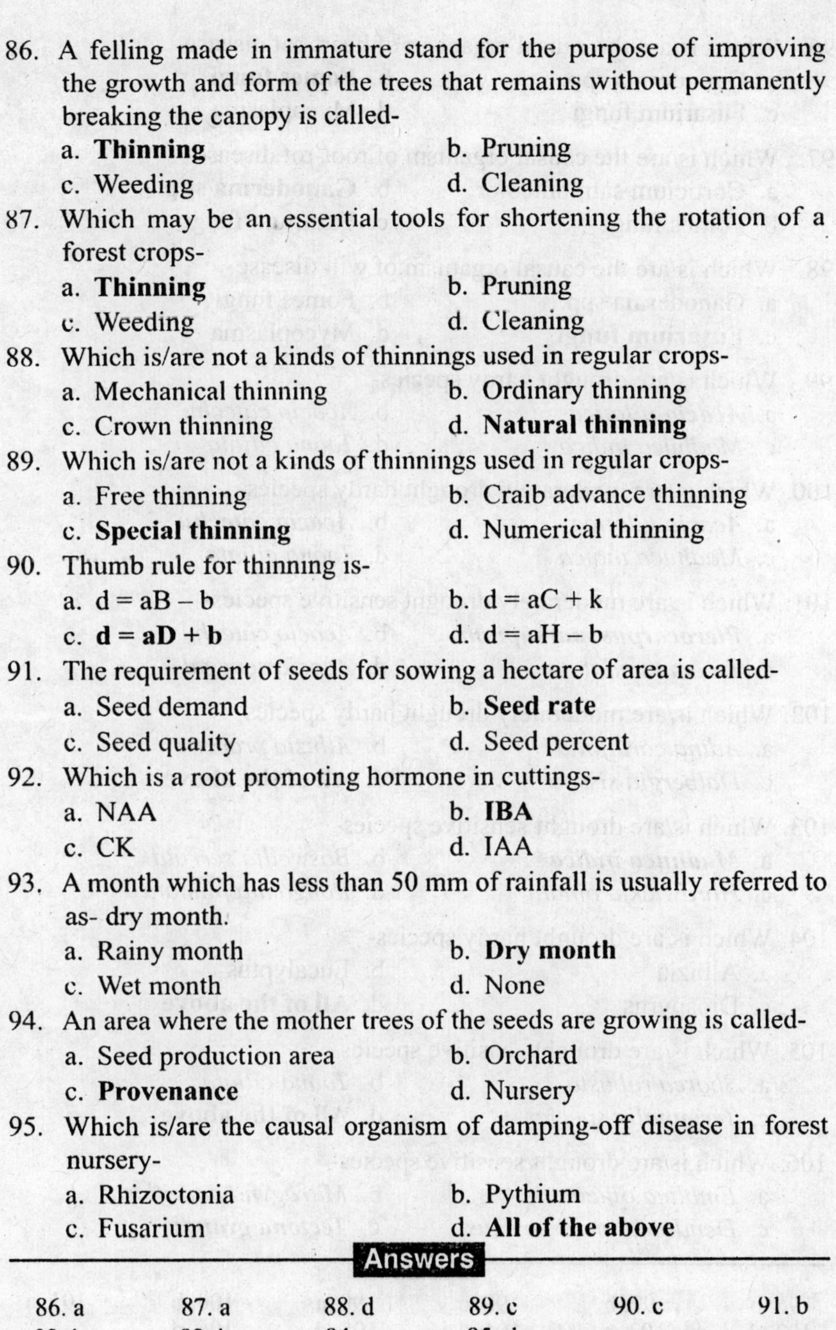

86. A felling made in immature stand for the purpose of improving the growth and form of the trees that remains without permanently breaking the canopy is called-
   a. **Thinning** b. Pruning
   c. Weeding d. Cleaning
87. Which may be an essential tools for shortening the rotation of a forest crops-
   a. **Thinning** b. Pruning
   c. Weeding d. Cleaning
88. Which is/are not a kinds of thinnings used in regular crops-
   a. Mechanical thinning b. Ordinary thinning
   c. Crown thinning d. **Natural thinning**
89. Which is/are not a kinds of thinnings used in regular crops-
   a. Free thinning b. Craib advance thinning
   c. **Special thinning** d. Numerical thinning
90. Thumb rule for thinning is-
   a. d = aB – b b. d = aC + k
   c. **d = aD + b** d. d = aH – b
91. The requirement of seeds for sowing a hectare of area is called-
   a. Seed demand b. **Seed rate**
   c. Seed quality d. Seed percent
92. Which is a root promoting hormone in cuttings-
   a. NAA b. **IBA**
   c. CK d. IAA
93. A month which has less than 50 mm of rainfall is usually referred to as- dry month.
   a. Rainy month b. **Dry month**
   c. Wet month d. None
94. An area where the mother trees of the seeds are growing is called-
   a. Seed production area b. Orchard
   c. **Provenance** d. Nursery
95. Which is/are the causal organism of damping-off disease in forest nursery-
   a. Rhizoctonia b. Pythium
   c. Fusarium d. **All of the above**

Answers

86. a	87. a	88. d	89. c	90. c	91. b
92. b	93. b	94. c	95. d		

96. Which is/are the causal organism of heart-rot disease-
    a. Ganoderma spp.
    b. **Fomes fungi**
    c. Fusarium fungi
    d. Mycoplasma

97. Which is/are the causal organism of root-rot disease-
    a. Corticium salmonicolor
    b. **Ganoderma spp.**
    b. Fomes fungi
    c. Fusarium fungi

98. Which is/are the causal organism of wilt disease-
    a. Ganoderma spp.
    b. Fomes fungi
    c. **Fusarium fungi**
    d. Mycoplasma

99. Which is/are drought hardy species-
    a. ***Acacia nilotica***
    b. *Acacia catechu*
    c. *Madhuca indica*
    d. *Toona ciliata*

100. Which is/are moderately drought hardy species-
    a. *Acacia nilotica*
    b. ***Acacia catechu***
    c. *Madhuca indica*
    d. *Toona ciliata*

101. Which is/are moderately drought sensitive species-
    a. ***Pterocarpus marsupium***
    b. *Acacia catechu*
    c. *Tecoma undulata*
    d. *Acacia senegal*

102. Which is/are moderately drought hardy species-
    a. *Adina cordifolia*
    b. *Albizia procera*
    c. *Dalbergia sissoo*
    d. **All of the above**

103. Which is/are drought sensitive species-
    a. ***Madhuca indica***
    b. *Boswellia serrata*
    c. *Hardwickia binata*
    d. *Pongamia pinnata*

104. Which is/are drought hardy species-
    a. Albizia
    b. Eucalyptus
    c. Diospyrus
    d. **All of the above**

105. Which is/are drought sensitive species-
    a. *Shorea robusta*
    b. *Toona ciliata*
    c. *Terminalia species*
    d. **All of the above**

106. Which is/are drought sensitive species-
    a. *Emblica officinalis*
    b. *Mitragyna parviflora*
    c. *Dendrocalamus strictus*
    c. ***Tectona grandis***

Answers

96. b	97. b	98. c	99. a	100. b	101. a
102. d	103. a	104. d	105. d	106. d	

107. Which is/are drought hardy species-
a. *Azadirachta indica* b. *Tecoma undulata*
c. *Albizia procera* d. **All of the above**

108. Desert soil with high per cent of sand & salts is formed in which climate-
a. **Hot & dry area** b. Hot-humid & hot wet
c. Colder regions d. Alpine region

109. Grey-brown, red and lateritic soil is formed in which climate-
a. Hot & dry area b. **Hot-humid & hot wet**
c. Colder regions d. Alpine region

110. Podsol soil is formed in which climate-
a. Hot & dry area b. Hot-humid & hot wet
c. **Colder regions** d. Desert region

111. Conical shape crown is found in which species-
a. **Pinus** b. Silver fir
c. Neem d. Babul

112. Cylindrical shape crown is found in which species-
a. Deodar b. **Spruce**
c. Imli d. *Prosopis juliflora*

113. Spherical shape crown is found in which species-
a. Eucalyptus b. **Mahua**
c. Babul d. Spruce

114. Broad & flat topped crown is found in which species-
a. ***Accacia planifrons*** b. *Prosopis juliflora*
c. *Azadirachta indica* d. *Acacia catechu*

115. Broom shape crown is found in which species-
a. Pinus b. Silver fir
c. Neem d. **Babul**

116. Frondose shape crown is found in which species-
a. *Accacia planifrons* b. ***Prosopis juliflora***
c. *Azadirachta indica* d. *Acacia catechu*

117. Conical shape crown is found in which species-
a. **Deodar** b. Spruce
c. Imli d. *Prosopis juliflora*

Answers

107. d	108. a	109. b	110. c	111. a	112. b
113. b	114. a	115. d	116. b	117. a	

118. Cylindrical shape crown is found in which species-
a. *Accacia planifrons* b. *Prosopis juliflora*
c. *Azadirachta indica* d. ***Eucalyptus spp.***

119. Which is the Glover formula for thinning-
a. **d = D** b. d = 1.5 D
c. d = 2 D d. d= 3 D

120. Which is the Laurie formula for thinning-
a. d = D b. **d = 1.5 D**
c. d = 2 D d. d = 3 D

121. Which is the Sagreiya formula for thinning-
a. d = D b. d = 1.5 D
c. **d = 2 D** d. d = 3 D

122. Which is the Howard formula for thinning-
a. d = D b. d = 1.5 D
c. d = 2 D d. **d = 3 D**

123. Glover formula is used for thinning in which species-
a. **Deodar** b. Irregular sal crops
c. Sissoo d. Irregular teak crops

124. Laurie formula is used for thinning in which species-
a. Deodar b. **Irregular sal crops**
c. Sissoo d. Irregular teak crops

125. Sagreiya formula is used for thinning in which species-
a. Deodar b. Irregular sal crops
c. **Sissoo** d. Irregular teak crops

126. Howard formula is used for thinning in which species-
a. Deodar b. Irregular sal crops
c. Sissoo d. **Irregular teak crops**

127. Ordinary thinning also called-
a. German thinning b. Low thinning
c. Thinning from the below d. **All of the above**

128. In ordinary thinning trees are removed from which class-
a. Higher class b. Upper class
c. **Lower class** d. Medium class

Answers

118. d	119. a	120. b	121. c	122. d	123. a
124. b	125. c	126. d	127. d	128. c	

129. Ordinary thinning is most commonly used in which species-
a. **Light demander species**
b. Shade demander species
c. Shade bearer species
d. All of the above

130. Ordinary thinning is classified into how many grades-
a. Four
b. **Five**
c. Six
d. Eight

131. Which is not a grade of ordinary thinning-
a. A
b. E
c. B
d. **F**

132. Removal of dead, dying, diseased and suppressed trees of classes-III, IV and V is called-
a. **Light thinning**
b. Moderate thinning
c. Heavy thinning
d. Very Heavy thinning

133. Removal of dead, dying, diseased and suppressed and an occasional very defective dominant I (c) trees of classes-I (d), II (b), III, IV and IV is called-
a. Extremely Heavy thinning
b. **Moderate thinning**
c. Light thinning
d. Heavy thinning

134. Removal of all classes of trees of D grades and more dominants of I (a) without making permanent gaps in canopy of Classes-I (a), I (b), I (c), I (d) II, III, IV and V is called-
a. **Extremely Heavy thinning**
b. Moderate thinning
c. Light thinning
d. Heavy thinning

135. All classes of trees of B grades and removal of remaining dominated and defective codominants without making lasting gaps in canopy of Classes-I (b), I (c), I (d) II, III, IV and V is called-
a. Light thinning
b. Moderate thinning
c. **Heavy thinning**
d. Very Heavy thinning

136. All classes of trees of C grades and some goods dominants of some I (a) without making lasting gaps in canopy of Classes-I (b), I (c), I (d) II, III, IV and V is called-
a. **Very Heavy thinning**
b. Moderate thinning
c. Light thinning
d. Heavy thinning

Answers

129. a	130. b	131. d	132. a	133. b	134. a
135. c	136. a				

137. Light thinning also called as-
a. **A grade** b. C grade
c. D grade d. B grade

138. Extremely Heavy thinning also called as-
a. A grade b. C grade
c. D grade d. **E grade**

139. Which is/are a strong coppicer-
a. ***Acacia catechu*** b. *Pterocarpus marsupium*
c. *Acacia nilotica* d. *Pinus spp.*

140. Which is/are a fair coppicer-
a. *Albizia spp.* b. ***Terminalia spp.***
c. *Adina cordifolia* d. *Cedrus deodara*

141. Which is/are a poor coppicer-
a. ***Madhuca indica*** b. *Anogeissus latifolia*
c. *Manilkara hexandra* d. *Abies pindrew*

142. Which is/are a non coppicer-
a. *Butea monosperma* b. *Hardwickia binata*
c. *Bombax ceiba* d. ***Picea smithiana***

143. Which is/are a poor coppicer-
a. *Acacia nilotica* b. *Adina cordifolia*
c. *Madhuca indica* d. **All the above**

144. Which is/are a non coppicer-
a. *Pinus spp.* b. *Cedrus deodara*
c. *Abies pindrew* d. **All the above**

145. Which is/are a fair coppicer-
a. *Pterocarpus marsupium* b. *Quercus leucotrichophora*
c. *Manilkara hexandra* d. **All**

146. Which is/are a strong coppicer-
a. *Albizia spp.* b. *Anogeissus latifolia*
c. *Butea monosperma* d. **All the above**

147. Which is/are a strong coppicer-
a. *Dalbergia sissoo* b. *Diospyrus melanoxylon*
c. *Ougenia oojensis* d. **All of the above**

Answers

137. a	138. d	139. a	140. b	141. a	142. d
143. d	144. d	145. d	146. d	147. d	

148. Which is/are a strong coppicer-
a. ***Shorea robusta*** b. *Pterocarpus marsupium*
c. *Acacia nilotica* d. *Pinus spp.*

149. Which is/are a fair coppicer-
a. *Albizia spp.* b. ***Hardwickia binata***
c. *Adina cordifolia* d. *Cedrus deodara*

150. Which is/are a strong coppicer-
a. *Acacia nilotica* b. *Adina cordifolia*
c. *Madhuca indica* d. ***Tectona grandis***

151. Which is/are a strong coppicer-
a. *Pinus spp.* b. *Cedrus deodara*
c. ***Salix spp.*** d. *Abies pindrew*

152. Which is/are a strong coppicer-
a. *Acacia nilotica* b. ***Eucalyptus spp.***
c. *Madhuca indica* d. *Adina cordifolia*

153. Which shape beds are preferred than other shapes in nursery-
a. Square b. Triangular
c. Cuboidal d. **Rectangular**

154. Raised beds are prepared at the height above the ground level-
a. 5 – 10cm b. **10 – 15cm**
c. 30 – 50cm d. 10 – 100 cm

155. Sunken beds are prepared at the depth from the ground level-
a. 10 – 15cm b. **10 – 30cm**
c. 30 – 40cm d. 20 – 40 cm

156. Raised beds are suitable for which climatic conditions-
a. Normal rainfall area b. Dry area
c. **Heavy rainfall area** d. Low rainfall area

157. Sunken beds are suitable for which climatic conditions-
a. Normal rainfall area b. **Dry area**
c. Heavy rainfall area d. Low rainfall area

158. Level beds are suitable for which climatic conditions-
a. **Normal rainfall area** b. Dry area
c. Heavy rainfall area d. Low rainfall area

Answers

148. d	149. b	150. d	151. c	152. b	153. d
154. b	155. b	156. c	157. b	158. a	

159. The soil in beds should be dug out to a depth upto- 40 to 50cm
  a. 10 to 20cm b. 10 to 40cm
  c. 20 to 60cm d. **40 to 50cm**

160. Ratio of Sand: Soil: FYM for soil filling polybags in nursery-
  a. 1 : 1 : 1 b. 1: 2: 1
  c. **1 : 3 : 1** d. 3 : 1 : 1

161. Which is not a standard ploythene bags size used in nursery-
  a. 10 cm × 20 cm b. 15 cm × 20 cm
  c. **10 cm × 25 cm** d. 20 cm × 30 cm

162. Which size of polythene bag is used in nursery for seed germination-
  a. **10 cm × 20 cm** b. 15 cm × 20 cm
  c. 10 cm × 25 cm d. 20 cm × 30 cm

163. Which method is used to sowing small seed-
  a. Broadcasting b. Dibbling sowing
  c. **Drill sowing** d. Harrowing

164. Seed requirement in broadcasting methods as compare to required seed is-
  a. 2– 4 times more b. 2– 6 times more
  c. **3–4 times more** d. Equal amount

165. Seed requirement in dibbling sowing method as compare to required seed is-
  a. 2– 3 times more b. 2–4 times more
  c. 3–4 times more d. **Equal amount**

166. Seed requirement in drill sowing method as compare to required seed is-
  a. **2 times more** b. 3 times more
  c. 4 times more d. equal amount

167. Depth of seed sowing depends on which parameter-
  a. Area b. Soil
  c. **Seed diameter** d. Seed viability

168. Minute seed should be sown at the depth-
  a. 1.5 to 2.5 mm b. 2.5 to 4.0 mm
  c. **2.5 to 5.0 mm** d. 2.5 to 8 mm

Answers

159. d	160. c	161. c	162. a	163. c	164. c
165. d	166. a	167. c	168. c		

169. Which formula is used to calculate quantity of seed (W)-

a. Quantity of seeds (W) = $\frac{A \times P}{D \times N} \times 100$

b. Quantity of seeds (W) = $\frac{A + D}{P \times N} \times 100$

c. Quantity of seeds (W) = $\frac{A - D}{P \times N} \times 100$

d. **Quantity of seeds (W) = $\frac{A \times D}{P \times N} \times 100$**

170. Best time of seed sowing in nursery is-
a. March to May　　b. Feb to June
c. April to August　　d. **March to June**

171. Stump planting is common practice in which species-
a. Bamboo　　b. Sissoo
c. Teak　　d. **Both b and c**

172. How many year old seedling of teak and sissoo are suitable for stump planting-
a. **1.0 year**　　b. 1.5 years
c. 2 years　　d. 2.5 years

173. One year old seedling of teak and sissoo attaining thickness are suitable for stump planting-
a. 1.0-1.5 cm　　b. **1.5-2.0 cm**
c. 2.5-3.0 cm　　d. 3.5-4.0 cm

174. The process of moving of plant from one nursery bed to another for better root and shoot growth is called-
a. Transplanting　　b. Pricking out
c. Lining out　　d. **All of the above**

175. The plants which have been pricked out in the nursery are called-
a. Seedlings　　b. Sapling
c. Pole　　d. **Transplants**

Answers

169. d　170. d　171. d　172. a　173. b　174. d
175. d

176. Which species has viability period of two weeks-
   a. *Dipterocarpus spp.* b. *Shorea robusta*
   c. *Hopea odorata* d. **All of the above**

177. Which species has viability period of three months -
   a. *Chloroxylon spp.* b. *Mangifera indica*
   d. *Podocarpus spp.* d. **All of the above**

178. Which species has viability period of one year-
   a. Artocarpus b. Toona
   c. Mesua d. **All of the above**

179. Which species has viability period of two or more than two years-
   a. Accacia spp. b. Albizia spp.
   c. Cassia spp. d. **All of the above**

180. Which species has viability period of one year-
   a. ***Tectona grandis*** b. *Azadirachta indica*
   c. *Podocarpus spp.* d. *Prosopis spp.*

181. Seeds per kilogram is determined by counting-
   a. Four replicates of 100 seeds b. Five replicates of 500 seeds
   c. Six replicates of 100 seeds d. **Five replicates of 100 seeds**

182. Viability test required how many replicates-
   a. Four replicates of 100 seeds
   b. Four replicates of 400 seeds
   c. Five replicates of 100 seeds
   d. Five replicates of 500 seeds

183. The total number of seeds that germinate in the test plus the number of sound seeds remaining ungerminated at the end of the test is called-
   a. Germination energy b. Germination potential
   c. **Germinative capacity** d. Germination

184. The percentage of seed in a sample that have germinated in a test upto the time when the rate of germination reaches its peak is called-
   a. **Germination energy** b. Germination potential
   c. Germinative capacity d. Germination

Answers

176. d	177. d	178. d	179. d	180. a	181. d
182. a	183. c	184. a			

185. Tetrazolium test is performed to test-
   a. Seed germination b. **Seed viability**
   c. Germinative energy d. Germinative capacity
186. Which indicator is used in tetrazolium test (TZ)-
   a. 1, 2, 3- triphenyl-tetrazolium chloride/bromide
   b. 1, 3, 5- triphenyl-tetrazolium chloride/bromide
   c. 3, 2, 5- triphenyl-tetrazolium chloride/bromide
   d. **2, 3, 5- triphenyl-tetrazolium chloride/bromide**
187. During tetrazolium test live seed takes a colour-
   a. White b. Green
   c. **Red** d. Black
188. Marking of planting spots or soil-working spots with the help of sticks is called-
   a. Earthing up b. **Staking**
   c. Plant spotting d. Spacing
189. The distance between plant to plant in a plantation or in a crop is called-
   a. Earthing up b. Staking
   c. Plant spotting d. **Spacing**
190. An operation under which forked or multiple stems are reduced to a single stem to improve the form of the planted tree is called-
   a. Pruning b. Thinning
   c. **Singling** d. Staking
191. An operation in which competing plants of the same or similar species are removed to provide proper growing space is called-
   a. Thinning b. Weeding
   c. Spacing d. **Re-spacing**
192. A species which grows naturally in the country or in a region is called-
   a. Exotic species b. **Indigenous species**
   c. Wild species d. Domestic species
193. A species which is grown outside the limits of its natural range is called-
   a. **Exotic species** b. Indigenous species
   c. Wild species d. Domestic species

Answers

185. b	186. d	187. c	188. b	189. d	190. c
191. d	192. b	193. a			

194. Ordinary pits for planting is suitable for which soil-
 a. **Clayey loam** b. Sandy soil
 c. Very sandy d. Laterite soil

195. Saucer shaped pits for planting is suitable for which soil-
 a. Clayey loam b. **Sandy soil**
 c. Very sandy d. Black soil

196. Ring pits for planting is suitable for which soil-
 a. Clayey loam b. Sandy soil
 c. **Very sandy** d. Alluvial soil

197. Ordinary pits is suitable for which climatic condition-
 a. **All rainfall classes** b. Medium rainfall
 c. Less rainfall d. Dry conditions

198. Saucer shaped pits is suitable for which climatic condition-
 a. All rainfall classes b. **Medium rainfall**
 c. Less rainfall d. Dry conditions

199. Ring pits is suitable for which climatic condition-
 a. All rainfall classes b. Medium rainfall
 c. **Less rainfall** d. Dry conditions

200. Which formula is used to calculate number of plant (N) in line planting plantation-
 a. $\mathbf{N = \dfrac{100\times 100}{\text{distance of plants in lines (X)} \times \text{distance between the lines (Y)}}}$

 b. $N = \dfrac{100\times 100}{\text{Square of planting distance (X)}}$

 c. $N = \dfrac{100\times 100\times 1.155}{\text{Square of planting distance (X)}}$

 d. $N = \dfrac{2\times 100\times 100}{\text{Square of planting distance (X)}}$

Answers

194. a	195. b	196. c	197. a	198. b	199. c
200. a					

201. Which formula is used to calculate number of plant (N) in square planting plantation-

a. $N = \dfrac{100 \times 100}{\text{planting distance (X)}}$

b. $N = \dfrac{\mathbf{100 \times 100}}{\textbf{Square of planting distance (X)}}$

c. $N = \dfrac{2 \times 100 \times 100}{\text{Square of planting distance (X)}}$

d. $N = \dfrac{100 \times 100 \times 1.155}{\textbf{Square of planting distance (X)}}$

202. Which formula is used to calculate number of plant (N) in triangular planting plantation-

a. $N = \dfrac{\mathbf{100 \times 100 \times 1.155}}{\textbf{Square of planting distance (X)}}$

b. $N = \dfrac{100 \times 100 \times 2.155}{\text{Square of planting distance (X)}}$

c. $N = \dfrac{2 \times 100 \times 100}{\text{Square of planting distance (X)}}$

d. $N = \dfrac{1.155 \times 100 \times 100}{\text{planting distance (X)}}$

Answers

201. b 202. a

203. Which formula is used to calculate number of plant (N) in quincunx planting plantation-

a. $N = \dfrac{\mathbf{2 \times 100 \times 100}}{\textbf{Square of planting distance (X)}}$

b. $N = \dfrac{4 \times 100 \times 100}{\text{Square of planting distance (X)}}$

c. $N = \dfrac{1.155 \times 100 \times 100}{\text{planting distance (X)}}$

d. $N = \dfrac{100 \times 100}{\text{Square of planting distance (X)}}$

204. Wire fencing has how many strands-

a. Two strands
b. Three strands
c. **Four strands**
d. Six strands

205. Standard height of stone wall fencing is-

a. 1.0 m
b. **1.25 m**
c. 1.5m
d. 2.0 m

206. *Casuarina equisetifolia* is a native of which country-

a. **Indonesia**
b. Queensland
c. Central America
d. India

207. *Eucalyptus globulus* is a native of which country-

a. India
b. North America
c. **Australia**
d. China

208. *Cinnamomum camphora* is a native of which country-

a. Burma
b. Europe
c. **China**
d. Brazil

209. *Hevea brasiliensis* is a native of which country-

a. Japan
b. Central America
c. Russia
d. **Brazil**

Answers

203. a 204. c 205. b 206. a 207. c 208. c
209. d

210. *Salix alba* is a native of which country-
   a. Australia b. **China**
   c. Indonesia d. Mexico

211. Champion and Seth (1968) classified Indian forests int0-
   a. 5 major groups and 15 type groups
   b. 6 major groups and 16 type groups
   c. **5 major groups and 16 type groups**
   d. 6 major groups and 18 type groups

212. Tropical forest major group has how many types group-
   a. Three b. Five
   c. Six d. **Seven**

213. Montane sub-tropical forests major group has how many types group-
   a. **Three** b. Five
   c. Four d. Five

214. Montane temperate forests major group has how many types group-
   a. Two b. **Three**
   c. Four d. Five

215. Sub-alpine forests major group has how many types group-
   a. **One** b. Three
   c. Four d. Five

216. Alpine scrub major group has how many types group-
   a. One b. **Two**
   c. Three d. Four

217. Which is/are not a types groups of montane temperate forests major group-
   a. Montane wet temperate forests b. Himalayan moist forests
   c. Himalayan dry forests d. **Dry evergreen forests**

218. Which is/are not a types groups of montane sub-tropical forests major group-
   a. Subtropical broad leaved forests
   b. Subtropical pine forests
   c. **Littoral and swampy forests**
   d. Subtropical dry evergreen forests

Answers

210. b	211. c	212. d	213. a	214. b	215. a
216. b	217. d	218. c			

219. Which is/are not a types groups of tropical forests major group-
a. Moist deciduous forests
b. Semi- evergreen forests
c. **Himalayan moist forests**
d. Dry deciduous forests

220. Which is/are not a types groups of tropical forests major group-
a. **Himalayan dry forests**
b. Littoral and swampy forests
c. Thorn forests
d. Dry evergreen forests

221. Mean annual temperature in the tropical forest is-
a. **> 24°C**
b. 17° - 24°C
c. 7° - 17°C
d. < 7°C

222. Mean annual temperature in the montane sub-tropical forests is-
a. > 24°C
b. **17° - 24°C**
c. 7° - 17°C
d. < 7°C

223. Mean annual temperature in the montane temperate forests is-
a. > 24°C
b. 17° - 24°C
c. **7° - 17°C**
d. < 7°C

224. Mean annual temperature in the sub-alpine forests is-
a. > 24°C
b. 17° - 24°C
c. 7° - 17°C
d. **< 7°C**

225. Mean annual rainfall in the wet forest occurs-
a. **> 2500 mm**
b. 1250 - 2500 mm
c. 750 - 1250 mm
d. < 750 mm

226. Mean annual rainfall in the moist and semi moist forests–
a. > 2500 mm
b. **1250 - 2500 mm**
c. 750 - 1250 mm
d. < 750 mm

227. Mean annual rainfall in the dry forests–
a. > 2500 mm
b. 1250 - 2500 mm
c. **750 - 1250 mm**
d. < 750 mm

228. Mean annual rainfall in the desert forests –
a. > 2500 mm
b. 1250 - 2500 mm
c. 750 - 1250 mm
d. **< 750 mm**

229. Master seed year in *Abies pindrew*-
a. **10** year
b. 12 year
c. 15 year
d. 18 year

Answers

219. c	220. a	221. a	222. b	223. c	224. d
225. a	226. b	227. c	228. d	229. a	

230. Master seed year in *Acacia catechu-*
   a. **1-2** year   b. 5 year
   c. 7 year   d. 10 year

231. Master seed year in *Cedrus deodara-*
   a. 1-2 year   b. **4-5 year**
   c. 8 year   d. 10 year

232. Master seed year in *Cupressus torulosa-*
   a. 2-3 year   b. 4-5 year
   c. **7-8 year**   d. 10 year

233. Master seed year in *Picea smithiana-*
   a. 3-5 year   b. 4-5 year
   c. 7-8 year   d. **5-6 year**

234. Master seed year in *Pinus roxburghii-*
   a. 2 year   b. **4-5 year**
   c. 8 year   d. 10 year

235. Master seed year in *Pinus wallichiana-*
   a. **2-3 year**   b. 4-5 year
   c. 7-8 year   d. 10 year

236. Master seed year in *Shorea robusta-*
   a. **3-5 year**   b. 4-5 year
   c. 7-8 year   d. 5 year

237. Time of flowering in *Abies pindrew-*
   a. Sept-October   b. July-August
   c. Mar-November   d. **None**

238. Time of seed/cone collection in *Abies pindrew*, *Cedrus deodara* & *Picea smithiana-*
   a. Sept-October   b. July-August
   c. **Oct-November**   d. Mar-June

239. Time of seed collection in *Acacia catechu-*
   a. **Jan – February**   b. Nov- March
   c. Mar-June   d. May-August

Answers

230. a	231. b	232. c	233. d	234. b	235. a
236. a	237. d	238. c	239. a		

240. Time of seed collection and time of flowering in *Adina cordifolia* respectively-
   a. **April-June and June-August**
   b. Feb-March and April-May
   c. Sept-November and Jan – Feb

241. Time of flowering and seed collection in *Ailanthus excelsa* respectively-
   a. June-September and Dec-March
   b. Sept-November and Feb-March
   c. **Feb-March and April-May**
   d. Sept-November and Jan – February

242. Time of flowering and seed collection in *Albizia procera* respectively-
   a. June-September and Dec-March
   b. **Sept-November and Feb-March**
   c. May and June-December
   d. Sept-November and Jan – February

243. Time of flowering and seed collection in *Anogeissus latifolia* respectively-
   a. Feb-April and May-June
   b. Jan-March and April-May
   c. March-May and June-July
   d. **June-September and Dec-March**

244. Time of flowering and seed collection in *Azadirachta indica* respectively-
   a. Jan-March and April-May
   b. **March-May and June-July**
   c. Feb-April and May-June
   d. May and June-December

245. Time of flowering and seed collection in *Boswellia serrata* respectively-
   a. **Feb-April and May-June**
   b. May and June-December
   c. August-September and Jan-February
   d. Jan-March and April-May

Answers

240. a 241. c 242. b 243. d 244. b 245. a

246. Time of flowering and seed collection in *Casuarina equisetifolia* respectively-
   a. August-September and Jan-February
   b. Jan-March and April-May
   c. **May and June-December**
   d. Feb-April and May-June

247. Time of flowering and seed collection in *Dalbergia sissoo* respectively-
   a. August-September and Jan-February
   b. **April-May and Dec-February**
   c. March-April and May-June
   d. Feb-April and May-June

248. Time of flowering and seed collection in *Dalbergia latifolia* respectively-
   a. **August-September and Jan-February**
   b. March-April and May-June
   c. Feb-April and May-June
   d. April-May and Dec-February

249. Time of flowering and seed collection in *Dendrocalamus strictus* respectively-
   a. Feb-March and Dec-February
   b. Feb-April and May-June
   c. **March-April and May-June**
   d. August-September and Jan-February

250. Time of flowering and seed collection in *Pterocarpus marsupium* respectively-
   a. April-June and Aug-September
   b. Sept-December and March-April
   c. March-April and June-July
   d. **Sept-October and Dec-April**

251. Time of flowering and seed collection in *Quercus species* respectively-
   a. **April-June and Aug-September**
   b. March-April and June-July
   c. August-September and Jan-February
   d. July-August and Nov-January

Answers

246. c   247. b   248. a   249. c   250. d   251. a

252. Time of flowering and seed collection in *Santalum album* respectively-
   a. March-April and June-July
   b. July-August and Nov-January
   c. **Sept-December and March-April**
   d. August-September and Jan-February

253. Time of flowering and seed collection in *Shorea robusta* respectively-
   a. August-September and Jan-February
   b. July-August and Nov-January
   c. **March-April and June-July**
   d. March-April and May-June

254. Time of flowering and seed collection in *Tectona grandis* respectively-
   a. **July-August and Nov-January**
   b. August-September and Jan-February
   c. March-April and June-July
   d. March-April and Nov-January

255. Which is/are drought hardy species-
   a. *Ailanthus excelsa*　　b. *Azadirachta indica*
   c. *Bombax ceiba*　　d. **All of the above**

256. Which is/are frost hardy species-
   a. *Acacia catechu*　　b. *Anogeissus latifolia*
   c. *Anogeissus pendula*　　d. **All of the above**

257. Which is/are frost tender species-
   a. *Acacia catechu*　　b. *Anogeissus latifolia*
   c. *Anogeissus pendula*　　d. ***Abies pindrew***

258. Which is/are drought hardy species-
   a. *Boswellia serrata*　　b. *Dalbergia latifolia*
   c. *Diospyros melanoxylon*　　d. **All**

259. Which is/are drought hardy species-
   a. *Prosopis juliflora*　　b. *Schleichera oleosa*
   c. *Syzygium cumini*　　d. **All of the above**

260. Which is/are drought sensitive and frost tender-
   a. *Abies pindrew*　　b. *Albizia lebbeck*
   c. *Casuarina equisetifolia*　　d. **All**

Answers

252. c	253. c	254. a	255. d	256. d	257. d
258. d	259. d	260. d			

261. Which is/are drought sensitive and frost tender-
   a. *Eucalyptus hybrid* b. *Populus ciliata*
   c. *Pterocarpus marsupium* d. **All of the above**

262. Which is/are drought sensitive and frost tender-
   a. *Picea smithiana* b. *Tectona grandis*
   c. *Terminalia spp.* d. **All of the above**

263. Which is/are drought sensitive and frost hardy species-
   a. *Ailanthus excelsa* b. ***Anogeissus latifolia***
   c. *Azadirachta indica* d. *Bombax ceiba*

264. Which is/are drought sensitive and frost hardy species-
   a. *Ailanthus excelsa* b. *Anogeissus latifolia*
   c. ***Cedrus deodara*** d. *Bombax ceiba*

265. Which is/are drought sensitive and frost hardy species-
   a. *Dendrocalamus strictus* b. *Madhuca indica*
   c. *Pinus roxburghii* d. **All of the above**

266. Which is/are drought sensitive and frost hardy species-
   a. *Pinus wallichiana* b. *Quercus species*
   c. *Shorea robusta* d. **All of the above**

267. Which is/are frost tender species-
   a. *Anogeissus pendula* b. *Dalbergia sissoo*
   c. ***Ailanthus excelsa*** d. *Schleichera oleosa*

268. Which is/are frost hardy species-
   a. *Azadirachta indica* b. ***Anogeissus pendula***
   c. *Bombax ceiba* d. *Prosopis juliflora*

269. Which is/are frost tender species-
   a. ***Bombax ceiba*** b. *Dalbergia sissoo*
   c. *Schleichera oleosa* d. *Diospyros melanoxylon*

270. Which is/are frost tender species-
   a. *Azadirachta indica* b. *Boswellia serrata*
   c. *Prosopis juliflora* d. **All of the above**

Answers

261. d	262. d	263. b	264. c	265. d	266. d
267. c	268. b	269. a	270. d		

CHAPTER 12

Agroforestry

1. A collective name for a and use system and technology whereby woody perennials as deliberately used on the same land management unit an agriculture crops and/or animals in same form of spatial arrangement or temporal sequence-
   a. Forestry b. Social forestry
   c. **Agroforestry** d. Farm forestry
2. Term agroforestry was coined in the year-
   a. 1920 b. 1952
   c. 1976 d. **1977**
3. Full abbreviation of ICRAF-
   a. International Council for Research in Agroforestry
   b. Indian Council for Research in Agroforestry
   c. International Council for Research in Agriculture
   d. Indian Council for Research in Agriculture
4. ICRAF came into existence in 1977-
   a. 1906 b. 1957
   c. **1977** d. 1985
5. The National Commission on Agriculture (NCA) emphasised agroforestry education in which five year plan-
   a. Fifth five-year plan period (1985-90)
   b. Sixth five-year plan period (1985-90)
   c. **Seventh five-year plan period (1985-90)**
   d. Ninth five-year plan period (1985-90)
6. First agroforestry seminar in India was organised in the year-
   a. 1906 b. 1957
   c. 1977 d. **1985**

Answers

1. c 2. d 3. a 4. c 5. c 6. d

7. First agroforestry seminar in India was organised at-
   a. **Imphal**   b. Pantanagar
   c. New Delhi   d. Gujarat
8. International seminar on "Agroforestry for Rural Need" was organised in the year-
   a. 1926   b. 1947
   c. 1958   d. **1987**
9. First of all Post-graduation education in agroforestry was started in-
   a. Pussa Agricultural Institute, New Delhi
   b. **Dr. Y.S. Parmar University of Horticulture & Forestry, Solan**
   c. Gujarat Agricultural University, Gujarat
   d. Forest Research Institute, Dehra Dun
10. Gujarat Agricultural University has started a P.G programme in agroforestry at the Aspee College of Horticulture & Forestry in the year-
    a. 1952   b. 1976
    c. 1985   d. **1993**
11. Benefits from agroforestry is classified into how many categories-
    a. Two   b. **Three**
    c. Five   d. Six
12. Which is not a categories of agroforestry benefits-
    a. Environmental benefits   b. Economic benefits
    c. **Ecological benefits**   d. Social benefits
13. Allelochemicals produced by the plants are-
    a. Primary plant metabolites
    b. **Secondary plants metabolites**
    c. Tertiary plant metabolites
    d. Not plant products
14. Allelochemicals are escaped or released from the plant parts by-
    a. Volatilization   b. Leaching
    c. Exudation   d. **All of the above**
15. Which is a method or way of releasing allelochemicals from plant parts-
    a. Weathering   b. **Decomposition**
    c. Mineralization   d. Nutrient pumping

Answers

7. a	8. d	9. b	10. d	11. b	12. c
13. b	14. d	15. b			

16. The negative effect of trees on the intercrops can be reduced by-
   a. Planting in E-W direction
   b. Planting in N-W direction
   c. Planting in W-S direction
   d. Planting in N-E direction

17. Allelopathy is the best examples for which-
   a. Lichen
   b. Mychorrhiza
   c. Commensalism
   d. **Amensalism**

18. Home garden also called as-
   a. Hedge row intercropping
   b. One-tier system
   c. Two-tier system
   d. **Multi-tier system**

19. Which is/are the fundamental attributes of all agroforestry systems-
   a. Production
   b. Conservation
   c. Protection
   d. **Both a and c**

20. The direct or indirect effects of one plant to another through the production of chemical inhibitors that are released into the environment is called-
   a. Amensalism
   b. Parasitism
   c. **Allelopathy**
   d. Commensalism

21. Allelochemicals are mostly produced by which plant parts-
   a. Roots
   b. **Leaves**
   c. Stems
   d. Fruits

22. The species interaction due to chemical influences is also called as-
   a. Allelochemistry
   b. Phytochemical ecology
   c. Allelobiology
   d. **All of the above**

23. The process of taking nutrients from deeper soil profile and depositing them on the surface layer is referred-
   a. Leaching
   b. Nutrient immobilization
   c. **Nutrient pumping**
   d. Mineralization

24. The process by which complex organic substances are broken down to simple inorganic elements is called as-
   a. Leaching
   b. Nutrient immobilization
   c. Nutrient pumping
   d. **Mineralization**

Answers

16. a	17. d	18. d	19. d	20. c	21. b
22. d	23. c	24. d			

25. The practices of rotational cropping helps in improving and maintaining-
    a. Soil fertility
    b. Soil productivity
    c. Water conservation
    d. **All of the above**
26. The process of cycling of nutrients from soil to the plants and back to the soil is called as-
    a. Decomposition
    b. **Nutrient cycling**
    c. Mineralization
    d. Leaching
27. India has been classified into how many agroecological regions-
    a. 5
    b. **8**
    c. 14
    d. 15
28. Annual precipitation is received maximum in which agroecological regions-
    a. Humid Bengal-Assam Basin
    b. Sub Humid Sutlej-Ganga Alluvial Plains
    c. Humid To Semi-Arid Western Ghats And Karnataka Plateau
    d. Sub Humid to Humid Eastern & South-Eastern Uplands
29. Annual precipitation is received minimum in which agroecological regions-
    a. Semi-Arid Lava Plateau And Central Highlands
    b. **Arid Western Plains**
    c. Humid To Semi-Arid Western Ghats And Karnataka Plateau
    d. Humid Eastern Himalayan Region And Bay Islands
30. Shifting cultivation is most common practices in which agroecological regions-
    a. Sub Humid to Humid Eastern & South-Eastern Uplands
    b. Semi-Arid Lava Plateau And Central Highlands
    c. **Humid Eastern Himalayan Region And Bay Islands**
    d. Sub Humid Sutlej-Ganga Alluvial Plains
31. According to Nair agroforestry system can be classified into how many categories-
    a. Three
    b. **Four**
    c. Five
    d. Six

Answers

25. d 26. b 27. b 28. a 29. b 30. c
31. b

32. Agroforestry system classification was given by Nair in the year-
    a. 1958 b. 1976
    c. **1987** d. 1993

33. Which is not a key features of D & D design-
    a. Flexibility b. **Productivity**
    c. Speed d. Repetition

34. The procedures of agroforestry D & D are usually done in which ways-
    a. Macro D b. Micro D
    c. Mini D d. **Both a and b**

35. Macro D & D covers-
    a. Entire ecological zone within a country
    b. Chosen ecological zone within a state
    c. Entire ecological zone within a state
    d. Chosen ecological zone within a district

36. Micro D & D covers-
    a. Entire ecological zone within a state
    b. Chosen ecological zone within a district
    c. Entire ecological zone within a country
    d. **Chosen ecological zone within a state**

37. A family of procedure for the diagnosis of land management problems & potentials and the design of agroforestry solution is called-
    a. Macro D b. Micro D
    c. **Agroforestry D & D** d. Land amelioration

38. Full form of D & D is-
    a. Design and Diagnosis
    b. **Diagnosis & Design**
    c. Design & Development
    d. Development & Diagnosis

39. Which is/are not a main criteria for good agroforestry design-
    a. Productivity b. Sustainability
    c. Adoptability d. **Flexibility**

Answers

32. c	33. b	34. d	35. a	36. d	37. c
38. b	39. d				

40. On the structural basis agroforestry systems can be classified into how many categories-
    a. **Two**
    b. Three
    c. Four
    d. Five

41. The crops-land left without crops for periods ranging from one season to several years is called-
    a. Pasture land
    b. **Fallow land**
    c. Waste land
    d. Treated land

42. The main objective of fallows is-
    a. To recover depleted soil moisture
    b. To recover soil productivity
    c. **To recover depleted soil nutrients**
    d. To enhance microbial activity

43. Alley cropping also known as-
    a. Taungya
    b. Shifting cultivation
    c. Jhum Cultivation
    d. **Hedgerow intercropping**

44. A method of planting in which rows of trees are interspersed with rows of crops is called-
    a. Taungya
    b. Social forestry
    c. Farm forestry
    d. **Alley cropping**

45. Which soil nutrient is mostly provided to plants in alley cropping practices-
    a. **Nitrogen**
    b. Potassium
    c. Sulphur
    d. Phosphorus

46. Ideally hedgerows should be positioned or grown in which direction-
    a. North-east direction
    b. West-south direction
    c. **East-west direction**
    d. South-east direction

47. Spacing between rows to rows in alley cropping-
    a. 2-4 meter
    b. **4-8 meter**
    c. 4-6 meter
    d. 3-6 meter

48. Spacing between trees within rows in alley cropping-
    a. **0.25-2.0 meter**
    b. 0.5-2.0 meter
    c. 1-2.0 meter
    d. 1.5-2.5 meter

Answers

40. a	41. b	42. c	43. d	44. d	45. a
46. c	47. b	48. a			

49. Which tree species is most commonly used for alley cropping-
    a. **Legumes tree species** b. Non-leguminous
    c. Thorny species d. Bushy tree species

50. A belt/blocks of trees or shrubs established at right angles to the prevailing wind is called-
    a. Wind break b. Energy plantation
    c. **Shelterbelts** d. Block plantation

51. Ideal width of shelterbelts is-
    a. 25 m b. **50 m**
    c. 75 cm d. 100 cm

52. The ratio of height and width for shelterbelts should be-
    a. **1 : 10** b. 10 : 1
    c. 1 : 20 d. 10 : 25

53. A typical shelter belt has a shape-
    a. Quadrangles shape b. **Triangular shape**
    c. Square shape d. Rectangular shape

54. The minimum length of shelterbelts should be about its height-
    a. 10 times b. 20 times
    c. **25 times** d. 50 times

55. The strips of the trees and/or shrub planted to protect fields, homes, canals or other areas from wind and blowing soil or sand is called-
    a. **Wind break** b. Energy plantation
    c. Shelterbelts d. Block plantation

56. The wind-breaks reduce the wind velocity upto- 25 to 75 per cent of the open wind speed
    a. 10 to 25 per cent of the open wind speed
    b. 25 to 50 per cent of the open wind speed
    c. 50 to 75 per cent of the open wind speed
    d. **25 to 75 per cent of the open wind speed**

57. Wind-breaks should be planted at an angles to the wind from which protection is needed-
    a. 30° b. 45°
    c. 60° d. **90°**

Answers

49. a	50. c	51. b	52. a	53. b	54. c
55. a	56. d	57. d			

58. For better protection against environmental factors wind break should be planted in which direction-
a. East-west b. South-east
c. North-west d. **North-south**

59. For providing better shade to growing crops wind breaks should be planted in which direction-
a. **East-west** b. South-east
c. North-west d. North-south

60. A 20 m tall wind-break provides protection from……m on the upwind side and…….m on the downwind side respectively-
a. 50 m and 100 m b. **100 m and 500 m**
c. 500 m and 100 m d. 100 m and 200 m

61. The minimum length of wind-breaks should be about the mature height of the trees-
a. 10 times b. **12 times**
c. 15 times d. 25 times

62. Wind-breaks of how many rows are most effective-
a. **3 to 5** b. 4 to 5
c. 2 to 5 d. 4 to 6

63. Which species is more effective for wind breaks-
a. *Acacia nilotica* b. *Cedrus deodara*
c. ***Casuarina equisetifolia*** d. *Tectona grandis*

64. The production of woody plants combined with pasture is referred as-
a. Agri-horticulture b. **Silvopastural**
c. Agri-silviculture d. Agri-silvi-pastoral

65. The production of fruits plants combined with crops is referred as-
a. **Agri-horticulture** b. Silvopastural
c. Agri-silviculture d. Agri-silvi-pastoral

66. The production of woody plants & pasture combined with agricultural crops is referred as-
a. Agri-horticulture b. Silvopastural
c. Agri-silviculture d. **Agri-silvi-pastoral**

Answers

58. d	59. a	60. b	61. b	62. a	63. c
64. b	65. a	66. d			

67. The production of woody plants combined with agricultural crops is referred as-
   a. Agri-horticulture b. Silvopastural
   c. **Agri-silviculture** d. Agri-silvi-pastoral

68. A farming system that combines physical, social and economic functions on the area of land around the family home is called-
   a. Truck gardening b. Home garden
   c. Multitier system d. **Both b and c**

69. Home garden is most commonly practiced in which region-
   a. Temperate b. **Humid tropical**
   c. Tropical d. Arid

70. Main crop in home garden is-
   a. Cucumber b. **Coconut**
   c. Teak d. Sandal

71. In India every homestead has around land for personal production-
   a. 0.1 to 0.2 ha b. **0.2 to 0.5 ha**
   c. 0.5 to 1.0 ha d. 1.0 to 2.0 ha

72. Home garden has usually how many vertical canopy strata-
   a. **3-4** b. 3-5
   c. 2-4 d. 3-6

73. In home garden lowermost layer is dominated by-
   a. **Vegetables** b. Fruit plants
   c. Grasses d. Timber trees

74. In home garden intermediate layer is dominated by-
   a. Vegetables b. **Fruit plants**
   c. Grasses d. Timber trees

75. In home garden uppermost layer is dominated by-
   a. Vegetables b. Fruit plants
   c. Grasses d. **Timber trees**

76. Most common agroforestry system in arid region is-
   a. Agri-horticulture b. Silvopastural
   c. **Agro-silviculture** d. Agri-silvi-pastoral

Answers

67. c	68. d	69. b	70. b	71. b	72. a
73. a	74. b	75. d	76. c		

77. Coppicing and pollarding are forms of-
    a. Regeneration
    b. Tending operation
    c. **Heading back**
    d. Thinning
78. Closed nutrient cycles are known to operate in which forests-
    a. Deciduous forests
    b. Tropical forests
    c. Evergreen forests
    d. **Evergreen natural forests**
79. Which is/are a common example of the zonal pattern-
    a. Home garden
    b. **Alley cropping**
    c. Kitchen garden
    d. Silvopastural system
80. A geohydrological unit or a place land that drains at a common point is called-
    a. Hydrological cycle
    b. **Watershed**
    c. Catchment
    d. Waterhole
81. The drainage area of the catchment is-
    a. **> 1 lakh ha**
    b. < 1 lakh ha
    c. 1 lakh ha
    d. 0.5 lakh ha
82. The drainage area of the sub-catchment is-
    a. > 20000 ha
    b. < 20000 ha
    c. < 40000 ha
    d. **> 40000 ha**
83. The drainage area of the sub-watershed is-
    a. 100-200 ha
    b. 500-1000 ha
    c. 1000-2000 ha
    d. **2000-4000 ha**
84. The drainage area of the mini-watershed is-
    a. 100-200 ha
    b. 200-1000 ha
    c. **400-2000 ha**
    d. 2000-4000 ha
85. The drainage area of the micro-watershed is-
    a. < 200 ha
    b. > 500 ha
    c. **< 400**
    d. > 400 ha
86. Loyal timber of the poor is-
    a. Teak
    b. Sal
    c. Populus
    d. **Bamboo**
87. Queen of timber is-
    a. **Teak**
    b. Sal
    c. Sandal
    d. Bamboo

Answers

77. c	78. d	79. b	80. b	81. a	82. d
83. d	84. c	85. c	86. d	87. a	

CHAPTER 13

Social Forestry

1. The word Social forestry was coined by-

   a. **Westoby**   b. K. M. Munshi

   c. Brandis   d. Mohit Husain

2. Which was the first country to embark on a major community reforestation programme-

   a. Japan   b. Germany

   c. India   d. **China**

3. Social forestry was first adopted successfully in which Indian state-

   a. West Bengal   b. **Gujarat**

   c. Uttar Pradesh   d. Uttarakhand

4. Forestry outside the conventional forests which primarily aims at providing continuous flow of goods and services for the benefit of people called-

   a. Agroforestry   b. **Social forestry**

   c. Plantation forestry   d. People forestry

5. Afforestation in the post-independence period can be divided in to-

   a. Two phases   b. **Three phases**

   c. Four phases   d. Five phases

6. Farm forestry was started in the year-

   a. 1920   b. 1958

   c. **1970**   d. 1976

7. The practice of forestry in all its aspects in and the around the farms or village lands integrated with other farm operations is called-

   a. **Farm forestry**   b. Extension forestry

   c. Mixed forestry   d. Recreation forestry

Answers

1. a   2. d   3. b   4. b   5. b   6. c
7. a

8. The practice of forestry in areas devoid of tree growth and other vegetation situated in places away from the conventional forest areas with the object of increasing the area under tree growth is called-
   a. Farm forestry b. **Extension forestry**
   c. Mixed forestry d. Recreation forestry
9. The practice of forestry for raising fodder grass with scattered fodder trees, fruit trees and fuel wood trees on suitable wastelands, panchayat lands and village commons is called-
   a. Farm forestry b. Extension forestry
   c. **Mixed forestry** d. Recreation forestry
10. The first taungya plantation in India was raised in which region-
    a. Gujarat b. West Bengal
    c. **North Bengal** d. Meghalaya
11. The practice of forestry with the object of raising flowering trees and shrubs mainly to serve as recreation forests for the urban and rural population is called-
    a. Farm forestry b. Extension forestry
    c. Mixed forestry d. **Recreation forestry**
12. The practice of forestry with the object of developing or maintaining a forest of high scenic value is called-
    a. Farm forestry b. Extension forestry
    c. **Aesthetic forestry** d. Recreation forestry
13. The practice of cutting down trees and setting them or fire and raising crops on the resulting ash called-
    a. Taungya b. **Jhum cultivation**
    c. Hill cultivation d. All of the above
14. Jhuming cultivation also known as-or shifting cultivation or slash and burn cultivation
    a. Shifting cultivation b. Slash and burn cultivation
    c. **Both a and b** d. None
15. Shifting cultivation in Assam, Tripura, and Arunachal Pradesh known as-
    a. **Jhum** b. Podu
    c. Penda d. Poonam

Answers

8. b	9. c	10. c	11. d	12. c	13. b
14. c	15. a				

16. The National Forest Policy of 1988 was issued on-
a. 7th January
b. 7th March
c. 7th November
d. **7th December**

17. Social forestry has how many components-
a. Two
b. Three
c. **Four**
d. Six

18. Which is/are not a component of social forestry-
a. Farm forestry
b. Extension forestry
c. **Aesthetic forestry**
d. Recreation forestry

19. Shifting cultivation is practised extensively in which region-
a. **North-eastern**
b. Southern
c. Western
d. Northern

20. The taungya is a Burmese word coined in Burma in the year-
a. **1850**
b. 1890
c. 2000
d. 1976

21. Shifting cultivation in Odisha is known as-
a. Jhum
b. **Podu**
c. Penda
d. Poonam

22. World Metereological Organization (WMO) was formed in the year-
a. **1950**
b. 1976
c. 1987
d. 1998

23. Shifting cultivation in Madhya Pradesh is known as-
a. Jhum
b. Podu
c. **Penda**
d. Poonam

24. Headquarter of WMO is situated in-
a. Nairobi
b. Pune
c. **Geneva**
d. Rome

25. Shifting cultivation in Tamil Nadu is known as-
a. Penda
b. Dhai
c. Jhum
d. **Kumri**

26. In taungya meaning of 'taung.' is-
a. Forest
b. Cultivation
c. Shifting
d. **Hill**

Answers

16. d	17. c	18. c	19. a	20. a	21. b
22. a	23. c	24. c	25. d	26. d	

27. Shifting cultivation in Kerala is known as-
    a. Penda b. **Poonam**
    c. Jhum d. Dhai

28. Jhuming cycle (in years) in Arunchal Pradesh is-
    a. 4-5 b. 6-8
    c. 5-10 d. **1-17**

29. Jhuming cycle (in years) in Meghalaya, Mizoram & Tripura is-
    a. **4-5** b. 6-8
    c. 5-10 d. 1-17

30. Term taungya is-
    a. Latin b. Greek
    c. **Burmese** d. French

31. Shifting cultivation in Thailand is known as-
    a. Jumar b. **Tam-ray**
    c. Hanumo d. Chena

32. In taungya meaning of 'ya' is-
    a. Forest b. **Cultivation**
    c. Shifting d. Hill

33. Shifting cultivation in Philippines is known as-
    a. Karen b. Ray
    c. **Hanumo** d. Chena

34. Gregarious flowering is usually followed by the death of-
    a. Rhizomes b. **Clumps**
    c. Culms d. Shoots

35. Shifting cultivation in Sri Lanka is known as-
    a. Jumar b. Tam-ray
    c. Hanumo d. **Chena**

36. Taungya system also called as-
    a. Jhum cultivation b. Shifting cultivation
    c. Jumar d. **Hill cultivation**

37. Bamboo also known as-
    a. Green gold b. Friend of the people
    c. Cradle of coffin timber d. **All**

Answers

27. b	28. d	29. a	30. c	31. b	32. b
33. c	34. b	35. d	36. d	37. d	

38. Taungya system in East Africa is called as-
    a. **Shamba**   b. Jhum
    c. Jumar   d. Hanumo

39. Australian wattle is-
    a. *Acacia catechu*   b. *Acacia senegal*
    c. ***Acacia auriculiformis***   d. *Acacia nilotica*

40. Taungya was first introduced by-
    a. Westoby   b. K. M. Munshi
    c. **Brandis**   d. Mohit Husain

41. Taungya was first introduced in India in the year-
    a. 1850   b. **1890**
    c. 1896   d. 2000

42. The first taungya plantation in India was raised in the year-
    a. 1850   b. 1890
    c. **1896**   d. 2000

43. Taungya is now applied to the method of raising forest plantations in combination with field crops and also known as-
    a. **Agri-silviculture**   b. Horti-silviculture
    c. Jumar   d. Kumri

44. Which is/are not a type of taungya-
    a. Departmental taungya   b. Leased taungya
    c. Village taungya   d. **Auction taungya**

45. What is the duration of leased taungya-
    a. **2 year**   b. 4 years
    c. 6 years   d. 10 years

46. Gregarious flowering occurs in-
    a. Teak   b. Sal
    c. **Bamboo**   d. Neem

47. What is the duration of departmental taungya-
    a. 2 year   b. 2-4 years
    c. **2-5 years**   d. 10 years

48. In village taungya each family has about-
    a. 0.4-0.8 ha   b. **0.8-1.6 ha**
    c. 1.6-2.0 ha   d. 2.0-2.5 ha

Answers

38. a	39. c	40. c	41. b	42. c	43. a
44. d	45. a	46. c	47. c	48. b	

49. Joint forest management (JFM) was started in the year-
    a. **1990** b. 2000
    c. 2002 d. 2006
50. National Wasteland Development Board (NWDB) was formed in the year-
    a. 1920 b. 1956
    c. **1985** d. 1996
51. Joint forest management (JFM) was firstly started in which state-
    a. Odisha b. Rajasthan
    c. **West Bengal** d. Gujarat
52. Forest products contributes to world gross domestic product about-
    a. **1%** b. 2%
    c. 3% d. 5%

Answers

49. a 50. c 51. c 52. a

CHAPTER 14

Genetics and Tree Improvement

1. Term genetics was coined by-
   a. Mendel b. Johnson
   c. **Bateson** d. Darwin

2. Term allele was coined by-
   a. Mendel b. Johnson
   c. Nilsson-Ehle d. **Bateson**

3. One mutant gene – one metabolic block hypothesis was given by-
   a. **A.Garrod** b. T.H.Morgan
   c. Nilsson-Ehle d. Mendel

4. One gene one enzyme hypothesis was given by-
   a. A.Garrod b. T.H.Morgan
   c. Nilsson-Ehle d. **Beadle & Tatum**

5. Collective study of heredity and variations is called-
   a. Mutation b. **Genetics**
   c. Tree improvement d. Ecology

6. Father of genetic is-
   a. **Mendel** b. Johnson
   c. Nilsson-Ehle d. Bateson

7. Father of tree improvement is-
   a. Mendel b. **Bruce Zobel**
   c. Tolbert d. Baker

Answers

1. c 2. d 3. a 4. d 5. b 6. a
7. b

8. Transmission of genetic characters from parents to offsprings is called-
   a. Genetics   b. Mutation
   c. Transformation   d. **Heredity**
9. Individuals of same species have some differences is called-
   a. **Variation**   b. selection
   c. Mutation   d. Heredity
10. Alternative form of a gene which are located on same position on homologous chromosome-
    a. Gene   b. Chromatid
    c. **Allele**   d. Centromere
11. The external and morphological appearance of an organism for a particular character is called-
    a. **Phenotype**   b. Genotype
    c. Variations   d. Phenology
12. The genetic constitute or genetic make-up of an organism for a particular character is called-
    a. Phenotype   b. **Genotype**
    c. Variations   d. Phenology
13. Superiority of offspring over its parents is called-
    a. Hybrid vigour   b. Heterosis
    c. **Both a and b**   d. Mutation
14. Loss of hybrid vigour due to inbreeding is called as-
    a. Mutation   b. Inbreeding effect
    b. **Inbreeding depression**   d. Selection
15. Theory of linkage was given by-
    a. **T.H.Morgan**   b. Correns
    c. Johannsen   d. Nilsson-Ehle
16. Gregor Johann Mendel was born in-
    a. 1820   b. **1822**
    c. 1895   d. 1922
17. Gregor Johann Mendel was died in-
    a. 1822   b. 1856
    c. **1884**   d. 1910

Answers

8. d	9. a	10. c	11. a	12. b	13. c
14. b	15. a	16. b	17. c		

18. Who has rediscovered the work of Mendel-
    a. Carl Correns (Germany)
    b. Hugo de Varies (Holland)
    c. Tschemark (Austria)
    d. **All of the above**
19. Mendel started his experiment on which plant-
    a. **Pea**
    b. Mung
    c. Bean
    d. Mustard
20. Mendel studied how many characters in peal plant-
    a. 3 characters
    b. **7 characters**
    c. 9 characters
    d. 14 characters
21. Gene which controls more than one character is called-
    a. Multiple gene
    b. **Pleiotropic gene**
    c. Oligotrophic gene
    d. Polytrophic gene
22. Removal of stamen from the bisexual flowers at juvenile stage called-
    a. Budding
    b. Grafting
    c. Cutting
    d. **Emasculation**
23. Term factor was given-
    a. E.Baur
    b. Bateson & Punnett
    c. Waldeyer
    d. **Johannsen**
24. Checker Board method was first time used by-
    a. Bateson & Punnett
    b. Mendel
    c. **Punnet**
    d. Johnson
25. A cross in which F_1 individuals are crossed with either recessive or dominant parents is-
    a. Test cross
    b. **Back cross**
    c. Out cross
    d. Reciprocal cross
26. Chromosomal theory of inheritance was given by-
    a. Henking
    b. Mc Clung
    c. **Walter Sutton & Theodor Boveri**
    d. Wilson & Stevens
27. When F_1 individuals is crossed with dominant parents then it is called-
    a. Test cross
    b. Back cross
    c. **Out cross**
    d. Reciprocal cross

Answers

18. d	19. a	20. b	21. b	22. d	23. d
24. c	25. b	26. c	27. c		

28. When F_1 progeny is crossed with recessive parents then it is called-
    a. **Test cross**   b. Back cross
    c. Out cross   d. Reciprocal cross

29. Which cross is used to find out the genotype of dominant individuals-
    a. Out cross   b. Reciprocal cross
    c. **Test cross**   d. Back cross

30. When two parents are used in two experiments in such a way that in one experiment "A" is used as male parent and "B" is used as female parent, while in other experiment vice-versa. Such type of a set of experiments is called-
    a. Back cross   b. **Reciprocal cross**
    c. Out cross   d. Test cross

31. More than two alternative forms of same gene called as-
    a. **Multiple allele**   b. Pleiotropic gene
    c. Oligotrophic gene   d. Polytrophic gene

32. Gene which cause death of individual when it comes in homozygous condition called-
    a. Epistatic gene   b. Recessive gene
    c. **Lethal gene**   d. Dominant gene

33. When a gene prevents the expression of another non-allelic gene this phenomena called-
    a. Pleiotropy   b. Mutation
    c. Selection   d. **Epistasis**

34. Collective inheritance of the character is called-
    a. Hereditary   b. **Linkage**
    c. Chiasma   d. Epistasis

35. Linkage first time was seen in *Lathyrus odoratus* by-
    a. **Bateson & Punnett**   b. Mendel
    c. Hugo de Varies   d. T. H. Morgan

36. Eugenics is a Greek word which mean-
    a. Abnormal born   b. **Well born**
    c. Hybrid   d. Cybrid

Answers

28. a	29. c	30. b	31. a	32. c	33. d
34. b	35. a	36. b			

37. Term phenotype and genotype was coined by-
    a. **Johannsen** b. Bateson
    c. Correns d. E.Baur
38. Sum total of genes in a reproductive gametes of a population is-
    a. Gene flow b. Genetic gain
    c. **Gene pool** d. Gene frequency
39. Migration of gene from one population to another population by cross fertilization is called-
    a. **Gene flow** b. Genetic gain
    c. Gene pool d. Gene frequency
40. The existence within the population of disadvantageous allele in a heterozygous genotype is known as-
    a. Gene flow b. Genetic gain
    c. Gene pool d. **Genetic load**
41. A proportion of different alleles of a gene in a population is called-
    a. Genetic gain b. Genetic load
    c. Gene pool d. **Gene frequency**
42. Incomplete dominance theory was given by-
    a. **Correns** b. Bateson & Punnett
    c. Mc Clung d. C.B. Bridges
43. Coupling and repulsion theory was given by-
    a. Wilson & Stevens b. Sir Francis Galton
    c. **Bateson & Punnett** d. Hugo de Varies
44. Which was the first plant in which complete genome was sequenced-
    a. *Neurospora spp.* b. *Pisum sativum*
    c. *Lamarkiana oenothora* d. ***Arabidopsis thaliana***
45. A store house of clones of known DNA fragments, genes, gene maps, seeds, spores, frozen sperms or eggs or embryos is called-
    a. Poly house b. **Gene bank**
    c. Cryopreservation d. Green house
46. Collection of many of the desired genes of DNA fragments maintained in clones of bacterial or some other cells is called-
    a. **Gene library** b. Gene pool
    c. Gene bank d. Cryopreservation

Answers

37. a	38. c	39. a	40. d	41. d	42. a
43. c	44. d	45. b	46. a		

47. A cell division which produced genetically identical cells which are similar to parent cell-
a. Amitosis
b. Mitosis
c. Meiosis
d. **Both a and b**

48. Mitosis occurs in which cells of an organism-
a. Reproductive cells
b. Ovary
c. **Somatic cells**
d. All of the above

49. Replication of nuclear DNA and synthesis of histone protein takes place in-
a. G_1 phase
b. G_2 phase
c. **S phase**
d. M phase

50. While synthesis of different types of RNA & proteins takes place in-
a. **G_1 phase**
b. G_2 phase
c. S phase
d. M phase

51. Which is a correct sequence of cell cycle-
a. $G_1 \rightarrow M \rightarrow G_2 \rightarrow S$
b. $G_1 \rightarrow G_2 \rightarrow S \rightarrow M$
c. **$G_1 \rightarrow S \rightarrow G_2 \rightarrow M$**
d. $G_1 \rightarrow G_2 \rightarrow M \rightarrow S$

52. Which is not a phase of mitosis division-
a. Prophase
b. Metaphase
c. Anaphase
d. **Leptotene**

53. Genic balance theory was given by-
a. Waldeyer
b. Strasburger
c. **C.B. Bridges**
d. Mendel

54. Longest phase of mitosis is-
a. **Prophase**
b. Metaphase
c. Anaphase
d. Telophase

55. Father of eugenics is-
a. **Sir Francis Galton**
b. Montgomery
c. Wilson & Stevens
d. McClintock

56. In which phase of mitosis golgi complexes, ER, nucleolus and nuclear membrane disappear-
a. **Prophase**
b. Anaphase
c. Metaphase
d. Telophase

Answers

47. d	48. c	49. c	50. a	51. c	52. d
53. c	54. a	55. a	56. a		

57. Term mitosis was given by-
a. Remake b. **Flemming**
c. Farmer & Moore d. Strasburger

58. Morphological studies of chromosomes is most easily done in which phase of mitosis-
a. Prophase b. Anaphase
c. **Metaphase** d. Telophase

59. Shortest phase of mitosis-
a. Prophase b. **Anaphase**
c. Metaphase d. Telophase

60. The production of four haploid cells which are dissimilar to parent cell is occurs in-
a. Amitosis b. Fission
c. **Meiosis** d. Mitosis

61. In which cell division nucleus divides twice and division of chromosome occur only once-
a. Amitosis b. Fission
c. **Meiosis** d. Mitosis

62. Prophase-I of meiosis is divided into-
a. Leptotene, zygotene, pacyhtene and diplotene
b. Leptotene, zygotene, pacyhtene, diplotene and anaphase
c. Leptotene, zygotene, pacyhtene, telophase and diakinesis
d. **Leptotene, zygotene, pacyhtene. diplotene and diakinesis**

63. Term chromosome was coined by-
a. **Waldeyer** b. Strasburger
c. C.B. Bridges d. Mendel

64. Term allosome & heterosome was given by-
a. Sir Francis Galton b. **Montgomery**
c. Wilson & Stevens d. McClintock

65. Chromosomal theory of sex determination was given by-
a. Strasburger b. Jacob & Monad
c. Rudolf Virchow d. **Wilson & Stevens**

Answers

57. b	58. c	59. b	60. c	61. c	62. d
63. a	64. b	65. d			

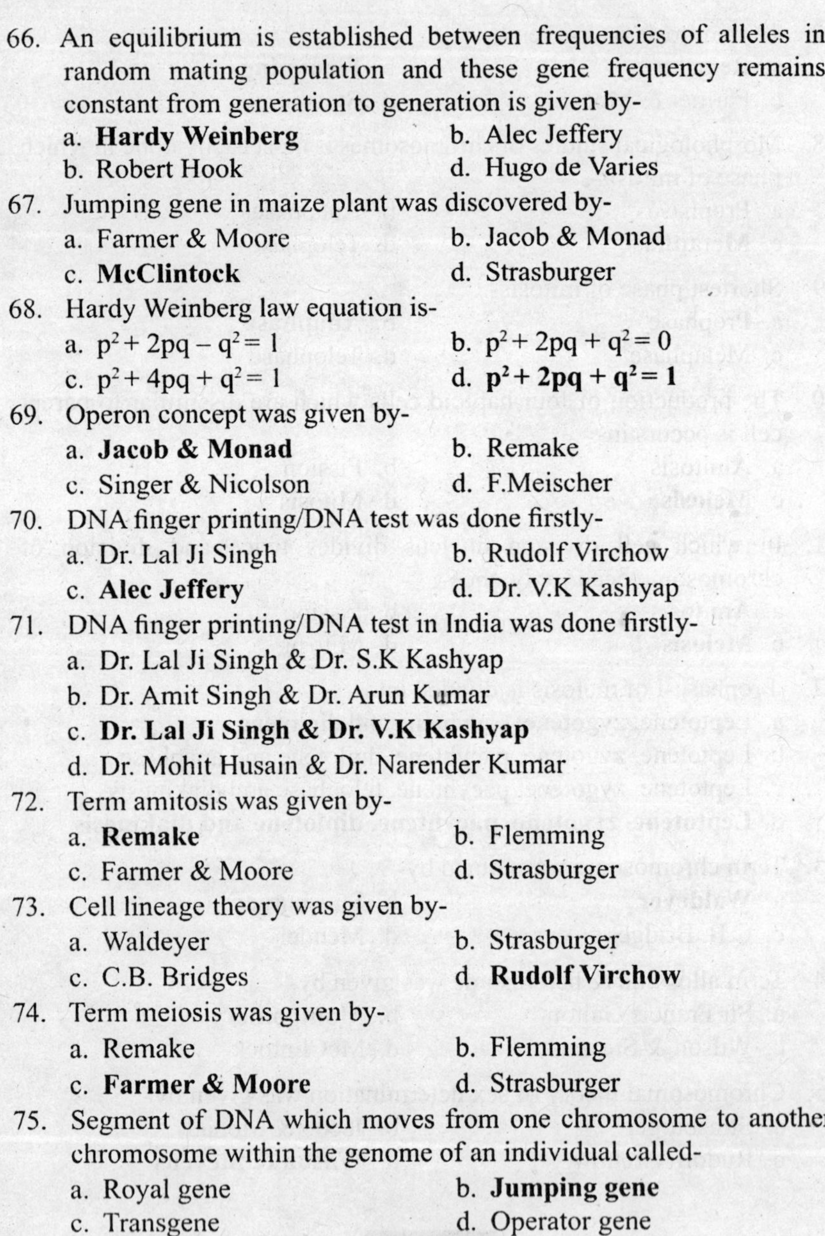

66. An equilibrium is established between frequencies of alleles in random mating population and these gene frequency remains constant from generation to generation is given by-
    a. **Hardy Weinberg** b. Alec Jeffery
    b. Robert Hook d. Hugo de Varies
67. Jumping gene in maize plant was discovered by-
    a. Farmer & Moore b. Jacob & Monad
    c. **McClintock** d. Strasburger
68. Hardy Weinberg law equation is-
    a. $p^2 + 2pq - q^2 = 1$ b. $p^2 + 2pq + q^2 = 0$
    c. $p^2 + 4pq + q^2 = 1$ d. $\mathbf{p^2 + 2pq + q^2 = 1}$
69. Operon concept was given by-
    a. **Jacob & Monad** b. Remake
    c. Singer & Nicolson d. F.Meischer
70. DNA finger printing/DNA test was done firstly-
    a. Dr. Lal Ji Singh b. Rudolf Virchow
    c. **Alec Jeffery** d. Dr. V.K Kashyap
71. DNA finger printing/DNA test in India was done firstly-
    a. Dr. Lal Ji Singh & Dr. S.K Kashyap
    b. Dr. Amit Singh & Dr. Arun Kumar
    c. **Dr. Lal Ji Singh & Dr. V.K Kashyap**
    d. Dr. Mohit Husain & Dr. Narender Kumar
72. Term amitosis was given by-
    a. **Remake** b. Flemming
    c. Farmer & Moore d. Strasburger
73. Cell lineage theory was given by-
    a. Waldeyer b. Strasburger
    c. C.B. Bridges d. **Rudolf Virchow**
74. Term meiosis was given by-
    a. Remake b. Flemming
    c. **Farmer & Moore** d. Strasburger
75. Segment of DNA which moves from one chromosome to another chromosome within the genome of an individual called-
    a. Royal gene b. **Jumping gene**
    c. Transgene d. Operator gene

Answers

66. a	67. c	68. d	69. a	70. c	71. c
72. a	73. d	74. c	75. b		

76. A part of genetic material or DNA, which acts as single regulated unit having one or more structural genes called-
a. **Operon** b. Genetic code
c. Cistron d. Codon

77. Term protoplasm was given by-
a. Bloor b. F.Meischer
c. **J.E.Purkinje** d. George Gamow

78. Phenotypic and genotypic ratio in monohybrid cross is respectively-
a. 1 : 2 : 1 and 3 : 1 b. **3 : 1 and 1 : 2 : 1**
c. 9 : 3 : 3 : 1 and 1 : 1 : 1 : 1 d. 1 : 2 : 1 and 1 : 3 : 1

79. Term nucleic acid was given by-
a. Crick b. Watson & Crick
c. Zacharis d. **Altman**

80. Phenotypic and genotypic ratio in dihybrid cross is respectively-
a. 1 : 1 : 1 : 1 and 1 : 1 : 1 : 1
b. 1 : 2 : 2 : 4 : 1 : 2 : 1 : 2 : 1 and 9 : 3 : 3 : 1
c. **9 : 3 : 3 : 1 and 1 : 2 : 2 : 4 : 1 : 2 : 1 : 2 :1**
d. 1 : 1 : 1 : 1 and 9 : 3 : 3 : 1

81. Nucleic acid discovered was discovered by-
a. Nirenberg b. Fischer
c. Khorana d. **Meischer**

82. Term DNA was coined by-
a. Crick b. Watson & Crick
c. **Zacharis** d. Altman

83. Phenotypic and genotypic ratio in monohybrid test cross is respectively-
a. 1 : 1 and 1 : 2 : 1 b. 1 : 1 : 1 : 1 and 1 : 1 : 1 : 1
c. 1 : 2 : 1 and 1 : 3 : 1 d. **1 : 1 and 1 : 1**

84. DNA was discovered by-
a. Nirenberg b. Fischer
c. Khorana d. **Meischer**

Answers

76. a	77. c	78. b	79. d	80. c	81. d
82. c	83. d	84. d			

85. Phenotypic and genotypic ratio in dihybrid test cross is respectively-
   a. 1 : 1 : 1 : 1 and 1 : 1 : 1 : 1
   b. 9 : 3 : 3 : 1 and 1 : 2 : 2 : 4 : 1 : 2 : 1 : 2 : 1
   c. 1 : 1 : 1 : 1 and 9 : 3 : 3 : 1
   d. 1 : 2 : 1 and 3 : 1

86. Total gene complement inherited haploid sets of chromosomes in an individual called as-
   a. Gene pool b. **Genome**
   c. Codon d. Cistron

87. First animal clone was developed in which institute-
   a. Franklin Institute of Scotland in 1987
   b. **Roslin Institute of Scotland in 1997**
   c. Veterinary institute of Italy in 1989
   d. Indian Institute of avian in Bareilly 2000

88. First clone animal was-
   a. Mice b. Dog
   c. Goat d. **Sheep**

89. Fluid mosaic model was given by-
   a. Strasburger b. Robert Hook
   c. **Singer & Nicolson** d. Rudolf Virchow

90. The name of first clone animal was-
   a. Molly b. Polly
   c. **Dolly** d. Jolly

91. Term genetic code was coined by-
   a. Nirenberg b. Mathai
   c. Khorana d. **Gamow**

92. Genetic code was discovered by-
   a. George Gamow
   b. **Nirenberg, Mathai & Khorana**
   c. Robert Brown
   d. Crick

Answers

85. a	86. b	87. b	88. d	89. c	90. c
91. d	92. b				

93. Central Dogma was given by-
    a. George Gamow
    b. Nirenberg, Mathai & Khorana
    c. Robert Brown
    d. **Crick**

94. Term mitochondria was coined by-
    a. Robert Hook
    b. Robert Brown
    c. **Benda**
    d. Strasburger

95. Bouquet stage, longest & thinest chromosomes are seen in which stage of meiosis-
    a. Prophase
    b. **Leptotene**
    c. Metaphase
    d. Telophase

96. Synapsis or synaptonemal complex is seen in which stage-
    a. Diplotene
    b. **Zygotcnc**
    c. Leptotene
    d. Pacyhtene

97. Term cytoplasm was coined by-
    a. Robert Hook
    b. Robert Brown
    c. Benda
    d. **Strasburger**

98. Tetrad condition is seen in which stage-
    a. Diplotene
    b. Zygotene
    c. Leptotene
    d. **Pacyhtene**

99. Chiasmata visible (crossing over) in the stage-
    a. Diakinesis
    b. **Diplotene**
    c. Pacyhtene
    d. Zygotene

100. Terminalisation of chiasmata is occurs in which stage-
    a. Diplotene
    b. Zygotene
    c. Leptotene
    d. **Diakinesis**

101. Any successful tree improvement programme has-
    a. One aspect
    b. **Two aspects**
    c. Three aspects
    d. Four aspects

102. Commonwealth Forestry Institute (CFI) is situated in-
    a. Rome
    b. Geneva
    c. **Oxford**
    d. Queensland

Answers

93. d	94. c	95. b	96. b	97. d	98. d
99. b	100. d	101. b	102. c		

103. Barner has introduced the concept of race in the year-
a. 1921 b. **1966**
c. 1975 d. 1989

104. Exotic forestry is most wide spread in-
a. Arid region b. Temperate region
c. Alpine region **d. Tropical region**

105. Division of centromere occurs in-
a. Prophase b. Metaphase
c. **Anaphase** d. Telophase

106. The membrane surrounding cell vacuole is called-
a. **Tonoplast** b. Cell wall
c. Plasma membrane d. Cell membrane

107. The diagrammatic representation of chromosomes is called-
a. Karyotype b. Holotype
c. Isotype d. **Idiogram**

108. Which is not found in animal cell-
a. Nucleus b. Golgi bodies
c. **Chloroplast** d. Mitochondria

109. Principal constituents of chromosomes are-
a. DNA b. RNA
c. Protein d. **Both a and c**

110. Knife of DNA is-
a. Exonuclease b. **Endonuclease**
c. RNA polymerase d. Both b and c

111. Term nucleus was coined by-
a. Robert Hook b. **Robert Brown**
c. Meischer d. Strasburger

112. Cell was discovered by-
a. **Robert Hook** b. Robert Brown
c. Meischer d. Strasburger

113. Hardest material in animal kingdom is-
a. Dentine b. Bone
c. Teeth d. **Enamel**

Answers

103. b	104. d	105. c	106. a	107. d	108. c
109. d	110. b	111. b	112. a	113. d	

114. Hardest material in plant kingdom is-
   a. Seed b. **Sporopollenin**
   c. Wood d. Enamel
115. Simple carbohydrates which are soluble in water and sweet in taste are called-
   a. **Sugar** b. Mantose
   c. Sacrum d. Lipid
116. Which is known as grape or blood sugar-
   a. Maltose b. Galactose
   c. **Glucose** d. Sucrose
117. Which is known as fruit sugar-
   a. Maltose b. **Fructose**
   c. Glucose d. Sucrose
118. Which is known as brain sugar-
   a. **Galactose** b. Fructose
   c. Glucose d. Sucrose
119. Which is known as cane sugar or table sugar -
   a. Galactose b. Fructose
   c. Glucose d. **Sucrose**
120. Which is known as animal starch-
   a. **Glycogen** b. Fructose
   c. Glucose d. Sucrose
121. Amino acids which is not synthesis in our body called-
   a. Non-essential amino acid b. **Essential amino acid**
   c. Saturated amino acid d. None
122. Which is/are the example of essential amino acids-
   a. Threonine b. Valine
   c. Leucine d. **All of the above**
123. Amino acids which are synthesized in our body are called-
   a. **Non-essential amino acid** b. Essential amino acid
   c. Saturated amino acid d. None
124. Which is/are the example of essential amino acids-
   a. Tryptophan b. Arginine
   c. **Glycine** d. Histidine

Answers

114. b	115. a	116. c	117. b	118. a	119. d
120. a	121. b	122. d	123. a	124. c	

125. Total number of amino acid required by our body-
a. 8 b. 10
c. **20** d. 30

126. Simplest amino acid is-
a. Tryptophan b. Arginine
c. **Glycine** d. Histidine

127. Nucleic acids are the polymer of-
a. Amino acid b. **Nucleotides**
c. Nucleosides d. Nitrogenous

128. Nucleotides are made up of-
a. **N_2-base + Pentose sugar + Phosphate**
b. N_2-base + Hexose sugar + Phosphate
c. N_2-base + Pentose sugar + Sulphate
d. N_2-base + Hexose sugar + Chloride

129. Which is/are the pyrimidines base-
a. Cytocine b. Uracil
c. Thymine d. **All of the above**

130. Which is/are the purine base-
a. Adenine b. Guanine
c. Cytocine d. **Both a and b**

131. DNA have which sugar-
a. Ribose b. **Pentose**
c. Hexose d. Triose

132. RNA have which sugar-
a. **Ribose** b. Pentose
c. Hexose d. Triose

133. DNA is-
a. Deoxyribonucleic acid b. Ribonucleic acid
c. Deoxynucleic acid d. Deoxy tri ribonucleic acid

134. RNA is-
a. Deoxyribonucleic acid b. **Ribonucleic acid**
c. Deoxynucleic acid d. Ribo tri ribonucleic acid

Answers

125. c	126. c	127. b	128. a	129. d	130. d
131. b	132. a	133. a	134. b		

135. Both polynucleotide chains of DNA are-
   a. Complementary and antiparallel to each other
   b. Complementary and parallel to each other
   c. Supplementary and antiparallel to each other
   d. Parallel and antiparallel to each other

136. Adenine (A) binds to thymine (T) by-
   a. Two hydrogen bonds b. Three hydrogen bonds
   c. Four hydrogen bonds d. Five hydrogen bonds

137. Cytosine (C) binds to guanine (G) by-
   a. Two hydrogen bonds b. **Three hydrogen bonds**
   c. Four hydrogen bonds d. Five hydrogen bonds

138. Out of two strand of DNA only one strand participates in transcription is called-
   a. Antisense strand b. Non-coding strand
   c. Template strand d. **Both a and b**

139. The strand which do not participate in transcription is called-
   a. Sense strand b. Coding strand
   c. Template strand d. **All of the above**

140. Which is right handed DNA-
   a. Z-DNA b. **B-DNA**
   c. C-DNA d. A-DNA

141. Which is left handed DNA-
   a. **Z-DNA** b. B-DNA
   c. C-DNA d. A-DNA

142. Who has proved the genetic material is DNA-
   a. Oswald Avery, Colin Macleod and Mc Clinton
   b. Oswald Avery, Colin Paul and Maclyn Mccarty
   c. Oswald Tippo, Colin Macleod and Khorana
   d. **Oswald Avery, Colin Macleod and Maclyn Mccarty**

143. Semi-conservative mode of DNA replication was first proposed by-
   a. Oswald Avery, Colin Macleod and Maclyn Mccarty
   b. Meselson & Stahl
   c. **Watson & Crick**
   d. Mathai and Khorana

Answers

135. a	136. a	137. b	138. d	139. d	140. b
141. a	142. d	143. c			

144. Semi-conservative mode of DNA replication experimentally proved by-
   a. Meselson & Stahl
   b. Oswald Avery, Colin Macleod and Maclyn Mccarty
   c. Mathai and Khorana
   d. Watson & Crick

145. DNA replication takes place in-
   a. **5'→3'** b. 3'→5'
   c. 4'→5' d. 2'→5'

146. Transcription takes place in which direction-
   a. **5'→3'** b. 3'→5'
   c. 4'→5' d. 2'→5'

147. Which nitrogenous base is found only in RNA-
   a. Adenine b. Guanine
   c. Cytocine d. **Uracil**

148. RNA and DNA is respectively-
   a. Single stranded and double stranded
   b. Double stranded and single stranded
   c. Single stranded and single stranded
   d. Double stranded and Triple stranded

149. Formation of RNA over DNA template with the help of RNA polymerase enzyme is called-
   a. Translation b. **Transcription**
   c. Replication d. Transformation

150. Which enzyme is used in transcription-
   a. DNA polymerase b. **RNA polymerase**
   c. DNA ligase d. Endonuclease

151. Which enzyme is used in replication-
   a. **DNA polymerase** b. RNA polymerase
   c. DNA ligase d. Endonuclease

152. The relationship between the sequence of amino acids in a polypeptide chain and nucleotide sequence of nucleic acids is called-
   a. Operon b. Codon
   c. **Genetic code** d. Cistron

Answers

144. a	145. a	146. a	147. d	148. a	149. b
150. b	151. a	152. c			

153. Which is/are the stop codon-
a. UAA b. UAG
c. UGA d. **All of the above**

154. Which is/are the initiation codon-
a. AUG b. UAA
c. GUG d. **Both a and c**

155. The formation of m-RNA from DNA and then synthesis of protein from it, is known as-
a. **Central dogma** b. Translation
c. Transcription d. Replication

156. Formation of proteins from m-RNA is called-
a. Replication b. Transcription
c. **Translation** d. Central dogma

157. Central dogma includes-
a. Transcription and replication
b. **Transcription and translation**
c. Transcription, translation and replication
d. Translation and replication

158. Function of m-RNA is-
a. New polypeptide chain formation
b. Carrier of amino acids
c. Provide attachment site to t- RNA
d. All of the above

159. Formation of DNA from DNA with the help of DNA polymerase is called-
a. **Replication** b. Transcription
c. Translation d. Central dogma

160. Function of r-RNA is-
a. New polypeptide chain formation
b. Carrier of amino acids
c. **Provide attachment site to t- RNA**
d. All of the above

Answers

153. d	154. d	155. a	156. c	157. b	158. a
159. a	160. c				

161. Function of t-RNA is-
   a. New polypeptide chain formation
   b. **Carrier of amino acids**
   c. Provide attachment site to t- RNA
   d. All of the above
162. Cell theory was proposed by-
   a. Watson and crick
   b. Robert Hook and Robert Brown
   c. **Schleiden (Botanist) and Schwann (Zoologist)**
   d. Schleiden (Zoologist) and Schwann (Botanist)
163. Cell-membrane is-
   a. Semi-permeable b. **Selective permeable**
   c. Non permeable d. None
164. Endoplasmic Reticulum (ER) helps in-
   a. Respiration b. **Protein synthesis**
   c. Transpiration d. Photosynthesis
165. Rough ER helps in-
   a. Lipid synthesis b. **Protein synthesis**
   c. Respiration d. Photosynthesis
166. Smooth ER helps in-
   a. **Lipid synthesis** b. Protein synthesis
   c. Respiration d. Photosynthesis
167. Director of macromolecular traffic in cell is known as-
   a. Nucleus b. Ribosome
   c. Mitochondria d. **Golgi body**
168. Which cell organelle is filled with acid hydrolases enzyme-
   a. Chloroplast b. ER
   c. Ribosome d. **Lysosome**
169. Which cell organelle is also called as suicidal bags of cell-
   a. Golgi body b. Mitochondria
   c. **Lysosome** d. Ribosome
170. Power house of the cell is-
   a. Nucleus b. Ribosome
   c. **Mitochondria** d. Golgi body

Answers

161. b	162. c	163. b	164. b	165. b	166. a
167. d	168. d	169. c	170. c		

171. Red colour of chilli and tomato is due to red pigment called-
   a. Xanthene
   b. Chlorophyll
   c. Chromophyll
   d. **Lycopene**

172. Protein factory of cell is called-
   a. Nucleus
   b. **Ribosome**
   c. Mitochondria
   d. Golgi body

173. Which cell organelle is consider as controller or director of cell-
   a. **Nucleus**
   b. Ribosome
   c. Mitochondria
   d. Golgi body

174. Ribosomes are made up of-
   a. Proteins plus nucleic acids DNA
   b. **Proteins plus nucleic acids RNA**
   c. Lipid plus nucleic acids RNA
   d. Lipid plus nucleic acids DNA

175. Ribosome, Basal bodies and Centriole are-
   a. Single membrane cell organelles
   b. Double membrane cell organelles
   c. Triple membrane cell organelles
   d. **Membrane less cell organelles**

176. The trees produced from the seed of a parent tree are called its-
   a. Sibling
   b. **Progeny**
   c. Race
   d. Family

177. Group of populations that generally interbreed with one another and that intergrade more or less continuously are referred to as-
   a. Population
   b. Species
   c. **Race**
   d. Sibling

178. Individuals that are more closely related to each other than to other individuals in a population are called as-
   a. **Family**
   b. Species
   c. Race
   d. Sibling

179. A group of individuals within a family are referred to as-
   a. Progeny
   b. Species
   c. Race
   d. **Siblings**

Answers

171. d	172. b	173. a	174. b	175. d	176. b
177. c	178. a	179. d			

180. The group of related individuals when only one parent is common is called-

a. Siblings b. Full sib family
c. **Half sib family** d. Open-pollinated family

181. The group of related individuals when both parents are common is called-

a. Siblings b. **Full sib family**
c. Half sib family d. Open-pollinated family

182. The group of related individuals when one parent is common and other parent (s) are unknown is called-

a. Siblings b. Full sib family
c. Half sib family d. **Open-pollinated family**

183. Variation in a population is increased by-

a. Mutation b. Gene flow
c. Selection d. **Both a and b**

184. Variation in a population is decreased by-

a. Natural selection b. Gene flow
c. Genetic drift. d. **Both a and c**

185. The concept of ecotype was suggested by-

a. Reiter b. **Turreson**
c. Huxley d. Linnaeus

186. According to......ecotype genotypical response of a species to a particular habitat-

a. Huxley b. Odum
c. **Turreson** d. John Ray

187. The concept of cline was first given by-

a. **Huxley** b. Odum
c. **Turreson** d. John Ray

188. Father of Binomial nomenclature is-

a. Reiter b. Turreson
c. Huxley d. **Linnaeus**

189. A gradient measurable characteristic is called-

a. Progeny b. Sibling
c. Race d. **Cline**

Answers

180. c	181. b	182. d	183. d	184. d	185. b
186. c	187. a	188. d	189. d		

190. A population of individuals that has become adapted to a specific environment in which it has been planted is called-
a. Ecophenes
b. Ecotype
c. **Land race**
d. Cline

191. The average performance of the progeny of an individual when it is mated to a number of the other individuals in the population is called-
a. General combining ability (GCA)
b. Specific combining ability (SCA)
c. General competing ability (GCA)
d. Specific competing ability (SCA

192. As the average performance of the progeny of a cross between two specific parents that are different from what would be expected on the basis of their general combining abilities alone-
a. General combining ability (GCA)
b. **Specific combining ability (SCA)**
c. General competing ability (GCA)
d. Specific competing ability (SCA)

193. Specific combining ability (SCA) value can be-
a. Positive
b. Negative
c. **Both**
d. None

194. The breeding value of an individual is defined as twice its-
a. General combining ability (GCA)
b. Specific combining ability (SCA)
c. Genetic gain
d. Selection differential

195. A ratio indicating the degree to which parents pass their characteristics along to their offspring-
a. Heredity
b. Heterosis
c. **Heritability**
d. Genetic gain

196. Heritability is how many types-
a. **Two**
b. Three
c. Four
d. Five

Answers

190. c	191. a	192. b	193. c	194. a	195. c
196. a					

197. Heritability value can be-

a. Positive
b. Negative
c. Zero
d. **Both a and c**

198. The ratio of total genetic variation (V_G) in a population to the phenotypic variation (V_P)-

a. Broad-sense heritability (H^2)
b. Narrow-sense heritability (h^2)
c. Specific combining ability (SCA)
d. General combining ability (GCA)

199. The ratio of additive genetic variation (V_A) in a population to total variation (V_P)-

a. Broad-sense heritability (H^2)
b. **Narrow-sense heritability (h^2)**
c. General combining ability (GCA)
d. Specific combining ability (SCA)

200. The difference between the mean of the selected individuals (X_s) and the population mean (X) is called-

a. **Selection differential**
b. Genetic gain
c. Gene load
d. Heritability

201. The multiply product of narrow-sense heritability (h^2) and selection differential (S) is called-

a. Selection differential
b. **Genetic gain**
c. Gene load
d. Heritability

202. Choosing individuals solely on the basis of their phenotypes, without regards to any information about performance of offsprings or siblings is called-

a. **Mass selection**
b. Family selection
c. Sib selection
d. Tandem selection

203. The selection of entire family on the basis of their average phenotypic values is called-

a. Mass selection
b. **Family selection**
c. Sib selection
d. Tandem selection

204. Selection of individuals on the basis of the performance of their siblings not on their own performance is called-

a. Progeny testing
b. Within family selection
c. **Sib selection**
d. Tandem selection

Answers

197. d	198. a	199. b	200. a	201. b	202. a
203. b	204. c				

205. The selection of parent trees on the basis of performance of their progeny is called-
a. **Progeny testing** b. Within family selection
c. Sib selection d. Mass selection

206. Selection of individuals on the basis of their deviation from the family mean, and family values per se are given no weight when selection are made called-
a. Mass selection b. Family selection
c. Sib selection d. **Within family selection**

207. Selection on the basis of families followed by selection of individuals within families is called-
a. Mass selection
b. Family selection
c. **Family within family selection**
d. Sib selection

208. Which is/are the methods of multiple traits selection-
a. Tandem selection b. Independent culling
c. Selection index d. **All of the above**

209. The selection procedure that involves many cycles of selection and breeding is known as-
a. Mass selection b. Current selection
c. **Recurrent selection** d. Progeny selection

210. A tree that has been selected for grading because of its desirable phenotypic qualities but that has not yet been graded or tested called-
a. **Candidate tree** b. Elite tree
c. Check tree d. Comparison tree

211. Morphologically superior tree is called-
a. **Plus tree** b. Elite tree
c. Check tree d. Comparison tree

212. Trees that are located in the same stand having nearly same age and growing on the same or better site as the select tree and against which the select tree is graded called-
a. Candidate tree b. Elite tree
c. Plus tree d. **Comparison tree**

Answers

205. a	206. d	207. c	208. d	209. c	210. a
211. a	212. d				

213. Polyploidy is more common in which species-
  a. Softwood b. **Hardwood**
  c. Conifers d. None

214. Polyploidy suggested as a tree improvement tools by-
  a. Turreson b. Bruce Zobel
  c. **Gustafsson** d. Mendel

215. The best known polyploidy conifer is-
  a. Deodar b. Pinus
  c. **Redwood** d. Rosewood

216. The ultimate source of all variation is-
  a. Natural selection b. Crossing over
  c. Gene flow d. **Mutation**

217. Phenotypic variation includes-
  a. Non-additive variations b. Environmental variations
  c. Additive variations d. **All of the above**

218. Genetic variation includes-
  a. Additive variations b. Non- additive variations
  c. Environmental variations d. **Both a and b**

Answers

213. b 214. c 215. c 216. d 217. d 218. d

CHAPTER 15

Tree Seed Technology

1. The science of determining methods for collection, extraction, cleaning, grading, storage and testing of seeds and heir effects on seed quality is called-
   a. Seed pathology b. **Seed technology**
   c. Seed biology d. Silviculture
2. Fertilized mature ovule called-
   a. Fruit b. Flower
   c. **Seed** d. Embryosac
3. Fertilized mature ovary called-
   a. **Fruit** b. Flower
   c. Seed d. Embryosac
4. Seed is a link between how many generation-
   a. One generation b. **Two generation**
   c. Three generation d. Four generation
5. Which species fruits consider as seed-
   a. Sal b. Bamboo
   c. Teak d. **All of the above**
6. Teak seeds are-
   a. Endospermic b. Non-endospermic
   c. Ex-albuminous seed d. **Both b and c**
7. Morphological, physiological and functional changes that occurs from the time of fertilization until the matured seeds are ready for harvest is called-
   a. Seed development b. Seed vigour
   c. Seed quality d. **Seed maturation**

Answers

1. b 2. c 3. a 4. b 5. d 6. d
7. d

8. Moisture content in mature seed will be-
   a. 10-20% b. **15-20%**
   c. 15-30% d. 10-30%

9. The point of maximum dry weight of seed is called-
   a. Heat point b. Peak point
   c. Saturation point d. **Maturity**

10. Maturation time period of teak seed is-
    a. 60 – 120 days b. **120 – 200 days**
    c. 150 – 200 days d. 200 – 240 days

11. Common seed colour of forest trees is-
    a. Red b. Yellow
    c. **Brown** d. Black

12. Winged seeds are found in-
    a. Dipterocarpus b. Terminalia
    c. Pterocarpus d. **All of the above**

13. Seeds of most conifers are-
    a. **Winged** b. Crown
    c. Spherical d. Hairy

14. Hairy seeds are found in-
    a. Bombax b. Salix
    c. Populus d. **All of the above**

15. Fairy seed are the characteristics of which family-
    a. Compositae b. Dipterocarpaceae
    c. **Apocynaceae** d. Verbenaceae

16. Hard seed coat are the characteristics of-
    a. Ebenaceae b. Dipterocarpaceae
    c. Apocynaceae d. **Leguminaceae**

17. Indian Seed Act was formed in the year-
    a. 1920 b. **1966**
    c. 1968 d. 1983

18. Indian Seed Act has-
    a. 15 sections b. 20 sections
    c. **25 sections** d. 30 sections

Answers

8. b	9. d	10. b	11. c	12. d	13. a
14. d	15. c	16. d	17. b	18. c	

19. Seed Rules was formed in the year-
   a. 1906 b. 1966
   c. **1968** d. 1983

20. Seed order was formed in the year-1983-
   a. 1906 b. 1966
   c. 1968 d. **1983**

21. Nucleus seed is the progeny of-
   a. **Nucleus seed** b. Breeder seed
   c. Foundation seed d. Registered seed

22. Breeder seed is the progeny of-
   a. Breeder seed b. **Nucleus seed**
   c. Foundation seed d. Certified seed

23. Foundation seed is the progeny of-
   a. Nucleus seed b. **Breeder seed**
   c. Foundation seed d. Registered seed

24. Registered seed is the progeny of-
   a. Nucleus seed b. Breeder seed
   c. **Foundation seed** d. Certified seed

25. Certified seed is the progeny of-
   a. Nucleus seed b. Breeder seed
   c. Foundation seed d. **Registered seed**

26. Tag colour of nucleus seed is-
   a. Azar blue b. White
   c. Purple d. **None**

27. Tag colour of foundation seed is-
   a. Azar blue b. **White**
   c. Purple d. Pink

28. Tag colour of breeder seed is-
   a. **Yellow** b. White
   c. Purple d. Green

29. Tag colour of certified seed is-
   a. Golden yellow b. White
   c. **Purple** d. Yellow

Answers

19. c	20. d	21. a	22. b	23. b	24. c
25. d	26. d	27. b	28. a	29. c	

30. Tag colour of registered seed is-
    a. **Yellow** b. White
    c. Purple d. Green

31. Certification of seed is required in case of-
    a. Foundation seed b. Registered seed
    c. Certified seed d. **All of the above**

32. An orchard which is seedlings origin is called-
    a. **Seedling Seed Orchard** b. Clonal Seed Orchard
    c. Seed orchard d. All of the above

33. An orchard which is vegetative propagation origin is called-
    a. Seedling Seed Orchard b. **Clonal Seed Orchard**
    c. Seed orchard d. All of the above

34. Seeds which has life span more than 15 days called-
    a. Sound seed b. **Orthodox seed**
    c. Recalcitrant seed d. Viable seed

35. Seed which cannot be dried to moisture contents below 30 per cent without injury and are unable to tolerate freezing called-
    a. Sound seed b. Orthodox seed
    c. **Recalcitrant seed** d. Viable seed

36. Seed certification is not required in case of-
    a. Foundation seed b. Registered seed
    c. Certified seed d. **Nucleus seed**

37. Seed certification is not required in case of-
    a. **Breeder seed** b. Registered seed
    c. Certified seed d. Foundation seed

38. Which seed is genotypical as well as phenotypical 100% pure-
    a. Nucleus seed b. Breeder seed
    c. **Both a and b** d. Registered seed

39. Which seed is genotypical 100% but phenotypical 98% pure-
    a. Registered seed b. Certified seed
    c. Foundation seed d. **All of the above**

Answers

30. a	31. d	32. a	33. b	34. b	35. c
36. d	37. a	38. c	39. d		

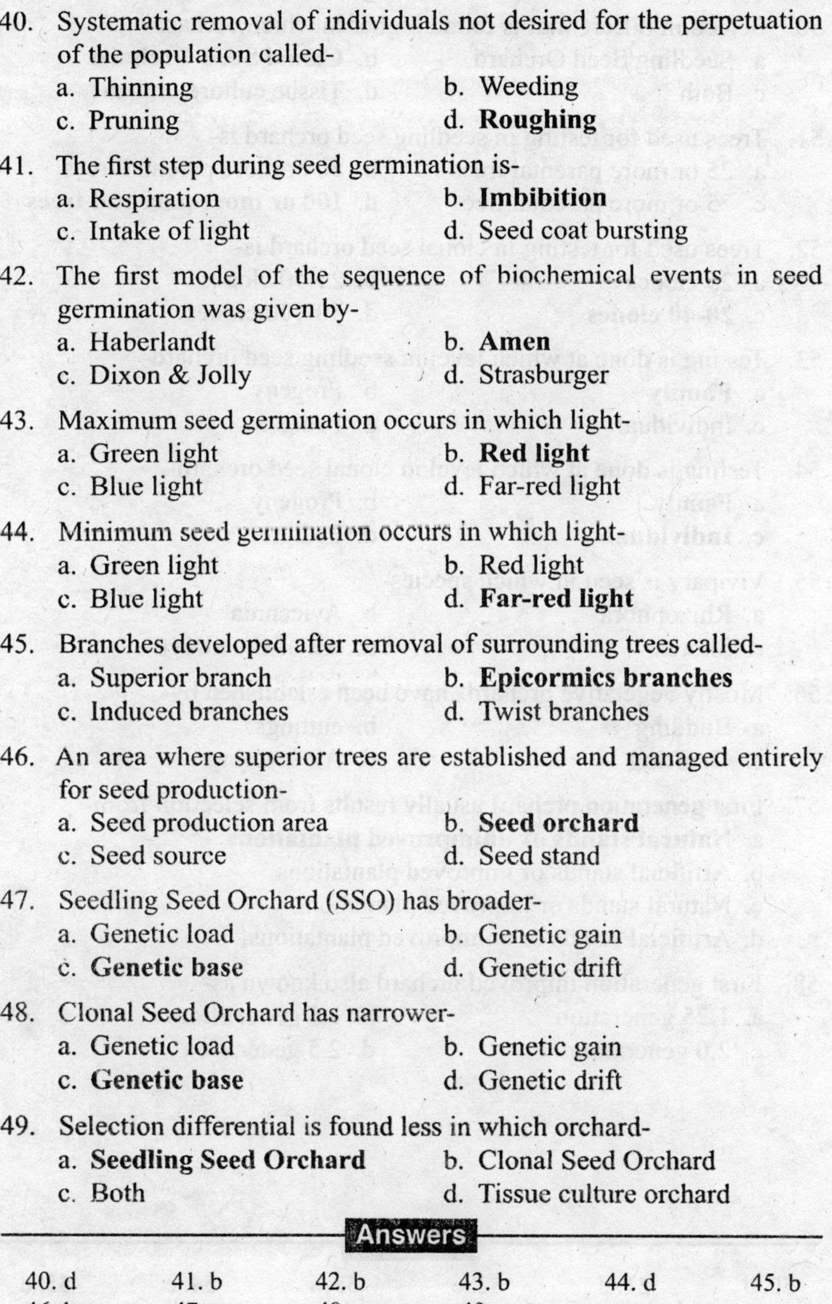

40. Systematic removal of individuals not desired for the perpetuation of the population called-
    a. Thinning b. Weeding
    c. Pruning d. **Roughing**
41. The first step during seed germination is-
    a. Respiration b. **Imbibition**
    c. Intake of light d. Seed coat bursting
42. The first model of the sequence of biochemical events in seed germination was given by-
    a. Haberlandt b. **Amen**
    c. Dixon & Jolly d. Strasburger
43. Maximum seed germination occurs in which light-
    a. Green light b. **Red light**
    c. Blue light d. Far-red light
44. Minimum seed germination occurs in which light-
    a. Green light b. Red light
    c. Blue light d. **Far-red light**
45. Branches developed after removal of surrounding trees called-
    a. Superior branch b. **Epicormics branches**
    c. Induced branches d. Twist branches
46. An area where superior trees are established and managed entirely for seed production-
    a. Seed production area b. **Seed orchard**
    c. Seed source d. Seed stand
47. Seedling Seed Orchard (SSO) has broader-
    a. Genetic load b. Genetic gain
    c. **Genetic base** d. Genetic drift
48. Clonal Seed Orchard has narrower-
    a. Genetic load b. Genetic gain
    c. **Genetic base** d. Genetic drift
49. Selection differential is found less in which orchard-
    a. **Seedling Seed Orchard** b. Clonal Seed Orchard
    c. Both d. Tissue culture orchard

Answers

40. d	41. b	42. b	43. b	44. d	45. b
46. b	47. c	48. c	49. a		

50. Selection differential is found higher in which orchard-
a. Seedling Seed Orchard b. **Clonal Seed Orchard**
c. Both d. Tissue culture orchard

51. Trees used for testing in seedling seed orchard is-
a. 25 or more parental trees b. 50 or more parental trees
c. 75 or more parental trees d. **100 or more parental trees**

52. Trees used for testing in clonal seed orchard is-
a. 25 clones b. 25-50 clones
c. **20-40 clones** d. 50-100 clones

53. Testing is done at which level in seedling seed orchard-
a. **Family** b. Progeny
c. Individual d. Pedigree

54. Testing is done at which level in clonal seed orchard-
a. Family b. Progeny
c. **Individual** d. Pedigree

55. Vivipary is seen in which species-
a. Rhizophora b. Avicennia
c. Shorea d. **All of the above**

56. Mostly vegetative orchards have been established by-
a. Budding b. cuttings
c. **Grafting** d. Air layering

57. First generation orchard usually results from selection from-
a. **Natural stands or unimproved plantations**
b. Artificial stands or improved plantations
c. Natural stands or improved plantations
d. Artificial stands or unimproved plantations

58. First generation improved orchard also known as-
a. 1.25 generation b. **1.5 generation**
c. 2.0 generation d. 2.5 generation

Answers

50. b	51. d	52. c	53. a	54. c	55. d
56. c	57. a	58. b			

CHAPTER 16

Soil Science

1. A dynamic body developed by the natural forces acting on natural materials and usually differentiated into horizons is called-
   a. Land — b. **Soil**
   c. Vegetation — d. Rocks
2. Soil has how much percent of solid components-
   a. 45% — b. 5%
   c. **50%** — d. 75%
3. Soil has how much percent of liquid components-
   a. 45% — b. 5%
   c. **50%** — d. 75%
4. A vertical section of soil shows how many five master horizons-
   a. 4 — b. **5**
   c. 6 — d. 8
5. Organic horizon of soil profile is-
   a. **O** — b. A
   c. B — d. C
6. 'A' horizon of soil is called-
   a. Washing out — b. Elluvial horizon
   c. Leaching horizon — d. **All of the above**
7. 'B' horizon of soil is called-
   a. Washing in — b. Illuvial horizon
   c. Leaching horizon — d. **Both a and b**
8. Earth shows how many layer in their internal structure-
   a. 2 — b. **3**
   c. 4 — d. 5

Answers

1. b	2. c	3. c	4. b	5. a	6. d
7. d	8. b				

9. Which is/are not an internal layer of earth-
   a. Core b. Mantle
   c. **Buffer** d. Crust

10. Density of earth is-
    a. **5.5 gm/c.c** b. 2.6-2.7 gm/c.c
    c. 2.9 gm/c.c d. 3.0 gm/c.c

11. Density of rock is-
    a. 5.5 gm/c.c b. **2.6-2.7 gm/c.c**
    c. 2.9 gm/c.c d. 3.0 gm/c.c

12. Mixture of two or more minerals called-
    a. Land b. Soil
    c. Ores d. **Rock**

13. Study of rocks called petrology -
    a. Pedology b. Paedogenesis
    c. **Petrology** d. Geology

14. Study of origin, classification and description of soil is called -
    a. **Pedology** b. Paedogenesis
    c. Petrology d. Geology

15. Vertical section through a soil is called-
    a. Pedology b. **Soil profile**
    c. Petrology d. Geology

16. A plus B horizons collectively called as-
    a. Top soil b. Organic soil
    c. **Solum** d. Fertile oil

17. Most important process in chemical weathering process is-
    a. Oxidation b. Reduction
    c. Both a and b d. **Hydrolysis**

18. The relative proportion of sand, silt and clay in the soil know as-
    a. **Soil texture** b. Soil structure
    c. Lustre d. Soil profile

19. Mass of unit volume of soil including the pore space is called as-
    a. Particle density b. Absolute density
    c. Real density d. **Bulk density**

Answers

9. c	10. a	11. b	12. d	13. c	14. a
15. b	16. c	17. d	18. a	19. d	

20. The weight of per unit volume of the solid portion of soil is called-
a. Particle density
b. Absolute density
c. Gain density
d. **All of the above**

21. CO_2 content in soil as compare to atmospheric CO_2 is more than-
a. Four times
b. Eight times
c. **Ten times**
d. Hundred times

22. Soil colour chart was given by-
a. Mitchell
b. P. K Roy
c. Walkey & Black
d. **Mansell**

23. Which is not a class of soil colour chart-
a. Hue
b. **Spectrum**
c. Value
d. Chroma

24. Movement of water into the soil is called as-
a. Precipitation
b. Percolation
c. Run-off
d. **Infiltration**

25. Hue in Mansell soil colour chart shows-
a. Soil fertility index
b. **Dominant spectral colour**
c. Purity of colour
d. Lightness or darkness of colour

26. Chroma in Mansell soil colour chart shows-
a. Soil fertility index
b. Dominant spectral colour
c. **Purity of colour**
d. Lightness or darkness of colour

27. Value in Mansell soil colour chart shows-
a. Soil fertility index
b. Dominant spectral colour
c. Purity of colour
d. **Lightness or darkness of colour**

28. Soil water retained at the pressure-
a. -0.13 bar
b. 0.33 bar % +15 bar
c. **0.33 bar % -15 bar**
d. 1.33 bar % +15 bar

Answers

20. d　21. c　22. d　23. b　24. d　25. b
26. c　27. d　28. c

29. Amount of water required to produce a unit quantity of dry weight material is called as-
   a. Required water
   b. **Water use efficiency**
   c. Capillary water
   d. Water use point
30. Capacity of the soil to change its shape under moist conditions is called as-
   a. Soil structure
   b. Soil texture
   c. **Plasticity**
   d. Buffering capacity
31. Molybdenum is not found in the soil-
   a. Alkali soil
   b. Basic soil
   c. Red soil
   d. **Acidic soil**
32. The power to resist a change in pH by the soil is called -
   a. Electrical conductivity
   b. Soil pH
   c. **Buffering capacity**
   d. Amelioration capacity
33. The process of removal of silica instead of sesquioxides to concentrate in the solum is called -
   a. Calcination
   b. Podsolization
   c. Humification
   d. **Laterisation**
34. Saline soils also called as-
   a. Sodic soils
   b. Solonetz
   c. Non-saline alkali
   d. **Solon chalk**
35. Saline soils also called as-
   a. Sodic soils
   b. Solonetz
   c. **White alkali soils**
   d. Non-saline alkali
36. Alkali soils also called as-
   a. Sodic soils
   b. Solonetz
   c. Non-saline alkali
   d. **All of the above**
37. Particle density or absolute density or gain density and bulk or apparent density are expressed-
   a. Mg/m^3
   b. gms/cm^3
   c. kg/m^3
   d. **Both a and b**
38. Particle density or absolute density or gain density has value-
   a. 1.65 gm/c.c
   b. 1.85 gm/c.c
   c. **2.65 gm/c.c**
   d. 2.85 gm/c.c

Answers

29. b	30. c	31. d	32. c	33. d	34. d
35. c	36. d	37. d	38. c		

39. Bulk density has value-
    a. **1.65 gm/c.c** b. 1.85 gm/c.c
    c. 2.65 gm/c.c d. 2.85 gm/c.c
40. Particle density is always than bulk density-
    a. **Higher** b. Lower
    c. Equal d. Equal or lower
41. Number of land suitability classes is-
    a. **2** b. 8
    c. 12 d. 15
42. Which is called as night soil manure-
    a. Humus b. Vermicompost
    c. Solonchak d. **Poudrette**
43. Gypsum is used for reclamation of -
    a. Sodic soils b. Solonetz
    c. Non-saline alkali d. **All of the above**
44. Which soil is light textured -
    a. Sandy soil b. Sandy loam soil
    c. **Clay soil** d. Clay loam soil
45. One gram of soil contains-
    a. 10 million microorganisms b. 12 million microorganisms
    c. **14 million microorganisms** d. 16 million microorganisms
46. Red soils is found in-
    a. Bihar b. Rajasthan
    c. Maharashtra d. **Tamil Nadu**
47. Fossils are found in-sedimentary rocks-
    a. **Sedimentary rock** b. Igneous rock
    c. Metamorphic rock d. None
48. The length of engineer chain is -
    a. 10 ft. b. 66 ft.
    c. 75 ft. d. **100 ft.**
49. The length of Gunter chain is -
    a. 10 ft b. **66 ft**
    c. 75 ft d. 100 ft

Answers

39. a	40. a	41. a	42. d	43. d	44. c
45. c	46. d	47. a	48. d	49. d	

50. The transfer of minerals from the top soil to sub soil through the water is called as-
a. Mineralization
b. Elluviation
c. **Leaching**
d. Podsolization

51. Which fertilizer fixed immediately after application in the soil-
a. N
b. **P**
c. K
d. S

52. Cation exchange capacity is expressed-
a. me/100 g
b. Mme/100 g
c. C mole/kg.
d. **Both a and c**

53. The arrangement of primary particles and their aggregates into a soil called as-
a. Soil texture
b. **Soil structure**
c. Plasticity
d. Soil profile

54. Kaolinite is an example of-
a. 2:1 non-expanding types
b. 2:1 expanding types
c. **1:1 expanding types**
d. 1:2 expanding types

55. Montmorillonite is an example of-
a. 2:1 non-expanding types
b. **2:1 expanding types**
c. 1:1 expanding types
d. 1:2 expanding types

56. Hydrous mica or illite is an example of-
a. **2:1 non-expanding types**
b. 2:1 expanding types
c. 1:1 expanding types
d. 1:2 expanding types

57. Chlorites is an example of-
a. 2:1 non-expanding types
b. **2:2 expanding types**
c. 1:1 expanding types
d. 1:2 expanding types

58. Amount of exchangeable cations per unit weight of dry soil called-
a. **Cation exchange capacity**
b. Anion exchange capacity
c. Plasticity
d. Buffering capacity

59. Soil survey has been categorised into -
a. 2
b. **4**
c. 5
d. 6

Answers

50. c	51. b	52. d	53. b	54. c	55. b
56. a	57. b	58. a	59. b		

60. Capacity of a soil to absorb or release anion under normal soil conditions is called-
    a. Cation exchange capacity b. **Anion exchange capacity**
    c. Plasticity d. Buffering capacity

61. Cadastral maps having scale 1:8,000 or 1:4,000 are used in-
    a. **Detailed soil survey**
    b. Reconnaissance soil survey
    c. Detailed survey-reconnaissance soil survey
    d. Semi-detailed survey

62. Topographical maps having scale 1:50,000 are used in-
    a. Detailed soil survey
    b. **Reconnaissance soil survey**
    c. Detailed survey-reconnaissance soil survey
    d. Semi-detailed survey

63. Soil formation equation was first given by -
    a. Peterson b. D. K. Das
    c. **Dokuchaiev** d. Jenny

64. Which is/are the active factor in soil formation equation-
    a. Biosphere b. Climate
    c. Parent material d. **Both a and b**

65. Which is/are the passive factor in soil formation equation-
    a. Relief b. Time
    c. Parent material d. **All of the above**

66. Which is soil formation equation-
    a. **S = f (cl, b, r, p, t)** b. S = f (cl, q, r, p, t)
    c. f = s (cl, b, r, v, t) d. S = cl (f, b, r, p, t)

67. Who has given stages of soil development or formation-
    a. Mansell & Jenny b. **Mohr & Baren**
    c. Dokuchaiev & Jenny d. Jenny

68. Which is/are not a soil formation stages-
    a. Senile b. Virile
    c. **Juvenile** d. All of the above

Answers

60. b	61. a	62. b	63. c	64. d	65. d
66. a	67. b	68. c			

69. Mohr and van Baren recognized how many stages of soil development -
a. Three
b. Four
c. **Five**
d. Six

70. Soil formation equation was explained given by in 1941 -
a. Peterson
b. D. K. Das
c. Dokuchaiev
d. **Jenny**

71. 7th Approximation and their derivatives proposed how many soil orders-
a. Eight
b. Ten
c. **Twelve**
d. Fifteen

72. Which soil order has maximum sub soil orders-
a. **Mollisol**
b. Gelisol
c. Aridosol
d. Vertisol

73. Which soil order has minimum sub soil orders-
a. Mollisol
b. **Gelisol**
c. Aridosol
d. Vertisol

74. Soil order Gelisol has symbol-
a. Ept
b. 1st
c. Oll
d. **el**

75. The foliar spray of urea have per cent biuret concentration-
a. 1.0
b. **1.5**
c. 2.5
d. 4.5

76. The process of accumulation insoluble lime salts mostly Ca & Mg in soil horizons called -
a. Mineralization
b. **Calcification**
c. Leaching
d. Podsolization

77. E-value is used for determination of available elements in soils-
a. K
b. N
c. **P**
d. S

78. The ratio of the weight of organic carbon to the weight of the total nitrogen in soil called-
a. **C : N ratio**
b. Humus
c. Ammonification
d. Organic carbon

Answers

69. c	70. d	71. c	72. a	73. b	74. d
75. b	76. b	77. c	78. a		

79. Colloidal particles tend to attract each other and coagulate like a bigger particle called-
a. Aggregation
b. Calcification
c. Laterisation
d. **Flocculation**

80. The process of accumulation of iron and aluminium oxides in soil horizons called-
a. Mineralization
b. Calcification
c. **Laterisation**
d. Podsolization

81. L-value is used to calculate supply value in soils-
a. K
b. N
c. **P**
d. S

82. Formation of inorganic substances from the organic substances by microbial decomposition-
a. **Mineralization**
b. Calcification
c. Laterisation
d. Humification

83. The downward movement of water through the soil called as-
a. Precipitation
b. **Percolation**
c. Run-off
d. Infiltration

84. A-value and Y-value concept was given by respectively-
a. **Fried & Dean (1952) and Larson (1952)**
b. Larson (1952) and Fried & Dean (1953)
c. Fried & Dean (1941) and Larson (1953)
d. Mohr (1941) & Baren (1953)

85. The detachment and transportation of soil mass from one place to another place called-
a. Mineralization
b. **Soil erosion**
c. Tillage
d. Decomposition

86. Which is/are not a types of soil erosion by wind-
a. **Raindrop splash erosion**
b. Saltation
c. Suspension
d. Surface creep

87. Which is/are not a types of soil erosion by water -
a. Sheet erosion
b. Rill erosion
c. Gully erosion
d. **Saltation**

Answers

79. d	80. c	81. c	82. a	83. b	84. a
85. b	86. a	87. d			

88. Soil splash caused by the impact of falling rain drops is called-
   a. **Raindrop splash erosion**
   b. Saltation
   c. Suspension
   d. Surface creep
89. Number of soil irrigability classes is-
   a. 2
   b. **5**
   c. 8
   d. 12
90. Removal of uniform layers of surface soil by the action of rainfall and run-off water called -
   a. Saltation
   b. Rill erosion
   c. Gully erosion
   d. **Sheet erosion**
91. Removal of surface soil by running water with the formation narrow shallow channels called-
   a. Saltation
   b. **Rill erosion**
   c. Gully erosion
   d. Sheet erosion
92. Removal of surface soil by running water with the formation of channels that cannot be smoothed out completely by normal agricultural operation or cultivation called -
   a. Saltation
   b. Rill erosion
   c. **Gully erosion**
   d. Sheet erosion
93. Which is the first stage of soil erosion by water -
   a. Sheet erosion
   b. Rill erosion
   c. Gully erosion
   d. **Raindrop splash erosion**
94. Saltation is a process of movement of soil particles having size-
   a. 0.01 to 0.02 mm
   b. 0.02 to 0.50 mm
   c. 0.1 to 0.05 mm
   d. **0.05 to 0.5 mm**
95. Suspension is a process of movement of soil particles having size-
   a. < 0.5mm
   b. > 0.1mm
   c. **< 0.1mm**
   d. > 0.5mm
96. Surface creep is a process of movement of soil particles having size-
   a. > 0.2mm
   b. **> 0.5mm**
   c. > 0.6mm
   d. > 0.7mm
97. Percent soil loss in saltation of total percent of wind erosion is-
   a. **50-70**
   b. 5-25
   c. 3-4
   d. 1-4

Answers

88. a	89. b	90. d	91. b	92. c	93. d
94. d	95. c	96. b	97. a		

98. Percent soil loss in surface creep of total percent of wind erosion is-
   a. 50-70 b. **5-25**
   c. 3-4 d. 1-4
99. Percent soil loss in suspension of total percent of wind erosion is-
   a. 50-70 b. 5-25
   c. **3-4** d. 1-4
100. The rate of soil erosion in India is-
   a. 1.64 t/ha b. **16.4 t/ha**
   c. 28.9 t/ha d. 56.3 t/ha
101. Which is an example of igneous rock-
   a. **Granite** b. Limestone
   c. Sand stone d. Slate
102. Which is not an example of sedimentary rock-
   a. **Basalt** b. Limestone
   c. Sand stone d. Dolomite
103. Which is not an example of metamorphic rock-
   a. Marble b. Quartz
   c. **Dolomite** d. Gneisis
104. Soil is a-
   a. Greek word b. **Latin word**
   c. French word d. Dutch word
105. Coarse sand soil particles having size-
   a. **> 0.2mm** b. > 0.5mm
   c. > 0.6mm d. > 0.7mm
106. Silt soil particles having size-
   a. > 2.0 b. **0.02-0.002 mm**
   c. 0.02-0.2 mm d. 0.002 mm
107. Clay soil particles having size-
   a. > 2.0 b. 0.02-0.002 mm
   c. 0.02-0.2 mm d. **< 0.002 mm**
108. Soil of India broadly classified into how many groups-
   a. 4 b. 6
   c. **8** d. 12

Answers

98. b	99. c	100. b	101. a	102. a	103. c
104. b	105. a	106. b	107. d	108. c	

109. Largest soil groups in India is-
   a. **Alluvial soils** b. Black soils
   c. Red soils d. Laterite soils
110. Maximum leached soil groups in India is-
   a. Alluvial soils b. Black soils
   c. Red soils d. **Laterite soils**
111. Which is called early soils groups in India is-
   a. Alluvial soils b. Black soils
   c. **Red soils** d. Laterite soils
112. Regur refers to soils groups in India is-
   a. Alluvial soils b. **Black soils**
   c. Red soils d. Laterite soils
113. Alluvial soils groups occupied how much area in India is-
   a. **143 mha** b. 55 mha
   c. 15 mha d. 25 mha
114. Black soils groups occupied how much area in India is-
   a. 143 mha b. **55 mha**
   c. 15 mha d. 25 mha
115. Red soils groups occupied how much area in India is-
   a. 143 mha b. 55 mha
   c. **15 mha** d. 25 mha
116. Laterite soils groups occupied how much area in India is-
   a. 143 mha b. 55 mha
   c. 15 mha d. **25 mha**
117. Smallest unit of soil classifications is-
   a. **Soil series** b. Soil class
   c. Soil order d. Soil sub-order
118. Organic matter is found highest in-
   a. Forest ecosystem b. Pond ecosystem
   c. River ecosystem d. **Grasslands**
119. Organic matter present in the soil can be calculated by-
   a. SOC = Organic carbon (OC) × 1.124
   b. SOC = Organic carbon (OC) × 1.324
   c. **SOC = Organic carbon (OC) × 1.724**
   d. SOC = Organic carbon (OC) × 1.924

Answers

109. a	110. d	111. c	112. b	113. d	114. b
115. c	116. d	117. a	118. d	119. c	

120. Value of Bemlen factor is-
    a. **1.724**
    b. 2.324
    c. 1.734
    d. 4.324

121. Highest acidic soils found in-
    a. Gujarat
    b. Maharashtra
    c. **West Bengal**
    d. Bihar

122. Soils having atleast how much organic matter called as organic soils-
    a. 10%
    b. 15%
    c. **20%**
    d. 25%

123. Indian soils have how much percent organic matter-
    a. **0.45**
    b. 0.55
    c. 1.0
    d. 1.5

124. Zero tillage practice was first started in-
    a. India
    b. Japan
    c. China
    d. **USA**

125. Flame photometer is used to determine-
    a. N and P
    b. N and K
    c. **Na and K**
    d. N, P and K

126. Electrical conductivity (EC) of saline soil is-
    a. > 4 d/sm
    b. ≮ 4 ds/m
    c. **>4 ds/m**
    d. < 6 dsm^{-1}

127. Exchangeable sodium percent (ESP) of saline soil is-
    a. > 4 d/sm
    b. **< 15**
    c. > 15
    d. < 6

128. pH of saline soil is-
    a. **< 8.5**
    b. > 8.5
    c. > 7.5
    d. < 6.5

129. Calcium ammonium nitrate (CAN) also known as -
    a. Indian fertilizer
    b. China clay
    c. **Kissan Khad**
    d. Desi Khad

Answers

120. a	121. c	122. c	123. a	124. d	125. c
126. c	127. b	128. a	129. c		

130. Nitrogen percent in urea, thiourea and urea formaldehyde is respectively
  a. 46% N, 46.8% and 30-42% N
  b. 42% N, 26.8% and 38-52% N
  c. 43% N, 46.8% and 38-52% N
  d. **46% N, 36.8% and 38-42% N**

131. Electrical conductivity (EC) of alkaline soils is-
  a. >4 d/sm
  b. **< 4 ds/m**
  c. >4 ds/m
  d. < 6 dsm^{-1}

132. Exchangeable sodium percent (ESP) of alkaline soils is-
  a. > 4 d/sm
  b. < 15
  c. **> 15**
  d. < 6

133. pH of alkaline soils is-
  a. < 8.5
  b. **> 8.5**
  c. > 7.5
  d. < 6.5

134. pH scale was given by-
  a. Mohit Husain
  b. Mansell
  c. **Sorenson**
  d. Walkey & Black

135. The C: N ratio for humus is -
  a. 12 : 1
  b. 10 : 20
  c. **10 : 1**
  d. 10:25

136. The C: N ratio for normal soil is -
  a. **12 : 1**
  b. 10 : 20
  c. 10 : 1
  d. 10 : 25

137. Single super phosphate (SSP) contain-
  a. 10% P_2O_5, 15% Ca and 10% S
  b. 14% P_2O_5, 17% Ca and 12% S
  c. **16% P_2O_5, 19% Ca and 12% S**
  d. 18% P_2O_5, 21% Ca and 14% S

138. DAP contains-
  a. **18% N, 46% P_2O_5**
  b. 46% N, 26% P_2O_5
  c. 10% N, 34% P_2O_5
  d. 17% N, 34% P_2O_5

Answers

130. d	131. b	132. c	133. b	134. a	135. c
136. a	137. c	138. a			

139. Nitrogen percent in ammonium nitrate is-
   a. 28%
   b. **33%**
   c. 47%
   d. 50%

140. MAP contains-
   a. 10% N, 38% P_2O_5
   b. **11% N, 48% P_2O_5**
   c. 17% N, 40% P_2O_5
   d. 18% N, 40% P_2O_5

141. Land capability classification (LCC) has-
   a. 6 classes
   b. **8 classes**
   c. 12 classes
   d. 15 classes

142. LCC soil class I has colour -
   a. **Light green**
   b. Yellow
   c. Red
   d. Blue

143. N, P and K content in FYM is respectively –
   a. **0.5 N%, 0.2 P% and 0.5 K%**
   b. 0.5 N%, 0.4 P% and 0.5 K%
   c. 0.2 N%, 0.5 P% and 0.5 K%
   d. 0.25 N%, 0.20 P% and 0.55 K%

144. LCC soil class II has colour -
   a. Light green
   b. **Yellow**
   c. Red
   d. Blue

145. N, P and K content in vermicompost is respectively –
   a. 0.5 N%, 0.2 P% and 0.5 K%
   b. **3.0 N%, 1.0 P% and 1.5 K%**
   c. 0.2 N%, 0.5 P% and 1.5 K%
   d. 0.5 N%, 0.2 P% and 0.5 K%

146. LCC soil class III has colour -
   a. Light green
   b. Yellow
   c. **Red**
   d. Blue

147. N, P and K content in night soil is respectively –
   a. 0.5 N%, 0.2 P% and 0.5 K%
   b. 3.0 N%, 1.0 P% and 1.5 K%
   c. 2.2 N%, 4.5 P% and 1.5 K%
   d. **5.5 N%, 4.0 P% and 2.0 K%**

Answers

139. b	140. b	141. b	142. a	143. a	144. b
145. b	146. c	147. d			

148. LCC soil class IV has colour -
a. Light green b. Yellow
c. Red d. **Blue**

149. Soil transported by water is called -
a. Eolian b. Glacial
c. **Alluvial** d. Colluvial

150. Land capability soil class suitable for wildlife & watershed-
a. Class IV b. Class VI
c. Class VII d. **Class VIII**

151. LCC soil class V has colour -
a. **Dark green** b. Orange
c. Brown d. Purple

152. Soil transported by gravity is called -
a. Eolian b. Glacial
c. Alluvial d. **Colluvial**

153. Land capability soil class suitable for pasture and grassland -
a. Class I b. Class II
c. **Class V** d. Class VII

154. Soil transported by wind is called -
a. **Eolian** b. Glacial
c. Alluvial d. Colluvial

155. Soil transported by ice is called -
a. Eolian b. **Glacial**
c. Alluvial d. Colluvial

156. Land capability soil class suitable for cultivation is/are -
a. Class I b. Class II
c. Class III d. **All of the above**

157. LCC soil class VI has colour -
a. Dark green b. **Orange**
c. Brown d. Purple

158. Land capability soil class not suitable for cultivation is/are -
a. Class VI b. Class VII
c. Class VIII d. **All of the above**

Answers

148. d	149. c	150. a	151. a	152. d	153. c
154. a	155. b	156. d	157. b	158. d	

159. LCC soil class VII has colour -
  a. Dark green
  b. Orange
  c. **Brown**
  d. Purple

160. LCC soil class VIII has colour -
  a. Dark green
  b. Orange
  c. Brown
  d. **Purple**

161. Loamy soil has -
  a. 50% sand + 50% clay
  b. 60% sand + 40% clay
  c. **70% sand + 30% clay**
  d. 80% sand + 20% clay

162. Sandy soil has -
  a. 65% sand + 35% clay
  b. 70% sand + 30% clay
  c. 80% sand + 20% clay
  d. **85% sand + 15% clay**

Answers

159. c 160. d 161. c 162. d

CHAPTER 17

Agricultural Extension

1. Community Development Programme (CDP) was started in the year-
   a. **1952** b. 1953
   c. 1960 d. 1962
2. National Extension Service (NES) was started in the year-
   a. 1952 b. **1953**
   c. 1960 d. 1962
3. Agriculture extension was first started in-
   a. **Bihar Agricultural College, Sabour**
   b. Rajasthan Agricultural College, Udaipur
   c. Gujarat Agricultural College, Junagarh
   d. Tamil Nadu Agricultural College, Coimbatore
4. Pachayati Raj was started in India-
   a. 1950 b. 1953
   c. **1959** d. 1962
5. Tribal Area Development Programme (TADP) was started in the year-
   a. 1952 b. 1953
   c. 1960 d. **1962**
6. Etawah pilot project was headed by-
   a. M. K Gandhi b. D. Benor
   c. **Albert Mayer** d. R.N Tagore
7. Tribal Area Development Programme (TADP) was started in the year-
   a. 1958 b. **1964**
   c. 1965 d. 1969

Answers

1. a 2. b 3. a 4. c 5. d 6. c
7. b

8. Father of organic farming is-
   a. John Ray
   b. D. Benor
   c. **Masanobu Fukuoka**
   d. Albert Mayer
9. Small Farmers Development Agency (SFDA) was started in the year-
   a. 1958
   b. 1964
   c. 1965
   d. **1969**
10. National Agricultural Science Museum located at-
   a. Kolkata
   b. Dehra Dun
   c. Mumbai
   d. **New Delhi**
11. National Agricultural Science Museum located at New Delhi was established in the year-
   a. 1958
   b. 1994
   c. **2004**
   d. 2006
12. Integrated Rural Development Programme (IRDP) was started in the year-
   a. 1952
   b. **1978**
   c. 1984
   d. 1993
13. National Rural Employment Programme (NREP) was started in the year-
   a. 1952
   b. 1978
   c. **1980**
   d. 1991
14. Training of Rural Youth for Self-employment (TRYSEM) was started in the year-
   a. 1958
   b. 1979
   c. **1980**
   d. 2000
15. T and V programme was introduced in the year-
   a. 1943
   b. **1974**
   c. 1986
   d. 2002
16. ICAR was established on the recommendation of-
   a. Lord Fukoka Commission
   b. Swaminathan Commission
   c. B. P. Pal Commission
   d. **Lord Linlithgow Commission**

Answers

8. c	9. d	10. d	11. c	12. b	13. c
14. c	15. b	16. d			

17. The year of technology is -
    a. 2000 b. **2002**
    c. 2006 d. 2008
18. The first KVK was established at-
    a. Bihar b. Rajasthan
    c. Pantanagar d. **Pondicherry**
19. The first KVK was established in the year-
    a. 1958 b. **1974**
    c. 1980 d. 1993
20. Rural Landless Employment Guarantee Programme (RLEGP) was started in the year-
    a. 1958 b. 1974
    c. **1983** d. 1993
21. Shantiniketan project was headed by-
    a. M. K Gandhi b. D. Benor
    c. Albert Mayer d. **R.N Tagore**
22. Lab to Land programme was started in the year-
    a. **1979** b. 1984
    c. 1987 d. 1991
23. Lab to Land programme was started by ICAR on-
    a. Silver jubilee b. **Golden jubilee**
    c. Diamond jubilee d. Platinum jubilee
24. Jawahar Rozgar Yojana (JRY) was started in the year-
    a. 1979 b. **1984**
    c. 1987 d. 1991
25. Indira Aawaas Yojana was started in the year-
    a. 1979 b. 1981
    c. **1986** d. 1991
26. Rajiv Gandhi National Drinking Water Mission was started in the year-
    a. 1979 b. 1981
    c. 1986 d. **1991**

Answers

17. b	18. d	19. b	20. c	21. d	22. a
23. b	24. b	25. c	26. d		

27. Swaran Jayanthi Gram Swarojgar Yojana was started in the year-
    a. 1979 b. **1999**
    c. 2001 d. 2005

28. Television broadcast for rural development in India was started in the year-
    a. **1977** b. 1985
    c. 1996 d. 2000

29. *Grow more food* campaign was stared in-
    a. 1928 b. **1948**
    c. 1964 d. 1979

30. T and V system of extension was started by -
    a. Spencer Hatch b. **D. Benor**
    c. Albert Mayer d. R.N Tagore

31. Sevagram project was started by-
    a. Spencer Hatch b. **M. K Gandhi**
    c. Albert Mayer d. R.N Tagore

32. National Academy of Agriculture Research Management was located in-
    a. New Delhi b. Bhopal
    c. **Hyderabad** d. Dehra Dun

33. The Royal Commission came in the year-
    a. 1921 b. **1928**
    c. 1939 d. 1952

34. The full form of ATMA is-
    a. Agriculture Technology Management Authority
    b. Advances Technology Management Authority
    c. Agriculture Technology Machinery Agency
    d. **Agriculture Technology Management Agency**

35. The Government of India set up planning Commission during the year-
    a. **1954** b. 1979
    c. 1989 d. 1995

Answers

27. b 28. a 29. b 30. b 31. b 32. c
33. b 34. d 35. a

36. NATP was funded from-
   a. RBI b. NAWARD
   c. JICA d. **World Bank**

37. Who was the first chairman of National Commission on farmers-
   a. **Sompal** b. M. K Gandhi
   c. Swaminathan d. R.N Tagore

38. Who was the present chairman of National Commission on farmers-
   a. Sompal b. M. K Gandhi
   c. **Swaminathan** d. R.N Tagore

39. *National Food for work* programme was launched on-
   a. 14th July 2002 b. 14th December 2003
   c. **14th November 2004** d. 2nd June 2005

40. Role of different agencies for village development was included in-
   a. Close diagram b. Pie diagram
   c. **Chapati diagram** d. Histogram

41. The Apex bank providing rural credit in Agriculture field is-
   a. RBI b. **NAWARD**
   c. Gramin Bank d. World Bank

42. The functional unit of rural society is-
   a. Land b. Block
   c. **Village** d. Agriculture

43. First Five year plan was stared during the year-
   a. 1948-53 b. **1951-56**
   c. 1953-58 d. 1950-55

44. First rural development programme was-
   a. **Sriniketan project** b. Sewagram Project
   c. Etawah pilot project d. Gurgaon project

45. The principle of *one village one society* was suggested by-
   a. Mohan Singh Mehta Committee
   b. Swaminathan Commission
   c. **Maclogen committee**
   d. Lord Linlithgow Commission

Answers

36. d	37. a	38. c	39. c	40. c	41. b
42. c	43. b	44. a	45. c		

46. Panchayati Raj was first started in-
    a. Bihar
    b. **Rajasthan**
    c. Pantanagar
    d. Pondicherry

47. KVK was recommended by-
    a. **Mohan Singh Mehta Committee**
    b. Swaminathan Commission
    c. Maclogen committee
    d. Lord Linlithgow Commission

48. The National Agricultural Insurance Scheme (NAIS) was stated in-
    a. 1948-53
    b. 1951-56
    c. 1953-58
    d. **1999-2000**

49. The full form of NARS is-
    a. National Agricultural Research Scientist
    b. National Agronomy Research System
    c. National Advances Research System
    d. **National Agricultural Research System**

50. IVLP stands for-
    a. Indian Village Linkage Programme
    b. Institutional Village Labours Programme
    c. Indian Village Labours Programme
    d. **Institutional Village Linkage Programme**

51. National Institute of Agricultural Marketing (NIAM) is located in-
    a. **Jaipur**
    b. New Delhi
    c. Bhopal
    d. Lucknow

52. Reserve Bank of India was established on-
    a. 1st May, 1930
    b. **1st April, 1935**
    c. 1st August, 1935
    d. 1st April, 1934

53. Full form of SIDBI is -
    a. **Small Industries Development Bank of India**
    b. State Industries Designing Bank of India
    c. Small Industries Development Bazaar of India
    d. State Investment Development Bank of India

Answers

46. b	47. a	48. d	49. d	50. d	51. a
52. b	53. a				

54. The first fully Indian Bank was-
  a. State Bank of India b. **Punjab National Bank**
  c. Bank of India d. Indian Bank

55. The time period of 11th Five year Plan is-
  a. 2000-2005 b. 2005-210
  c. **2007-2012** d. 1998-2003

56. Co-operative credit societies act was passed in India during-
  a. 1885 b. 1892
  c. 1910 d. **1912**

57. National Policy for Farmers was approved during-
  a. 1912 b. 2000
  c. **2007** d. 2009

58. The term Extension was first used in-
  a. India b. **U.K.**
  c. Japan d. China

59. The functional unit of rural society is-
  a. Land b. Block
  c. **Village** d. Agriculture

60. ISARD stands for-
  a. Institute for Small on Agricultural and Rural Development
  b. **Institute for Studies on Agricultural and Rural Development**
  c. Indian State for Agricultural and Rural Development
  d. Institute for Studies on Advancement and Rural Development

61. The first Kshetriya Gramin Bank was opened in India during-
  a. 1921 b. **1975**
  c. 1979 d. 1982

62. The first state that adopted two tier panchayat raj was-
  a. Kerala b. Gujarat
  c. Tamil Nadu d. **Karnataka**

63. The chairman for national development of council in India is-
  a. **Prime Minister** b. President
  c. Director research d. Chief Minister

Answers

54. b	55. c	56. d	57. c	57. b	59. c
60. b	61. b	62. d	63. a		

64. Nobel Prize for Micro Credit was given to-
    a. **Mohammad Yunus** b. M. K Gandhi
    c. Swaminathan d. R.N Tagore
65. Directorate of plant protection, quarantine and storage is located at-
    a. Kerala b. New Delhi
    c. Lucknow d. **Faridabad**
66. Model village concept was given by-
    a. **Daniel Hamilton** b. M. K Gandhi
    c. Swaminathan d. R.N Tagore
67. CAMPA stands for-
    a. Complete Afforestation Fund Managing and Promoting Authority
    b. Compensatory Afforestation Forest Management and Planning Agency
    c. **Compensatory Afforestation Fund Management and Planning Authority**
    d. Common Afforestation Forest Management and Planning Agency
68. Percent of innovators in a society to adopt a new research or innovation is-
    a. **2.5** b. 13.5
    c. 34 d. 16
69. Percent of laggards in a society to adopt a new research or innovation is-
    a. 2.5 b. 13.5
    c. 34 d. **16**
70. 69. Percent of early adopters in a society to adopt a new research or innovation is-
    a. 2.5 b. **13.5**
    c. 34 d. 16
71. The NPV value for any system should be-
    a. Negative b. **Positive**
    c. Zero d. Both a and b
72. Percent of early majority in a society to adopt a new research or innovation is-
    a. 2.5 b. 13.5
    c. **34** d. 16

Answers

64. a	65. d	66. a	67. c	68. a	69. d
70. b	71. b	72. c			

73. Percent of late majority in a society to adopt a new research or innovation is-
    a. 2.5 b. 13.5
    c. **34** d. 16
74. The adoption of an innovation usually follows a-
    a. **S-shaped curve** b. J-shaped curve
    c. V-shaped curve d. I-shaped curve
75. Gurgaon project was started in the year-
    a. **1920** b. 1921
    c. 1929 d. 1948
76. Gurgaon project was started by-
    a. Spencer Hatch b. M. K Gandhi
    c. **F.L. Bryne** d. R.N Tagore
77. Marathondom project was started in the year-
    a. 1920 b. **1921**
    c. 1929 d. 1948
78. Marathondom project was started by-
    a. **Spencer Hatch** b. M. K Gandhi
    c. F.L. Bryne d. R.N Tagore
79. Sewagram project was started in the year-
    a. 1920 b. 1921
    c. **1929** d. 1948
80. Etawa pilot project was started in the year-
    a. 1920 b. 1921
    c. 1929 d. **1948**
81. Firka Development was started by-
    a. **T. Prakashan** b. M. K Gandhi
    c. **F.L. Bryne** d. R.N Tagore
82. Shantiniketan project was started in the year-
    a. 1920 b. **1921**
    c. 1929 d. 1948
83. Majdoor Manjil project was started by-
    a. Spencer Hatch b. M. K Gandhi
    c. F.L. Bryne d. **S.K. Dey**

Answers

73. c	74. a	75. a	76. c	77. b	78. a
79. c	80. d	81. a	82. b	83. d	

84. Father of extension -
    a. Spencer Hatch
    b. M. K Gandhi
    c. **Leagnes**
    d. R.N Tagore
85. Extension is a-
    a. Dutch word
    b. French word
    c. **Latin word**
    d. Greek word
86. Extension meaning is-
    a. **Stretching**
    b. Extending
    c. Expanding
    d. Tension

Answers

84. c   85. c   86. a

CHAPTER 18

Agricultural Statistics and Economics

1. Mean, median and mode are same in-
   a. Binomial distribution
   b. Poisson distribution
   c. **Normal distribution**
   d. All of the above

2. Which statistical design is the best design for field experiments -
   a. **Randomized Block Design**
   b. Completely Randomized Block Design
   c. Latin Square Design
   d. Completely Randomized Design

3. The square of the standard deviation is known as-
   a. Mode
   b. **Variance**
   c. Standard variance
   d. Dispersion

4. The relationship between independent and dependent variables is computed by-
   a. **Regression coefficient**
   b. Correlation
   c. Standard variance
   d. Harmonic mean

5. The regression coefficient is independent of-
   a. Change in direction
   b. Distribution of charge
   c. **Change of origin**
   d. Change of charge

6. The value of chi square (X^2) is always-
   a. Negative
   b. **Positive**
   c. Both a and b
   d. Never Positive

Answers

1. c 2. a 3. b 4. a 5. c 6. b

7. The ranges of chi square (X^2) is-
   a. –1 to +1
   b. 0 to 1
   c. $-\infty$ to $+\infty$
   d. **0 to ∞**

8. The correlation coefficient is the geometric mean between-
   a. **Two regression coefficients**
   b. Regression coefficients
   c. Two progressive mean
   d. Three harmonic mean

9. The ranges of multiple correlation coefficient lies between-
   a. **–1 to +1**
   b. 0 to 1
   c. $-\infty$ to $+\infty$
   d. 0 to ∞

10. Which statistical design is the best design for laboratory experiment-
    a. Randomized Block Design
    b. Completely Randomized Block Design
    c. Latin Square Design
    d. **Completely Randomized Design**

11. Local control is not applied for the design-
    a. Randomized Block Design
    b. Completely Randomized Block Design
    c. Latin Square Design
    d. **Completely Randomized Design**

12. If any items of the series is zero the which central tendency become zero -
    a. Mode
    b. **Geometric mean**
    c. Standard variance
    d. Dispersion

13. The allocation of the treatments to the different experimental units in a random manner called-
    a. Treatments
    b. Local control
    c. **Randomization**
    d. Dispersion

14. Which is appropriate design when the fertility gradient of the field is in one direction-
    a. **Randomized Block Design**
    b. Completely Randomized Block Design
    c. Latin Square Design
    d. Completely Randomized Design

Answers

7. d	8. a	9. a	10. d	11. d	12. b
13. c	14. a				

15. Which is appropriate design when the fertility gradient of the field is in two direction-
    a. Randomized Block Design
    b. Completely Randomized Block Design
    c. **Latin Square Design**
    d. Completely Randomized Design

16. Father of statistics is-
    a. Meischer
    b. Adam Smith
    c. W. Student
    d. **R.A. Fisher**

17. The minimum error degree of freedom for RBD should be at least-
    a. 5
    b. 10
    c. **12**
    d. 15

18. The hypothesis of no difference is known as-
    a. **Null hypothesis**
    b. Fisher hypothesis
    c. Smith hypothesis
    d. None

19. The probability of committing type 1 error is known as-
    a. Test of significance
    b. Local control
    c. Randomization
    d. **Level of significance**

20. Which test is used to test of significance of the difference between two means z-test-
    a. t-test
    b. f-test
    c. **z-test**
    d. All of the above

21. Which test is applicable for comparison of two means from independent samples -
    a. **t-test**
    b. f-test
    c. z-test
    d. Chi square test

22. The ranges of regression coefficient lies between-
    a. -1 to $+1$
    b. 0 to 1
    c. $-\infty$ **to** $+\infty$
    d. 0 to ∞

23. In hypothesis testing the most common level of significance that is used by statisticians is-
    a. 0.5
    b. **0.05**
    c. 1.0
    d. 0.10

Answers

15. c	16. d	17. c	18. a	19. d	20. c
21. a	22. c	23. b			

24. t-test is applicable when the number of treatments are -
    a. **2**
    b. 3
    c. 5
    d. 10

25. Principles of Experimental design is given by-
    a. Meischer
    b. Adam Smith
    c. W. Student
    d. **R.A. Fisher**

26. Anova was first proposed by-
    a. Meischer
    b. Adam Smith
    c. W. Student
    d. **R.A. Fisher**

27. The simplest measure of dispersion is-
    a. Mean
    b. Mode
    c. **Range**
    d. Median

28. The statistical test is used to determine the goodness of fit is-
    a. t-test
    b. f-test
    c. z-test
    d. **Chi square test**

29. The curve of normal distribution is-
    a. Sigmoid
    b. **Bell shaped**
    c. J-shaped
    d. V-shaped

30. The correlation coefficient will be negative when-
    a. X increases and Y increase
    b. **X increases and Y decreases**
    c. X decrease and Y decreases
    d. X decrease and Y increase

31. Binomial distribution is called-
    a. Pie distribution
    b. Fisher distribution
    c. **Bernoulli's distribution**
    d. Poisson distribution

32. Chi square test is parametric test-
    a. **Parametric test**
    b. Non-parametric test
    c. Both
    d. None

33. Error degree of freedom for RBD is-
    a. $(n-1)$
    b. $\mathbf{(n-1)(r-1)}$
    c. $(n-1)+(r-1)$
    d. $(n+1)$

34. The Indian economy type is-
    a. Compound
    b. Variable
    c. Pure
    d. **Mixed**

Answers

24. a	25. d	26. d	27. c	28. d	29. b
30. b	31. c	32. a	33. b	34. d	

35. Most valid law in agriculture production is-
    a. **Law of diminishing returns**
    b. Law of marginal returns
    c. Law of marginal utility
    d. Law of Equi-marginal utility
36. The formula for rate of turnover (σ) is-
    a. σ = Total income /Total expenses
    b. **σ = Total expenses/Gross income**
    c. σ = Total liabilities /Total assets
    d. σ = Total assets/Total liabilities
37. National Income in India is estimated by-
    a. SSO b. **CSO**
    c. ISO d. SCO
38. Headquarter of SIDBI is situated in-
    a. Jaipur b. New Delhi
    c. Kolkata d. **Lucknow**
39. The main security guard of International Trade is-
    a. WIPCO b. CSO
    c. ISO d. **WTO**
40. NABARD was established in which plan-
    a. **6 Five year plan** b. 7 Five year plan
    c. 8 Five year plan d. 9 Five year plan
41. The chairman of 12th Finance Commission-
    a. Swaminathan b. Trilochan Mahopatra
    c. Arun Jaitley d. **Rangarajan**
42. Father of economics is-
    a. Meischer b. **Adam Smith**
    c. W. Student d. R.A. Fisher
43. VAT is a/an -
    a. Direct tax b. Utility tax
    c. **Indirect tax** d. Simple tax
44. Randomised block design is applicable when the number of treatments are -
    a. **20** b. 25
    c. 30 d. 50

Answers

35. a	36. b	37. b	38. d	39. d	40. a
41. d	42. b	43. c	44. a		

45. Latin square design is applicable when the number of treatments are -
a. 5-8
b. 10-15
c. 5-12
d. **Both a and c**

46. Chi-square is applicable when the sample size is/are-
a. **≥ 50**
b. >30
c. < 50
d. < 30

47. Chi-square was given by-
a. Gosset
b. R. A. Fisher
c. **Karl Pearson**
d. Meischer

48. z-test is applicable when the sample size is/are-
a. ≥ 50
b. **> 30**
c. < 50
d. < 30

49. z-test was given by-
a. Gosset
b. **R. A. Fisher**
c. Karl Pearson
d. Meischer

50. Student t-test is applicable when the sample size is/are-
a. ≥ 50
b. > 30
c. < 50
d. **< 30**

51. Student t-test was given by-
a. **Gosset**
b. R. A. Fisher
c. Karl Pearson
d. Meischer

52. Error degree of freedom for completely randomised design (CRD) is-
a. **(n – 1)**
b. (n – 1) (r – 1)
c. (n – 1) + (r – 1)
d. (n + 1)

53. Error degree of freedom for Latin square design is-
a. (n – 1)
b. **(n – 1) (n – 2)**
c. (n – 1) + (n – 2)
d. (n + 1)

54. Rejecting null hypothesis (H_0) when it is true called-
a. **Type error-I**
b. Type error-II
c. Type error-III
d. Type error-IV

55. Accepting null hypothesis (H_0) when it is false called-
a. Type error-I
b. **Type error-II**
c. Type error-III
d. Type error-IV

Answers

45. d	46. a	47. c	48. b	49. b	50. d
51. a	52. a	53. b	54. a	55. b	

56. Degree of freedom is-
a. (n + k)
b. **(n – k)**
c. (n/k)
d. (n × k)

57. The ranges of t-test lies between-
a. –1 to +1
b. 0 to 1
c. **–∞ to +∞**
d. 0 to ∞

58. The ranges of normal distribution lies between-
a. –1 to +1
b. 0 to 1
c. **–∞ to +∞**
d. 0 to ∞

59. The ranges of Poisson distribution lies between-
a. –1 to +1
b. **$0 \le x \le \infty$**
c. –∞ to +∞
d. 0 to ∞

60. The ranges of f-test lies between-
a. –1 to +1
b. 0 to 1
c. –∞ to +∞
d. **$0 \le x \le \infty$**

61. The ranges of probability lies between-
a. –1 to +1
b. **0 to 1**
c. –∞ to +∞
d. 0 to ∞

Answers

56. b 57. c 58. c 59. b 60. d 61. b

CHAPTER 19

Plant Physiology

1. The study of various activities and metabolism of plants is known as plant physiology-
   a. Plant anatomy b. Histology
   c. Botany d. **Plant physiology**
2. Father of plant physiology is-
   a. Odum b. J. C. Bose
   c. **Stephan Hales** d. Dixon & Jolley
3. Father of Indian plant physiology is-
   a. Watson & Crick b. **J. C. Bose**
   c. Stephan Hales d. Dixon & Jolley
4. The movement of molecules or atoms or ions of a materials from an area of higher concentration to an area of their lower concentration is -
   a. **Diffusion** b. Osmosis
   c. Plasmolysis d. Exosmosis
5. Which is the correct order of diffusion rate in medium-
   a. Gas < Liquid < Solid b. Gas = Liquid > Solid
   c. **Gas > Liquid > Solid** d. Gas < Liquid = Solid
6. The water moves into the cell during the osmosis is called -
   a. Exosmosis b. Osmosis
   c. **Endosmosis** d. Imbibition
7. Diffusion pressure of a pure solvent is-
   a. 756 atm b. 1036 atm
   c. **1236 atm** d. 1500 atm

Answers

1. d	2. c	3. b	4. a	5. c	6. c
7. c					

8. Exchanges of gases like CO_2, O_2 and distribution of hormones in plants takes place through-
   a. **Diffusion** b. Osmosis
   c. Plasmolysis d. Exosmosis

9. Osmosis was discovered by -
   a. **Abbe Nollet** b. Duterochat
   c. Stephan Hales d. Traube

10. The highest osmotic pressure is found in-
   a. **Halophytes** b. Xerophytes
   c. Hydrophytes d. Mesophytes

11. The water moves out the cell during the osmosis is called -
   a. **Exosmosis** b. Osmosis
   c. Endosmosis d. Imbibition

12. Membrane which is permeable for solvent and do not allow the passage of solute called-
   a. Permeable b. **Semi-permeable**
   c. Non-permeable d. None

13. The pressure developed in a solution when solution and water are separated by semi-permeable membrane called -
   a. Turgor pressure b. Wall pressure
   c. Diffusion pressure d. **Osmotic pressure**

14. The osmotic pressure of pure water is zero-
   a. Negative b. Positive
   c. **Zero** d. Hundred

15. The lowest osmotic pressure is found in-
   a. Halophytes b. Xerophytes
   c. **Hydrophytes** d. Mesophytes

16. Which is the correct order of osmotic pressure in plants-
   a. **Hydrophytes< mesophytes < xerophytes < halophytes**
   b. Hydrophytes >mesophytes > xerophytes > halophytes
   c. Halophytes < mesophytes = xerophytes < Hydrophytes
   d. Hydrophytes< xerophytes < mesophytes < halophytes

Answers

8. a	9. a	10. a	11. a	12. b	13. d
14. c	15. c	16. a			

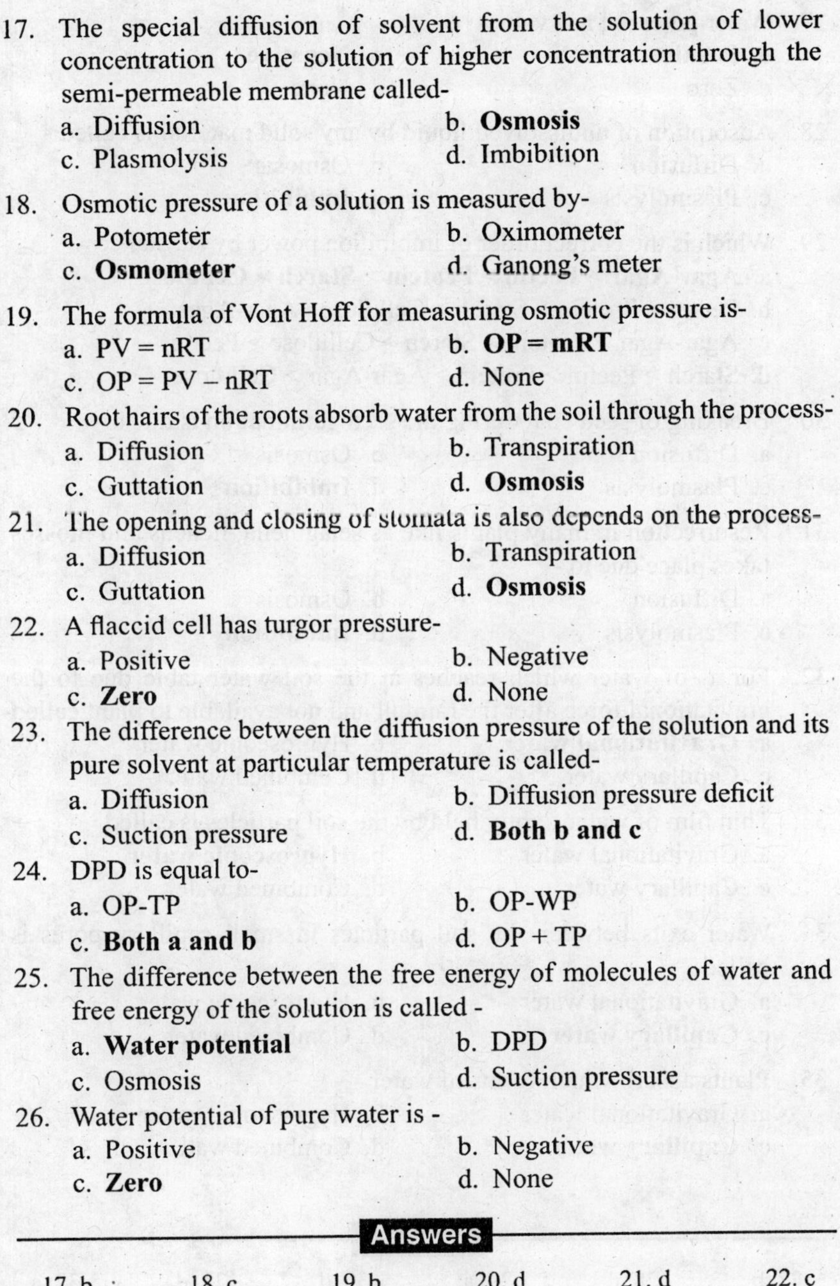

17. The special diffusion of solvent from the solution of lower concentration to the solution of higher concentration through the semi-permeable membrane called-
   a. Diffusion b. **Osmosis**
   c. Plasmolysis d. Imbibition
18. Osmotic pressure of a solution is measured by-
   a. Potometer b. Oximometer
   c. **Osmometer** d. Ganong's meter
19. The formula of Vont Hoff for measuring osmotic pressure is-
   a. PV = nRT b. **OP = mRT**
   c. OP = PV – nRT d. None
20. Root hairs of the roots absorb water from the soil through the process-
   a. Diffusion b. Transpiration
   c. Guttation d. **Osmosis**
21. The opening and closing of stomata is also depends on the process-
   a. Diffusion b. Transpiration
   c. Guttation d. **Osmosis**
22. A flaccid cell has turgor pressure-
   a. Positive b. Negative
   c. **Zero** d. None
23. The difference between the diffusion pressure of the solution and its pure solvent at particular temperature is called-
   a. Diffusion b. Diffusion pressure deficit
   c. Suction pressure d. **Both b and c**
24. DPD is equal to-
   a. OP-TP b. OP-WP
   c. **Both a and b** d. OP + TP
25. The difference between the free energy of molecules of water and free energy of the solution is called -
   a. **Water potential** b. DPD
   c. Osmosis d. Suction pressure
26. Water potential of pure water is -
   a. Positive b. Negative
   c. **Zero** d. None

Answers

17. b	18. c	19. b	20. d	21. d	22. c
23. d	24. c	25. a	26. c		

27. Water potential has value-
    a. Positive   b. **Negative**
    c. Zero   d. None
28. Adsorption of undissolved liquid by any solid material is called -
    a. Diffusion   b. Osmosis
    c. Plasmolysis   d. **Imbibition**
29. Which is the correct order of imbibition power by substances-
    a. **Agar-Agar > Pectin > Protein > Starch > Cellulose**
    b. Pectin > Protein > Starch > Cellulose> Agar-Agar
    c. Agar-Agar > Protein > Starch > Cellulose > Pectin
    d. Starch > Pectin > Protein > Agar-Agar > Cellulose
30. Breaking of seed coat during the seed germination is due to-
    a. Diffusion   b. Osmosis
    c. Plasmolysis   d. **Imbibition**
31. Resurrection in many plants like as selaginella, lichens and mosses takes place due to -
    a. Diffusion   b. Osmosis
    c. Plasmolysis   d. **Imbibition**
32. Forms of water which reaches at the soil water table due to the gravitational force after the rainfall and not available to plant called-
    a. **Gravitational water**   b. Hygroscopic water
    c. Capillary water   d. Combined water
33. Thin film of water tightly held by the soil particles is called-
    a. Gravitational water   b. **Hygroscopic water**
    c. Capillary water   d. Combined water
34. Water exits between the soil particles in small capillary pores is called-
    a. Gravitational water   b. Hygroscopic water
    c. **Capillary water**   d. Combined water
35. Plants absorb which forms of water -
    a. Gravitational water   b. Hygroscopic water
    c. **Capillary water**   d. Combined water

Answers

27. b	28. d	29. a	30. d	31. d	32. a
33. b	34. c	35. c			

36. Total amount of water present in the soil called as-
    a. Chresard　　b. Echard
    c. Field capacity　　d. **Holard**

37. The amount of water available to the plants is called-
    a. **Chresard**　　b. Echard
    c. Field capacity　　d. Holard

38. The amount of water not available to the plants is called-
    a. Chresard　　b. **Echard**
    c. Field capacity　　d. Holard

39. Maximum absorption of water takes place from-
    a. Meristematic zone　　b. **Root hair zone**
    c. Root cap zone　　d. Maturation zone

40. Maximum absorption of mineral takes place from-
    a. **Meristematic zone**　　b. Root hair zone
    c. Root cap zone　　d. Maturation zone

41. The movement of water from cell to cell through plasmodesmata is called-
    a. Symplast　　b. Apoplast
    c. Living pathway　　d. **Both a and c**

42. Osmotic active method of water absorption was given by-
    a. Munch　　b. Dixon & Jolley
    c. **Atkins & Priestley**　　d. Pfeffer

43. Non osmotic active method of water absorption was given by-
    a. Munch　　b. Dixon & Jolley
    c. Atkins & Priestley　　d. **Bennet-Clark**

44. Term active and passive absorption was proposed by-
    a. **Renner**　　b. Dixon & Jolley
    c. Atkins & Priestley　　d. Bennet-Clark

45. Upward movement of water against the gravitational force upto top parts of the plants is called-
    a. Guttation　　b. Transpiration
    c. Evaporation　　d. **Ascent of sap**

Answers

36. d	37. a	38. b	39. b	40. a	41. d
42. c	43. d	44. a	45. d		

46. Relay pump theory was proposed by-
    a. Priestley b. Dixon & Jolley
    c. J.C. Bose d. **Godlewski**
47. Translocation of food mainly occurs in the form of-
    a. Glucose b. Mannose
    c. **Sucrose** d. Maltose
48. Pulsation theory was proposed by-
    a. Priestley b. Boehm
    c. **J.C. Bose** d. Godlewski
49. Loss of water in the form of vapour, from the aerial parts of living plants is known as-
    a. Guttation b. **Transpiration**
    c. Evaporation d. Ascent of sap
50. Root pressure theory was proposed by-
    a. **Priestley** b. Boehm
    c. J.C. Bose d. Godlewski
51. Which theory is most accepted or universally accepted theory of ascent of sap-
    a. Root pressure theory
    b. **Transpiration pull & cohesion force theory**
    c. Pulsation theory
    d. Relay pump theory
52. Which theory is most accepted theory of food conduction in plants -
    a. Root pressure flow
    b. Transpiration pull & cohesion force theory
    c. **Mass flow hypothesis**
    d. Cytochrome pump theory
53. Transpiration pull & cohesion force theory was proposed by-
    a. Jamin b. Peter Mitchell
    c. Unger & Von Sachs d. **Dixon & Jolly**
54. Pressure flow or mass flow hypothesis of food translocation was given by-
    a. **Munch** b. Bennet & Clark
    c. D.E Koshland d. Vanden Honert

Answers

46. d	47. c	48. c	49. b	50. a	51. b
52. c	53. d	54. a			

55. The minimum transpiration is found in -
a. Mesophytes b. Hydrophyte
c. Xerophyte d. **Succulent xerophyte**

56. Capillary force theory was proposed by-
a. Priestley b. **Boehm**
c. J.C. Bose d. Godlewski

57. The maximum transpiration is found in -
a. **Mesophytes** b. Hydrophyte
c. Xerophyte d. **Succulent xerophyte**

58. Chain theory was proposed by-
a. **Jamin** b. Peter Mitchell
c. Unger & Von Sachs d. Dixon & Jolly

59. The transpiration is absent in -
a. Hydrophytes b. **Submerged hydrophyte**
c. Xerophyte d. Succulent xerophyte

60. Stomatal pore is surrounded by two specialised epidermal cells called as-
a. **Guard cells** b. Casparian cells
c. Accessory cells d. Stomatal cells

61. Imbibition force theory was proposed by-
a. Jamin b. Peter Mitchell
c. **Unger & Von Sachs** d. Dixon & Jolly

62. The structure of guard cell in monocots (Gramineae) is -
a. Kidney shape b. Bean shape
c. Sigmoid shape d. **Dumbled shape**

63. The amount of CO_2 in grams absorbed by 1 gram of chlorophyll in 1 hour is called as-
a. Photosynthetic value b. **Photosynthetic number**
c. Photosynthesis d. R. Q.

64. Cytochrome pump theory was proposed by-
a. Jamin b. Peter Mitchell
c. Unger & Von Sach d. **Lundegardh & Burstorm**

Answers

55. d	56. b	57. a	58. a	59. b	60. a
61. c	62. d	63. b	64. d		

65. Plants which are adapted to grow in high intensity of light is called-
    a. Sciophytes   b. **Heliophytes**
    c. Sporophytes   d. Oxalophytes

66. Carrier concept theory was proposed by-
    a. **Vanden Honert**   b. Bennet & Clark
    c. Emil Fischer   d. D.E Koshland

67. Plants which are adapted to grow in shade is called-
    a. **Sciophytes**   b. Heliophytes
    c. Sporophytes   d. Halophytes

68. When respiratory substrates are carbohydrates then respiration is known as-
    a. **Floating respiration**   b. Protoplasmic respiration
    c. Simple respiration   d. None

69. Efficiency of photosynthesis is-
    a. 25%   b. **30%**
    c. 35%   d. 42%

70. Protein- Lecithin theory was proposed by-
    a. Vanden Honert   b. **Bennet & Clark**
    c. Emil Fischer   d. D.E Koshland

71. When respiratory substrates are proteins then respiration is known as-
    a. Floating respiration   b. **Protoplasmic respiration**
    c. Simple respiration   d. None

72. Efficiency of respiration in plants is-
    a. 25%   b. 30%
    c. 35%   d. **42%**

73. Lock & key theory was proposed by-
    a. Vanden Honert   b. Bennet & Clark
    c. **Emil Fischer**   d. D.E Koshland

74. Induced fit theory was proposed by-
    a. Vanden Honert   b. Bennet & Clark
    c. Emil Fischer   d. **D.E Koshland**

Answers

65. b	66. a	67. a	68. a	69. b	70. b
71. b	72. d	73. c	74. d		

75. The plants grown in N_2-deficient places but are insectivorous habits called as-

a. Sciophytes b. **Insectivorous plants**
c. Sporophytes d. Halophytes

76. Final products of anaerobic respiration is-

a. CO_2 & H_2O b. **C_2H_5OH & CO_2**
c. C_2H_5OH & H_2O d. All of the above

77. Chemiosmotic theory was proposed by-

a. Liebig b. Bennet & Clark
c. Emil Fischer d. **Peter Mitchell**

78. ATP formed in aerobic respiration is-

a. **38** b. 36
c. 2 d. 30

79. Law of minimum was given by-

a. Hartwig b. **Liebig**
c. Moll d. Singer & Nicolson

80. ATP formed in anaerobic respiration is-

a. 38 b. 36
c. **2** d. 30

81. Law of limiting factors was given by-

a. Hartwig b. Liebig
c. Emil Fischer d. **Blackman**

82. Non-cyclic scheme in bacterial photosynthesis was given by-

a. Hartwig b. Liebig
c. **Olson** d. Blackman

83. Non green plants which are depends on dead organic matter called as-

a. Sciophytes b. Insectivorous plants
c. **Sporophytes** d. Halophytes

84. According to….transpiration is an essential evil-

a. Jamin b. Peter Mitchell
c. **Curtis** d. Dixon & Jolly

Answers

75. b	76. b	77. d	78. a	79. b	80. c
81. d	82. c	83. c	84. c		

85. According to.....transpiration is an unavoidable evil-
a. Blackman
b. Bennet & Clark
c. Emil Fischer
d. **Steward**

86. Which is/are not a types of transpiration-
a. Stomatal
b. Cuticular
c. Lenticular
d. **Folicular**

87. Stomatal transpiration percent of total transpiration in plants is-
a. **80-90%**
b. 9%
c. 5%
d. 0.1-1%

88. The outer wall of the guard cell is-
a. **Thin & elastic**
b. Thick & non-elastic
c. Thin & non-elastic
d. Thick & elastic

89. Which theory is most acceptable theory for stomatal movement-
a. **Active potassium (K^+) and hydrogen (H^+)**
b. Transpiration pull & cohesion force theory
c. Mass flow hypothesis
d. Passive potassium (K^+) and hydrogen (H^+)

90. Plant hormone which helps in closing of stomata is-
a. GA_3
b. Auxin
c. Cytokinin
d. **ABA**

91. Effect of low temperature on the initiation and development of flower called-
a. Cryopreservation
b. Photoperiodism
c. **Vernalization**
d. Anthesis

92. Vernalization phenomena was discovered by-
a. Muller
b. Allard Buchner
c. Munch
d. **Lysenko**

93. Term enzyme was given by-
a. Buchner
b. **Kuhne**
c. Michaelis & Menten
d. F.F.Blackman

94. First isolated enzyme zymase (from yeast) was discovered by-
a. **Buchner**
b. Kuhne
c. Lohman
d. Van Niel

Answers

85. d	86. d	87. a	88. a	89. a	90. d
91. c	92. d	93. b	94. a		

95. First discover ribozyme was-
a. L_{10} RNAase	b. L_{29} DNAase
c. **L_{19} RNAase**	d. L_{17} DNAase

96. 1 gram fat provide-
a. 2.8 K.Cal.	b. 4.8 K.Cal.
c. 8.8 K.Cal.	d. **9.8 K.Cal.**

97. Protein part of conjugated enzyme called-
a. Holoenzyme	b. **Apoenzyme**
c. Co-enzyme	d. Proenzyme

98. 1 gram protein provide-
a. 2.8 K.Cal.	b. **4.8 K.Cal.**
c. 8.8 K.Cal.	d. 9.8 K.Cal.

99. ATP was discovered by -
a. Buchner	b. Kuhne
c. **Lohman**	d. Van Niel

100. The concentration of substrate at which rate of reaction of that enzyme attains half of its maximum velocity called as-
a. Activation energy	b. **Km constant**
c. Rate of reaction	d. Y-value

101. Km constant concept for enzyme was given by-
a. **Michaelis & Menten**	b. Kuhne
c. Buchner	d. Lipman

102. In Photosynthesis process released oxygen comes from-
a. CO_2	b. **H_2O**
c. Chlorophyll	d. C_2H_5OH

103. In Photosynthesis process oxygen of glucose comes from -
a. **CO_2**	b. H_2O
c. Chlorophyll	d. C_2H_5OH

104. Photosynthesis a-
a. Photo-biochemical process	b. Anabolic process
c. Catabolic process	d. **All of the above**

Answers

95. c	96. d	97. b	98. b	99. c	100. b
101. a	102. b	103. a	104. d		

105. First true and oxygenic photosynthesis was started in-
a. **Cyanobacteria** b. Algae
c. Bryophytes d. Sulphur bacteria

106. Photosynthetic roots is/are found in-
a. Tinospora b. Nymphaea
c. Trapa d. **Both a and c**

107. Absorption spectrum of photosynthesis is-
a. Green b. Red
c. Yellow d. **Blue**

108. Action spectrum of photosynthesis is-
a. Green b. **Red**
c. Yellow d. Blue

109. Rate of photosynthesis is highest in which light-
a. Green b. Red
c. **White** d. Blue

110. Blackman discovered-
a. **Dark reaction** b. Light reaction
c. C_3-cycle d. C_4-cycle

111. Calvin cycle or C_3-cycle was discovered by-
a. Michaelis & Menten b. **Calvin & Benson**
c. Buchner d. Emerson & Arnold

112. Q_{10} for dark reaction and light reaction is respectively-
a. 1-3 and 1-2 b. 2-4 and 1-2
c. **2-3 and 1** d. 2-3 and 1-3

113. The concept of two photosystem was given by-
a. Michaelis & Menten b. Calvin & Benson
c. Park & Biggins d. **Emerson & Arnold**

114. Suddenly dropped down in quantum yield by giving only light greater than 680 nm wavelength called-
a. Green effect b. **Red drop**
c. Emerson effect d. Warburg effect

Answers

105. a	106. d	107. d	108. b	109. c	110. a
111. b	112. c	113. d	114. d		

115. Increase in quantum yield by giving light shorter or greater than 680 nm wavelength called-

a. Green effect | b. Red drop
c. **Emerson effect** | d. Warburg effect

116. The number of O_2 molecule evolved by one quantum of light in photosynthesis is called as-

a. **Quantum yield** | b. R. Q.
c. Quantum number | d. Octane number

117. Quantasome was discovered by-

a. Burgerstein | b. Calvin & Benson
c. **Park & Biggins** | d. Emerson & Arnold

118. Which is/are the universal pigment which are found in all O_2 liberating cells -

a. Chlorophyll-a | b. Chlorophyll-b
c. Carotene | d. **Both a and c**

119. Which is/are the antitranspirants substances-

a. Phenyl Mercuric Acetate | b. Abscisic acid
c. CO_2 | d. **All of the above**

120. Loss of water from the uninjured part or margin of leaves of the plant in the form of water droplets is called as-

a. **Guttation** | b. Transpiration
c. Evaporation | d. Ascent of sap

121. Guttation takes place through a special pore like structure called-

a. Stomata | b. Water stomata
c. Hydathode | d. **Both b and c**

122. Term guttation was coined by-

a. **Burgerstein** | b. Calvin & Benson
c. Park & Biggins | d. Emerson & Arnold

123. Fast flowing of liquid from the injured or cut parts of the plants is called as-

a. Guttation | b. Transpiration
c. **Exudation** | d. Ascent of sap

Answers

115. c	116. a	117. c	118. d	119. d	120. a
121. d	122. a	123. c			

124. Dropping of the soft parts of the pants due to loss of turgidity in their cells is called-

a. Guttation b. Transpiration
c. Evaporation d. **Wilting**

125. How many elements are considered as essential elements-

a. 15 b. 16
c. **17** d. 18

126. Criteria of essentiality of minerals was given by -

a. **Arnon** b. Champion & Seth
c. Haberlandt d. Emerson & Arnold

127. Dimorphic chloroplasts are present in the leaves of-

a. C_2- plants b. C_3-plants
c. **C_4-plants** d. C_5-plants

128. Kranz anatomy is present in the leaves of-

a. C_2- plants b. C_3-plants
c. **C_4-plants** d. C_5-plants

129. 129. In C_4-plants, C_3-Cycle occurs in-

a. **BS cells** b. Mesophyll cells
c. Guard cells d. Accessory cells

130. In C_4-plants, C_4-Cycle occurs in-

a. BS cells b. **Mesophyll cells**
c. Guard cells d. Accessory cells

131. First carboxylation in C_4-cycle occurs by-

a. PGA b. RNA polymerase
c. Rubisco d. **PEP case**

132. Concept of photorespiration was given by -

a. **Decker & Tio** b. Calvin & Benson
c. Haberlandt d. Emerson & Arnold

133. Rubisco act as oxygenase at-

a. **Higher concentration of O_2 & low CO_2 concentration**
b. Higher concentration of CO_2 & low O_2 concentration
c. Higher concentration of O_2 & High CO_2 concentration
d. Lower concentration of O_2 & low CO_2 concentration

Answers

124. d	125. c	126. a	127. c	128. c	129. a
130. b	131. d	132. a	133. a		

134. Photorespiration occurs in-
   a. Chloroplast
   b. Mitochondria
   c. Peroxisomes
   d. **All of the above**

135. Photorespiration occurs in-
   a. C_5-plants
   b. **C_3-plants**
   c. C_4-plants
   d. All plants

136. Intensity of light at which rate of photosynthesis becomes equal to rate of respiration in plants is known as-
   a. **Light compensation point**
   b. Null point
   c. Photo oxidation
   d. Extinction point

137. Net photosynthesis or net primary productivity at light compensation point is-
   a. Maximum
   b. Minimum
   c. **Zero**
   d. None

138. In aerobic respiration final electron acceptor is-
   a. CO_2
   b. O_2
   c. **Free oxygen**
   d. $NADPH^+$

139. Anaerobic respiration was first reported by-
   a. Buchner
   b. Kornberg
   c. Cruick Shank
   d. **Kostytchev**

140. Glycolysis occurs in-
   a. Matrix
   b. **Cytoplasm**
   c. Protoplasm
   d. Mitochondria

141. Glycolysis produces -
   a. 2 mole of pyruvic acid
   b. 2 ATP
   c. $2NADPH_2$
   d. **All of the above**

142. Glycolysis is also known as-
   a. **Embden, Meyerhof and Parnas (EMP) pathway**
   b. Endof, Meyer and Parnas (EMP) pathway
   c. Embden, Meyerhof and Parton (EMP) pathway
   d. Elson, Mendel and Parnas (EMP) pathway

143. A connecting link between glycolysis & Krebs-cycle is-
   a. ATP
   b. PEP case
   c. Rubisco
   d. **Acetyl Co-A**

Answers

134. d	135. b	136. a	137. c	138. a	139. d
140. b	141. d	142. a	143. d		

144. A common pathway between aerobic & anaerobic respiration is-
a. TCA b. **EMP**
c. ETC d. HMP

145. C_4-cycle was discovered by-
a. Decker & Tio b. Calvin & Benson
c. **H.A Kreb** d. Cruick Shank

146. C_4-cycle also known as-
a. Krebs-cycle b. Citric acid cycle
c. TCA d. **All of the above**

147. The enzyme of Krebs-cycle present in inner mitochondrial membrane is-
a. **Succinic dehydrogenase** b. OAA
c. Rubisco d. PEP case

148. Pentose phosphate pathway (PPP) is also called as-
a. Warburg-Dickens pathway b. HMP
c. TCA d. **Both a and b**

149. The minimum amount of oxygen at which aerobic respiration take place & anaerobic respiration become extinct is called as-
a. Compensation point b. Null point
c. Transition point d. **Extinction point**

150. Oxygen concentration at which aerobic respiration & anaerobic respiration take place simultaneously is called as-
a. Compensation point b. Null point
c. **Transition point** d. Extinction point

151. Glyoxylate cycle was discovered by-
a. Decker & Tio b. **Kornberg & Krebs**
c. H.A Kreb d. Cruick Shank

152. Glyoxylate cycle occurs in-
a. Glyoxisome b. Cytosol
c. Mitochondria d. **All of the above**

153. Fermentation process was discovered by-
a. Decker & Tio b. Kornberg & Krebs
c. H.A Kreb d. **Cruick Shank & Pasteur**

Answers

144. b	145. c	146. d	147. a	148. d	149. d
150. c	151. b	152. d	153. d		

154. Krebs-cycle occurs in-
   a. Matrix b. Cytoplasm
   c. Protoplasm d. **Mitochondria matrix**

155. The ratio of the volume of CO_2 released to the volume of O_2 taken in respiration is called-
   a. Respiration b. **R. Q.**
   c. Respiration efficiency d. None

156. Which is/are macronutrients-
   a. Fe b. Cu
   c. Zn d. **Ca**

157. Which is/are micronutrients-
   a. Ca b. Mg
   c. **Zn** d. P

158. Which is/are micronutrients-
   a. C b. H
   c. **Zn** d. O

159. Plants grown in moistened air with nutrients is called aeroponics-
   a. Hydroponics b. **Aeroponics**
   c. Tissue culture d. Bonsai

160. R. Q. value for carbohydrates--
   a. **1.0** b. 0.7
   c. > 1.0 d. 0

161. Hydroponics concept was given by-
   a. Cruick Shank & Pasteur b. Decker & Tio
   c. Kornberg d. **Geriack**

162. Inhibitory effect of high conc. of O_2 on photosynthesis is called as -
   a. **Warburg effect** b. Pasteur effect
   c. R.Q d. Krebs effect

163. The inhibition of anaerobic respiration by O_2 concentration is called as-
   a. Warburg effect b. **Pasteur effect**
   c. R.Q d. Krebs effect

164. R. Q. value for fat or oil-
   a. 1.0 b. **0.7**
   c. > 1.0 d. 0

Answers

154. d	155. b	156. d	157. c	158. c	159. b
160. a	161. d	162. a	163. b	164. b	

165. Term hormone was given by-
   a. Buchner   b. **Starling**
   c. Cruick Shank   d. Kostytchev
166. Term phytohormone was given-
   a. **Thieman**   b. Starling
   c. Cruick Shank   d. Kostytchev
167. R. Q. value for organic acid-
   a. 1.0   b. 0.7
   c. **> 1.0**   d. 0
168. High auxin and low cytokinin favour-
   a. **Root formation**   b. Shoot formation
   c. Apical meristem   d. Lateral growth
169. High Cytokinin and low auxin favour-
   a. Root formation   b. **Shoot formation**
   c. Apical meristem   d. Lateral growth
170. R. Q. value for CAM Plants is-
   a. 1.0   b. 0.7
   c. > 1.0   d. **0**
171. R. Q. value for protein plants is-
   a. 1.0   b. 0.7
   c. **0.9**   d. 0
172. Which is/are macronutrients-
   a. B   b. Cl
   c. Mn   d. **N**
173. The first natural cytokinin was identified & crystalized from immature corn grains by -
   a. Skoog   b. Starling
   c. **Letham**   d. Thieman
174. Which is/are micronutrients-
   a. **Mo**   b. K
   c. S   d. Mg
175. Which is root initiating hormone in cutting -
   a. NAA   b. **IBA**
   c. IAA   d. IPA

Answers

165. b	166. a	167. c	168. a	169. b	170. d
17[illegible]	172. d	173. c	174. a	175. b	

176. Which is natural auxin hormone-

a. NAA | b. IBA
c. **IAA** | d. IPA

177. Which is synthetic auxin-

a. **NAA** | b. Kinetin
c. IAA | d. Zeatin

178. The relative length of day and night for flowering in the plants called-

a. Photo veneration | b. **Photoperiodism**
c. Vernalization | d. Anthesis

179. Photoperiodism was first discovered by-

a. **Garner & Allard** | b. Haberlandt
c. Francis Galton | d. Dixon & Jolley

180. Photoperiodism was first discovered in which plant-

a. Pea | b. Mustard
c. Tomato | d. **Tobacco**

181. Garner & Allard classified the plants into how many categories-

a. **3** | b. 4
c. 5 | d. 6

182. R.Q is expressed as-

a. Volume of CO_2 consumed / volume of O_2 liberated
b. Volume of O_2 liberated/ volume of CO_2 consumed
c. Volume of O_2 liberated/ volume of CO_2 consumed
d. **Volume of CO_2 liberated/ volume of O_2 consumed**

183. Deficiency symptoms of mobile elements first appear in-

a. **Older plant parts** | b. Younger plant parts
c. Flower | d. Leaves

184. Deficiency symptoms of immobile elements first appear in-

a. Older plant parts | b. **Younger plant parts**
c. Flower | d. Leaves

185. Which is/are mobile element-

a. B | b. S
c. Ca | d. **Mg**

Answers

176. c	177. a	178. b	179. a	180. d	181. a
182. d	183. a	184. b	185. d		

186. Which is/are immobile element-
a. P b. Co
c. **K** d. N

187. Which is/are immobile element-
a. P b. Co
c. **K** d. N

188. Phytochrome pigment responsible for flowering is discovered by-
a. **Borthwick & Hendricks** b. Butler
c. Garner & Allard d. Thieman

189. Term phytochrome was given by-
a. Buchner b. Starling
c. **Butler** d. Thieman

190. Which is/are mobile elements-
a. N, S, Ca. Co and Fe b. **N, P, K and Mg**
c. B, S, Ca. Co and Fe d. N, P, K, Mg and S

191. Which is/are immobile elements-
a. N, S, Ca. Co and Fe b. N, P, K and Mg
c. **B, S, Ca. Co and Fe** d. N, P, K, Mg and S

192. Nitrogen in the plants is absorbed in the form-
a. NO^- b. NO^{-2}
c. $\mathbf{NO_3^-}$ d. N^{-3}

193. Whip tail of cauliflower in plants is caused due to deficiency-
a. S b. K
c. Zn d. **Mo**

194. Sulphur in the plants is absorbed in the form-
a. SO^- b. SO^{-2}
c. SO^{-3} d. $\mathbf{SO^{-4}}$

195. Iron in the plants is absorbed in the form-
a. Fe^{-2} b. $\mathbf{Fe^{+2}}$
c. Fe^{-3} d. Fe^{+3}

196. Which element play key role in stomatal movement & transpiration -
a. S b. **K**
c. Zn d. Mo

Answers

186. c	187. c	188. a	189. c	190. b	191. c
192. c	193. d	194. d	195. b	196. b	

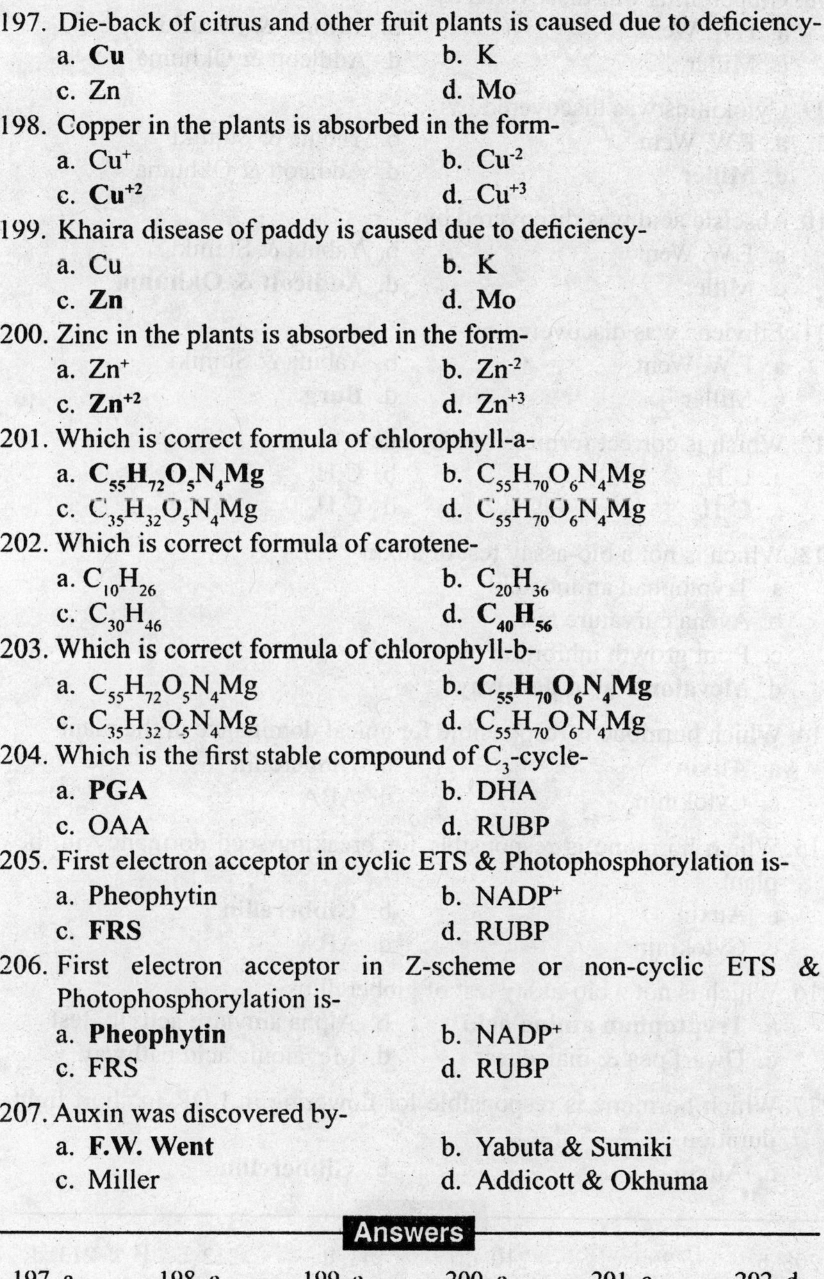

197. Die-back of citrus and other fruit plants is caused due to deficiency-
    a. **Cu** b. K
    c. Zn d. Mo
198. Copper in the plants is absorbed in the form-
    a. Cu^{+} b. Cu^{-2}
    c. **Cu^{+2}** d. Cu^{+3}
199. Khaira disease of paddy is caused due to deficiency-
    a. Cu b. K
    c. **Zn** d. Mo
200. Zinc in the plants is absorbed in the form-
    a. Zn^{+} b. Zn^{-2}
    c. **Zn^{+2}** d. Zn^{+3}
201. Which is correct formula of chlorophyll-a-
    a. **$C_{55}H_{72}O_5N_4Mg$** b. $C_{55}H_{70}O_6N_4Mg$
    c. $C_{35}H_{32}O_5N_4Mg$ d. $C_{55}H_{70}O_6N_4Mg$
202. Which is correct formula of carotene-
    a. $C_{10}H_{26}$ b. $C_{20}H_{36}$
    c. $C_{30}H_{46}$ d. **$C_{40}H_{56}$**
203. Which is correct formula of chlorophyll-b-
    a. $C_{55}H_{72}O_5N_4Mg$ b. **$C_{55}H_{70}O_6N_4Mg$**
    c. $C_{35}H_{32}O_5N_4Mg$ d. $C_{55}H_{70}O_6N_4Mg$
204. Which is the first stable compound of C_3-cycle-
    a. **PGA** b. DHA
    c. OAA d. RUBP
205. First electron acceptor in cyclic ETS & Photophosphorylation is-
    a. Pheophytin b. $NADP^{+}$
    c. **FRS** d. RUBP
206. First electron acceptor in Z-scheme or non-cyclic ETS & Photophosphorylation is-
    a. **Pheophytin** b. $NADP^{+}$
    c. FRS d. RUBP
207. Auxin was discovered by-
    a. **F.W. Went** b. Yabuta & Sumiki
    c. Miller d. Addicott & Okhuma

Answers

197. a	198. c	199. c	200. c	201. a	202. d
203. b	204. a	205. c	206. a	207. a	

208. Gibberellins was discovered by-
a. F.W. Went
b. **Yabuta & Sumiki**
c. Miller
d. Addicott & Okhuma

209. Cytokinins was discovered by-
a. F.W. Went
b. Yabuta & Sumiki
c. **Miller**
d. Addicott & Okhuma

210. Abscisic acid was discovered by-
a. F.W. Went
b. Yabuta & Sumiki
c. Miller
d. **Addicott & Okhuma**

211. Ethylene was discovered by-
a. F.W. Went
b. Yabuta & Sumiki
c. Miller
d. **Burg**

212. Which is correct formula of ethylene-
a. C_2H_2
b. C_2H_6
c. $\mathbf{C_2H_4}$
d. C_4H_6

213. Which is not a bio-assay test of auxin-
a. Tryptophan amino acid
b. Avena curvature test
c. Root growth inhibition test
d. **Mevalonic acid pathway**

214. Which hormone is responsible for apical dominance in the plant-
a. **Auxin**
b. Gibberellin
c. Cytokinin
d. ABA

215. Which hormone is responsible for breaking seed dormancy in the plant-
a. Auxin
b. **Gibberellin**
c. Cytokinin
d. ABA

216. Which is not a bio-assay test of gibberellins -
a. **Tryptophan amino acid**
b. Alpha amylase activity test
c. Dwarf pea & maize test
d. Mevalonic acid pathway

217. Which hormone is responsible for flowering in LDP, in short light duration-
a. Auxin
b. **Gibberellin**

Answers

208. b	209. c	210. d	211. d	212. c	213. d
214. a	215. b	216. a	217. b		

c. Cytokinin d. ABA

218. Which hormone is responsible for delay in senescence in plants-
a. Auxin b. Gibberellin
c. **Cytokinin** d. ABA

219. Which is not a bio-assay test of cytokinin-
a. Tobacco pith cell division test
b. **Alpha amylase activity test**
c. Chlorophyll preservation test
d. Soyabean and Radish cotyledon cell division test

220. Which hormone is responsible for flowering in short day plant-
a. Auxin b. Gibberellin
c. **Cytokinin** d. ABA

221. Which hormone is responsible for fruit ripening-
a. Auxin b. Gibberellin
c. Cytokinin d. **Ethylene**

222. Which hormone is responsible to induce abscission during adverse climatic conditions-
a. Auxin b. **ABA**
c. Cytokinin d. Ethylene

223. Which hormone is responsible for stomatal closing in plant-
a. Auxin b. Gibberellin
c. Cytokinin d. **ABA**

224. Which is the first stable compound of C_4-cycle-
a. PGA b. DHA
c. **OAA** d. RUBP

225. Which is the first stable compound of CAM-pathway -
a. PGA b. DHA
c. **OAA** d. RUBP

226. Which is the first stable compound of CAM-pathway was discovered by -
a. Calvin b. Hatch & Slack
c. Thiemann d. **Oleary & Rouhani**

Answers

218. c	219. b	220. c	221. d	222. d	223. d
224. c	225. c	226. d			

227. Which is/are the example of C_4- plants-

a. Sugarcane b. Maize

c. Amaranthus **d. Both a and b**

228. Which is/are the example of C_3- plants-

a. Sugarcane b. Maize

c. Wheat d. Chenopodium

229. Which is/are the example of C_3- plants-

a. Sugarcane b. Maize

c. Euphorbia **d. Rice**

230. Which is/are the example of CAM plants-

a. Agave b. Aloe

c. Bryophyllum **d. All of the above**

231. CO_2 compensation point C_3 plants-

a. 40-100 PPM b. 8-10 PPM

c. 50-100 PPM d. 20-50 PPM

232. CO_2 compensation point C_4 plants-

a. 40-100 PPM **b. 8-10 PPM**

c. 50-100 PPM d. 20-50 PPM

233. CO_2 compensation point CAM-plants-

a. 40-100 PPM b. 8-10 PPM

c. 50-100 PPM d. 20-50 PPM

234. Primary CO_2 acceptor in C_4 plants and CAM plants is-

a. RUBP b. PEP case

c. RUBP & PEP case d. PGA

235. Primary CO_2 acceptor in C_3 plants is-

a. RUBP b. PEP case

c. RUBP & PEP case d. PGA

236. Final products of aerobic respiration is-

a. CO_2 & H_2O b. C_2H_5OH & CO_2

c. C_2H_5OH & H_2O d. All of the above

237. Photosynthetic number or assimilatory number was given by-

a. Willstatter & Stoll b. Peter Mitchell

c. Unger & Von Sachs d. Dixon & Jolly

Answers

227. d	228. c	229. d	230. d	231. a	232. b
233. a	234. c	235. a	236. a	237. a	

CHAPTER 20

Memory based JRF Paper 2015

1. Cutting of trees into logs is called-
   a. Felling b. Seasoning
   c. Sawing d. **Bucking**
2. Science of determining past climate from trees primarily on the basis of annual rings is called-
   a. **Dendroclimatology** b. Dendrology
   c. Dendrochronology d. Silviculture
3. The classification of trees according to Champion and Seth is based on-
   a. Functional Basis b. Temperature basis
   c. **Ecosystem approach** d. None
4. Which is non-coppicer tree-
   a. Populus b. Eucalyptus
   c. Teak d. **Cedrus**
5. Piece of plant tissue used to initiate a culture is called-
   a. **Explant** b. Propagule
   c. Bud d. Stock
6. Which law is used to estimate the proportionate of trees in selective forests-
   a. Pressler b. Hoppus
   c. Biolley d. **De Liocourt'law**

Answers

1. d 2. a 3. c 4. d 5. a 6. d

7. Smallest permanent working plan unit in India is called-
   a. Block b. Sub-compartment
   c. circle d. **Compartment**
8. The non-random differential reproduction of genotype is called-.
   a. Selection b. **Mutation**
   c. Evaluation d. Breeding
9. Selection which provides rapid genetic gain while simultaneously maintaining the broad genetic base is- -
   a. Progeny selection b. Family selection
   c. **Mass selection** d. Individual selection
10. In paper making process, sizing is done to render-
    a. Bright b. Transpirants
    c. Pervious to ink d. **Impervious to ink**
11. A shelter belt in general protects the leeward area about-
    a. 10 times the height of shelterbelt
    b. **20 times the height of shelterbelt**
    c. 30 times the height of shelterbelt
    d. 35 times the height of shelterbelt
12. Which of the following is not a part of trinity of norms in forest management-
    a. Age- gradation
    b. Normal growing stock
    c. Normal increment
    d. **Site quality**
13. A systematic association between a fungus and the roots of a higher plant-
    a. **Mycorrhiza** b. Lichen
    c. Rhizobium d. Helotism
14. Boucherie process is used for the treatment of-
    a. **Green bamboos only** b. Hardwood
    c. Softwood d. All of the above
15. Soil, leaching or removal from upper horizon to lower horizon is called-
    a. Percolation b. Infiltration
    c. Eluviation d. **Illuviation**

Answers

7. d	8. b	9. c	10. d	11. b	12. d
13. a	14. a	15. d			

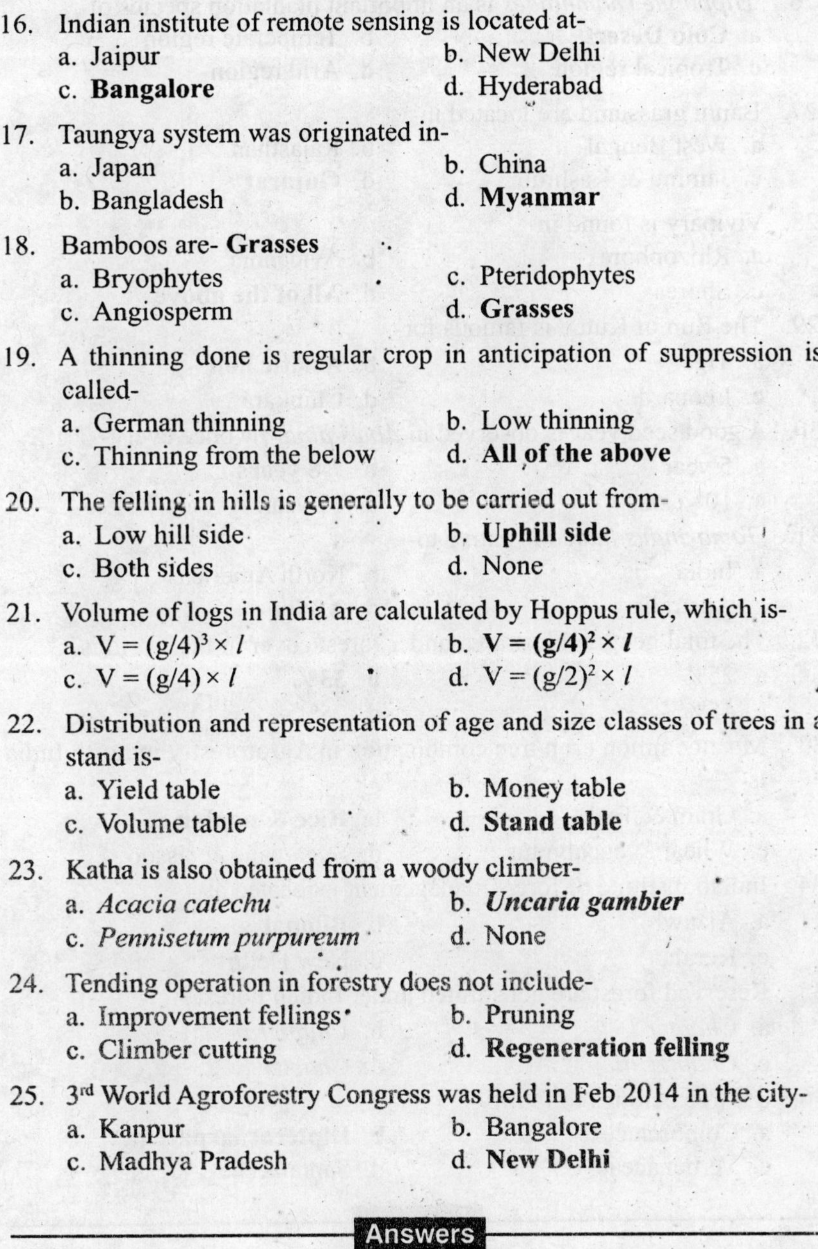

16. Indian institute of remote sensing is located at-
    a. Jaipur b. New Delhi
    c. **Bangalore** d. Hyderabad

17. Taungya system was originated in-
    a. Japan b. China
    b. Bangladesh d. **Myanmar**

18. Bamboos are- **Grasses**
    a. Bryophytes c. Pteridophytes
    c. Angiosperm d. **Grasses**

19. A thinning done is regular crop in anticipation of suppression is called-
    a. German thinning b. Low thinning
    c. Thinning from the below d. **All of the above**

20. The felling in hills is generally to be carried out from-
    a. Low hill side b. **Uphill side**
    c. Both sides d. None

21. Volume of logs in India are calculated by Hoppus rule, which is-
    a. $V = (g/4)^3 \times l$ b. $\mathbf{V = (g/4)^2 \times l}$
    c. $V = (g/4) \times l$ d. $V = (g/2)^2 \times l$

22. Distribution and representation of age and size classes of trees in a stand is-
    a. Yield table b. Money table
    c. Volume table d. **Stand table**

23. Katha is also obtained from a woody climber-
    a. *Acacia catechu* b. ***Uncaria gambier***
    c. *Pennisetum purpureum* d. None

24. Tending operation in forestry does not include-
    a. Improvement fellings b. Pruning
    c. Climber cutting d. **Regeneration felling**

25. 3rd World Agroforestry Congress was held in Feb 2014 in the city-
    a. Kanpur b. Bangalore
    c. Madhya Pradesh d. **New Delhi**

Answers

16. c	17. d	18. d	19. d	20. b	21. b
22. d	23. b	24. d	25. d		

26. *Hippophe rhamnoides* is an important plantation species of-
    a. **Cold Desert**
    b. Temperate region
    c. Tropical region
    d. Arid region
27. Banni grassland are located in-
    a. West Bengal
    b. Rajasthan
    c. Jammu & Kashmir
    d. **Gujarat**
28. Vivipary is found in-
    a. Rhizophora
    b. Avicennia
    c. Shorea
    d. **All of the above**
29. The Run of Kutch is famous for-
    a. Tiger
    b. **Asiatic lion**
    c. Leopard
    d. Chinkara
30. A good seed year is observed in *Abies pindrow* once every-
    a. 5 year
    b. **7-8 years**
    c. 10 years
    d. 12 year
31. *Tamarindus indica* is native to-
    a. India
    b. North America
    c. Australia
    d. **Africa**
32. The total geographical area under forest cover in the world is-
    a. 25%
    b. **33%**
    c. 45%
    d. 75%
33. Most common crop tree combination in Agroforestry in north India is-
    a. Gram & Teak
    b. **Rice & poplar**
    c. Wheat & eucalyptus
    d. Sugarcane & sissoo
34. Indian institute of forest management is located at-
    a. Aizawl
    b. **Bhopal**
    c. Kerala
    d. New Delhi
35. Reserved forest are constituted under Indian Forest Act-
    a. *Chapter i*
    b. ***Chapter ii***
    c. *Chapter iii*
    d. *Chapter iv*
36. Sal tree belongs to family-
    a. Combretaceae
    b. **Dipterocaarpaceae**
    c. Verbenaceae
    d. Santalaceae

Answers

26. a	27. d	28. d	29. b	30. b	31. d
32. b	33. b	34. b	35. b	36. b	

37. Scientific name of African elephant is-
    a. *Hystrix indica*
    b. ***Loxodonta africana***
    c. *Elephus maximus*
    d. *Elephus africana*
38. Sandal spike disease is caused by-
    a. Bacteria
    b. Fungi
    c. Virus
    d. **Mycoplasma**
39. REED abbreviation is-
    a. Reduced Emission from Daily plantation
    b. Radiation Emission from Deforestation
    c. **Reduced Emission from Deforestation and Degradation**
    d. Radiation Emission from Degradation forest
40. Diffusion treatment is applicable to-
    a. Green bamboo only
    b. **Green timber with water soluble wood**
    c. Green timber without water soluble wood
    d. All of the above
41. Champion and Seth classified the Indian forests into-
    a. 5 groups
    b. 10 groups
    c. **16 groups**
    d. 18 groups
42. Most of the commercially important oleoresin are obtained from-
    a. Acacia
    b. Camphiophora
    c. Wattle gum
    d. **Conifers**
43. Father of green revolution is-.
    a. **Norman Borlaug**
    b. Swaminathan
    c. Bruce Zobel
    d. Tolbert
44. In India forestry research h was initiated after establishment of FRI in the year-
    a. 1901
    b. **1906**
    c. 1919
    d. 1947
45. The maximum genetic gain is achieved through-
    a. **SSO**
    b. CSO
    c. ISO
    d. SCO

Answers

37. b	38. d	39. c	40. b	41. c	42. d
43. a	44. b	45. a			

46. Crown thinning is also called as-
    a. German thinning
    b. Ordinary thinning
    c. **French thinning**
    d. Natural thinning

47. Which among the following is an annual flowering bamboo-
    a. *Bambusa bamboos*
    b. *Bambusa vulgaris*
    c. ***Arundinaria wightiana***
    d. *Bambusa tudla*

48. A measure of the relative productivity of a site for a particular species is-
    a. Site productivity
    b. Site index
    c. Site
    d. **Site quality**

49. Removal of only certain spp. of high value of trees above a certain size and of certain spp. without full regard to silvicultural requirements is called-
    a. **Selective felling**
    b. Concentrated felling
    c. Improved felling
    d. Seeding felling

50. Number of plants required for 10 ha of plantation in which the plants are planted in a triangular pattern 5m will be-
    a. 1650
    b. 1850
    c. 3540
    d. **4620**

51. First cultivated crop in the world are-
    a. Tobacco & maize
    b. Potato & Onion
    c. **Wheat & barley**
    d. Rice & gram

52. Contribution of agriculture to GDP of India is-
    a. 5%
    b. 10%
    c. 12%
    d. **14%**

53. Teak defoliator is-
    a. ***Hyblea pupurea***
    b. *Hyblea shorea*
    c. *Hyblea tectonae*
    d. *Hyblea foligae*

54. D&D survey in agroforestry was proposed by-
    a. Nair
    b. Lundegardh
    c. **J.B Raintree**
    d. Swaminathan

55. The biological diversity act was enacted in the year-
    a. 1972
    b. 2000
    c. **2002**
    d. 2007

Answers

46. c	47. c	48. d	49. a	50. d	51. c
52. d	53. a	54. c	55. c		

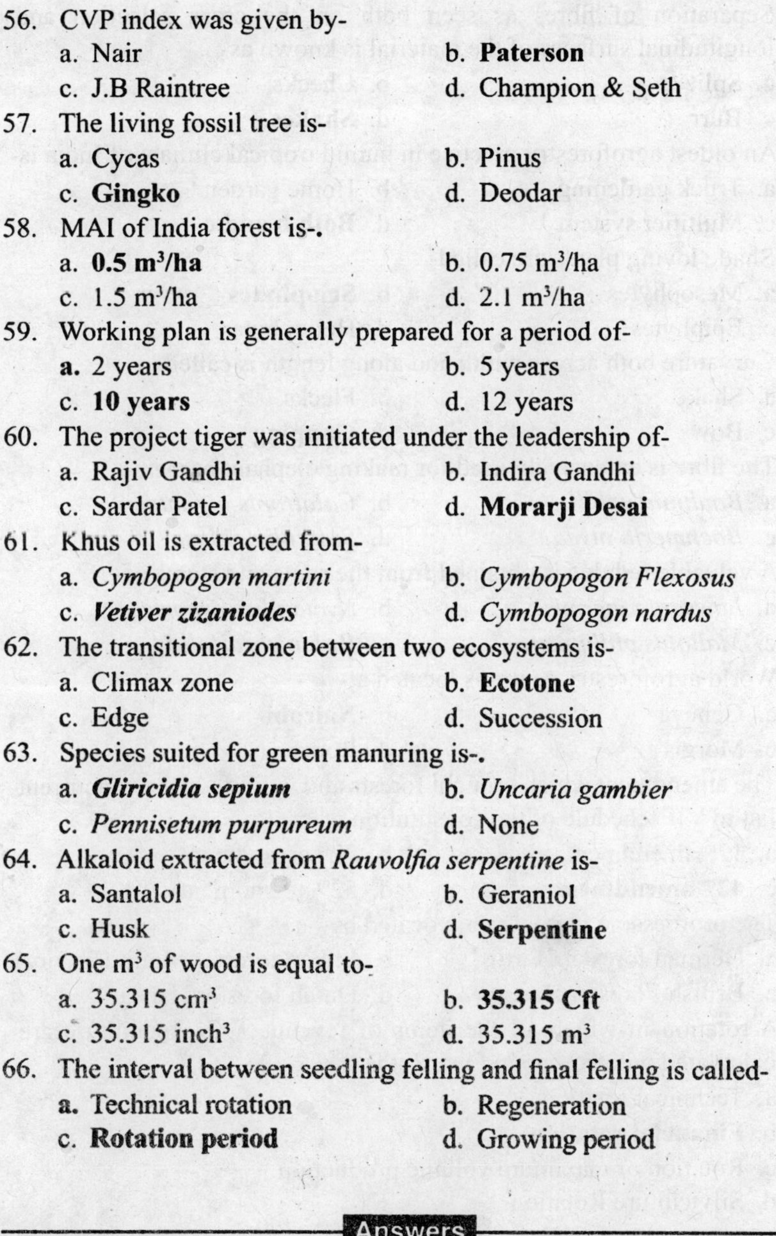

56. CVP index was given by-
    a. Nair b. **Paterson**
    c. J.B Raintree d. Champion & Seth
57. The living fossil tree is-
    a. Cycas b. Pinus
    c. **Gingko** d. Deodar
58. MAI of India forest is-.
    a. **0.5 m³/ha** b. 0.75 m³/ha
    c. 1.5 m³/ha d. 2.1 m³/ha
59. Working plan is generally prepared for a period of-
    a. 2 years b. 5 years
    c. **10 years** d. 12 years
60. The project tiger was initiated under the leadership of-
    a. Rajiv Gandhi b. Indira Gandhi
    c. Sardar Patel d. **Morarji Desai**
61. Khus oil is extracted from-
    a. *Cymbopogon martini* b. *Cymbopogon Flexosus*
    c. ***Vetiver zizaniodes*** d. *Cymbopogon nardus*
62. The transitional zone between two ecosystems is-
    a. Climax zone b. **Ecotone**
    c. Edge d. Succession
63. Species suited for green manuring is-.
    a. ***Gliricidia sepium*** b. *Uncaria gambier*
    c. *Pennisetum purpureum* d. None
64. Alkaloid extracted from *Rauvolfia serpentine* is-
    a. Santalol b. Geraniol
    c. Husk d. **Serpentine**
65. One m³ of wood is equal to-
    a. 35.315 cm³ b. **35.315 Cft**
    c. 35.315 inch³ d. 35.315 m³
66. The interval between seedling felling and final felling is called-
    a. Technical rotation b. Regeneration
    c. **Rotation period** d. Growing period

Answers

56. b	57. c	58. a	59. c	60. d	61. c
62. b	63. a	64. d	65. b	66. c	

67. Separation of fibres as seen both on the cross selection and longitudinal surfaces of the material is known as-
    a. Split b. Checks
    c. Burr d. **Shakes**
68. An oldest agroforestry practice in humid tropical climate of India is-
    a. Truck gardening b. Home garden
    c. Multitier system d. **Both b and c**
69. Shade loving plants are called-
    a. Mesophytes b. **Sciophytes**
    c. Epiphytes d. Heliophytes
70. Curvature both across width and along length is called-
    a. Shake b. Flecks
    c. Bow d. **Cupping**
71. The fibre is extensively used for making elephant harness-
    a. *Bauhinia vahlii* b. *Calotropis gigantean*
    c. *Boehmeria nivea* d. ***Sterculia villosa***
72. A valuable red dye is obtained from the roots and stems of-
    a. *Lawsonia inermis* b. *Nyctanthes arbortristis*
    c. *Mallotus philippensis* d. ***Rubia cordifolia***
73. World agroforestry centre is located at-
    a. Geneva b. **Nairobi**
    c. Morgis d. Rome
74. The amendment which brought forests and wildlife in the concurrent list in VII schedule of the constitution was-
    a. 12th amendment b. 22th amendment
    c. **42th amendment** d. 52th amendment
75. The progressive yield was advocated by-
    a. German forester Hartig b. Indian forester Mohit Husain
    c. British Forester Hartig d. Dutch forester Hartig
76. A rotation in which all the items of revenue and expenditure are calculated with compound interest is-
    a. Technical rotation
    b. **Financial rotation**
    c. Rotation of maximum volume production
    d. Silviculture Rotation

Answers

67. d	68. d	69. b	70. d	71. d	72. d
73. b	74. c	75. a	76. b		

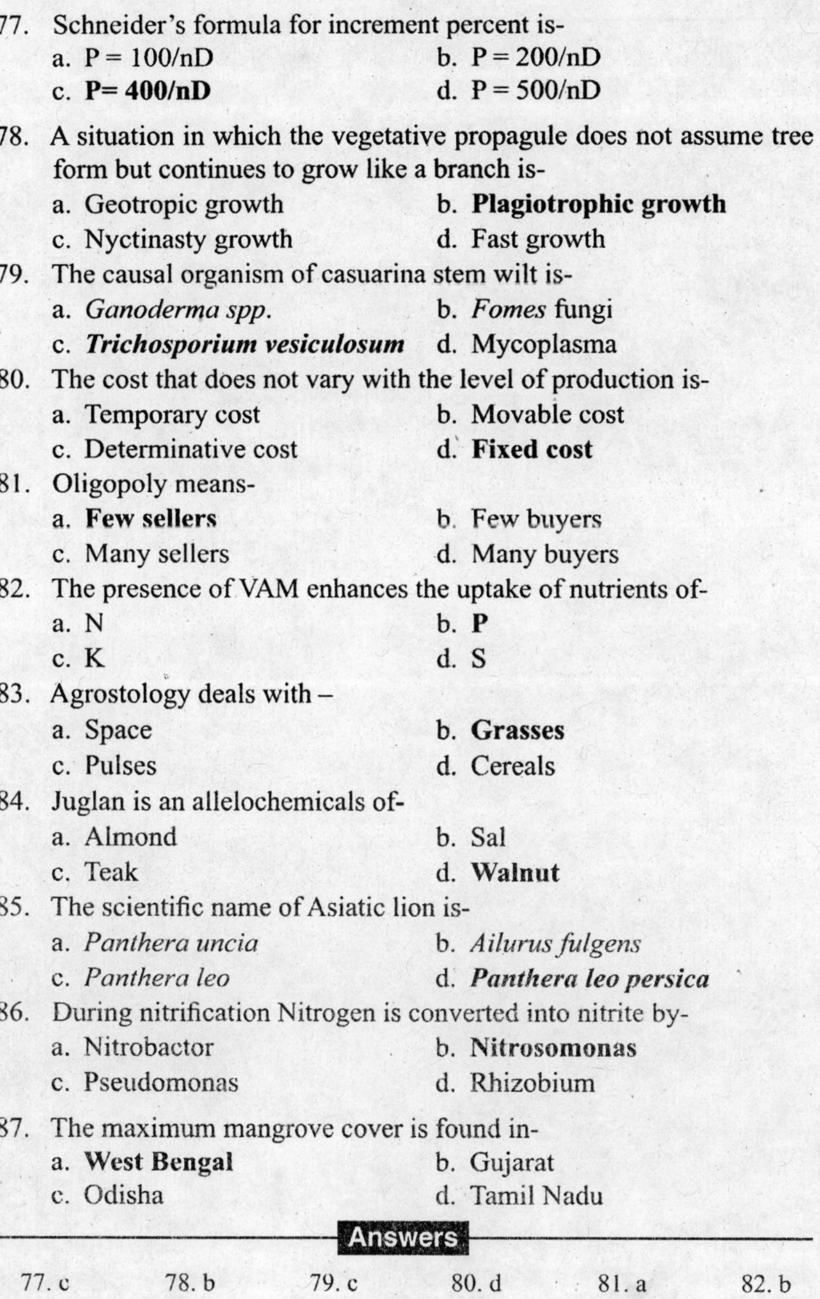

77. Schneider's formula for increment percent is-
    a. P = 100/nD b. P = 200/nD
    c. **P= 400/nD** d. P = 500/nD
78. A situation in which the vegetative propagule does not assume tree form but continues to grow like a branch is-
    a. Geotropic growth b. **Plagiotrophic growth**
    c. Nyctinasty growth d. Fast growth
79. The causal organism of casuarina stem wilt is-
    a. *Ganoderma spp.* b. *Fomes* fungi
    c. ***Trichosporium vesiculosum*** d. Mycoplasma
80. The cost that does not vary with the level of production is-
    a. Temporary cost b. Movable cost
    c. Determinative cost d. **Fixed cost**
81. Oligopoly means-
    a. **Few sellers** b. Few buyers
    c. Many sellers d. Many buyers
82. The presence of VAM enhances the uptake of nutrients of-
    a. N b. **P**
    c. K d. S
83. Agrostology deals with –
    a. Space b. **Grasses**
    c. Pulses d. Cereals
84. Juglan is an allelochemicals of-
    a. Almond b. Sal
    c. Teak d. **Walnut**
85. The scientific name of Asiatic lion is-
    a. *Panthera uncia* b. *Ailurus fulgens*
    c. *Panthera leo* d. ***Panthera leo persica***
86. During nitrification Nitrogen is converted into nitrite by-
    a. Nitrobactor b. **Nitrosomonas**
    c. Pseudomonas d. Rhizobium
87. The maximum mangrove cover is found in-
    a. **West Bengal** b. Gujarat
    c. Odisha d. Tamil Nadu

Answers

77. c 78. b 79. c 80. d 81. a 82. b
83. b 84. d 85. d 86. b 87. a

CHAPTER 21

Memory based JRF Paper 2016

1. Hybridoma technology is used in the production of-
   a. Hybrid plants
   b. Monoclonal serum
   c. **Monoclonal antibodies**
   d. Polyclonal antibodies

2. Spike disease in sandal is caused by-
   a. Fungi
   b. **Mycoplasma**
   c. Bacteria
   d. Insect

3. Forest conservation Act was formed in-
   a. 1927
   b. 1985
   c. 1827
   d. **1980**

4. Latest forest policy was formulated in the year-
   a. 1898
   b. **1984**
   c. 1894
   d. 2011

5. World environment day is celebrated on-
   a. 5th July
   b. 16th September
   c. 11th July
   d. **5th June**

6. Biodiesal plant is-
   a. **Jatropa**
   b. Azadirachta
   c. Madhuca
   d. Aloe

7. Offence and defence challan in the forest act comes under the section-
   a. Chapter **ix**
   b. Chapter x
   c. Chapter xi
   d. Chapter v

Answers

1. c 2. b 3. d 4. b 5. d 6. a
7. a

8. Transition zone between two communities called as-
   a. Climax zone
   b. Edge
   c. **Ecotone**
   d. Succession
9. Monoclimax theory was given by
   a. E. P. Odum
   b. Reiter
   c. A. G. Tansley
   d. **Clements**
10. In forestry D.B.H is measured at the height-
   a. 1.37 cm
   b. **1.37 m**
   c. 1.30 cm
   d. 1.30 m
11. In the slope area D.B.H is measured from-
   a. Low hill side
   b. Both sides
   b. **Uphill side**
   d. None
12. Which N_2 base is found only in RNA-
   a. Adenine
   c. Thymine
   c. Guanine
   d. **Uracil**
13. Which cell organelle have stroma and grana lamellae
   a. Nucleus
   b. Ribosomes
   c. **Chloroplast**
   d. Endoplasmic reticulum
14. Ozone depleting factor is
   a. Carbon di oxide
   b. Nitric acid
   c. Sulphuric acid
   d. **Chloro floro carbon**
15. Alternate source of fuel energy is
   a. CNG
   b. Propane
   c. Biogas
   d. **All of these**
16. Indian Lac Research Institute is situated in
   a. Jhansi (U.P)
   b. Bihar
   c. **Ranchi**
   d. Maharashtra
17. ICFRE was formed in the year
   a. **1901**
   b. 1929
   c. 1919
   d. 1927
18. Indian Institute of Forest management is situated in
   a. Jaipur
   b. Ranchi
   c. Dehra Dun
   d. **Bhopal**

Answers

8. c	9. d	10. b	11. b	12. d	13. c
14. d	15. d	16. c	17. a	18. d	

19. Indian Institute of Plantation management is situated in
a. Kerala
b. Tamil Nadu
c. **Bangalore**
d. Haryana

20. The shape of a typical shelter belt is
a. **Triangular**
b. Conical
c. Square
d. Cylindrical

21. Table based on one variable-
a. **Local volume table**
b. General volume table
c. Regional volume table
d. Yield table

22. One example of height measuring instrument-
a. Barometer
b. Hygrometer
c. **Spiegel Relaskop**
d. Dendrometer

23. The canopy range of dense forest is-
a. **Less than one**
b. Equal to one
c. More than one
d. None

24. Geo-stationary satellite revolved at a distance of
a. **800 m**
b. 6400 m
c. 1200 m
d. 22000 m

25. *Foris* is a-
a. Greek word
b. English word
c. **Latin word**
d. Sanskrit word

26. What are the factors that causes uprooting of seedlings
a. Wind
b. Temperature
c. Rainfall
d. **Both a and b**

27. Particle density is higher than
a. **Bulk density**
b. Real density
c. fluid density
d. Soil density

28. Soil having pH 3.2 is
a. **Acidic**
b. Basic
c. Alkaline
d. Acidic-alkaline

29. Transfer of pollen from anther to stigma of one flower to another flower of same plant
a. Autogamy
b. Geitenogamy
c. Allogamy
d. **Both a and c**

Answers

19. c	20. a	21. a	22. c	23. a	24. a
25. c	26. d	27. a	28. a	29. d	

30. Noise pollution is measured-
a. Hertz c. **Decibel**
b. Amplitude d. Centimetre

31. The RNA which carry information is called
a. m-RNA c. **t-RNA**
b. r-RNA d. hn-RNA

32. In tree damping off is caused by
a. **Fungi** c. Virus
b. Bacteria d. Earthworms

33. The name of teak defoliator is
a. ***Hyblea pupurea*** c. *Hyblea shorea*
b. *Hyblea tectonae* d. *Hyblea foligae*

34. Removal of vegetation from an area and development of newly vegetation is called
a. Afforestation c. Plantation
b. **Reforestation** d. Deforestation

35. The procedures of agroforestry D & D are usually done in which ways-
a. Macro D b. Micro D
c. Mini D d. **Both a and b**

36. Macro D & D covers-
a. **Entire ecological zone within a country**
b. Chosen ecological zone within a state
c. Entire ecological zone within a state
d. Chosen ecological zone within a district

37. A family of procedure for the diagnosis of land management problems & potentials and the design of agroforestry solution is called-
a. Macro D b. Micro D
c. **Agroforestry D & D** d. Land amelioration

38. Full form of D & D is-
a. Design and Diagnosis b. **Diagnosis & Design**
c. Design & Development d. Development & Diagnosis

Answers

30. c	31. c	32. a	33. a	34. b	35. d
36. a	37. c	38. b			

39. Which is/are not a main criteria for good agroforestry design-
a. Productivity
b. Sustainability
c. Adoptability
d. **Flexibility**

40. On the structural basis agroforestry systems can be classified into how many categories-
a. **Two**
b. Three
c. Four
d. Five

41. The crops-land left without crops for periods ranging from one season to several years is called-
a. Pasture land
b. **Fallow land**
c. Waste land
d. Treated land

42. The main objective of fallows is-
a. To recover depleted soil moisture
b. To recover soil productivity
c. **To recover depleted soil nutrients**
d. To enhance microbial activity

43. Alley cropping also known as-
a. Taungya
b. Shifting cultivation
c. Jhum Cultivation
d. **Hedgerow intercropping**

44. A method of planting in which rows of trees are interspersed with rows of crops is called-
a. Taungya
b. Social forestry
c. Farm forestry
d. **Alley cropping**

45. Which soil nutrient is mostly provided to plants in alley cropping practices-
a. **Nitrogen**
b. Potassium
c. Sulphur
d. Phosphorus

46. Gregarious flowering occurs in-
a. Teak
b. Sal
c. **Bamboo**
d. Neem

47. Spacing between rows to rows in alley cropping-
a. 2-4 meter
b. **4-8 meter**
c. 4-6 meter
d. 3-6 meter

Answers

39. d	40. a	41. b	42. c	43. d	44. d
45. a	46. c	47. b			

48. Spacing between trees within rows in alley cropping-
    a. **0.25-2.0 meter** b. 0.5-2.0 meter
    c. 1-2.0 meter d. 1.5-2.5 meter
49. Which tree species is most commonly used for alley cropping-
    a. **Legumes tree species** b. Non-leguminous
    c. Thorny species d. Bushy tree species
50. A belt/blocks of trees or shrubs established at right angles to the prevailing wind is called-
    a. Wind break b. Energy plantation
    c. **Shelterbelts** d. Block plantation
51. Ideal width of shelterbelts is-
    a. 25 m b. **50 m**
    c. 75 cm d. 100 cm
52. The ratio of height and width for shelterbelts should be-
    a. **1 : 10** b. 10 : 1
    c. 1 : 20 d. 10 : 25
53. A typical shelter belt has a shape-
    a. Quadrangles shape b. **Triangular shape**
    c. Square shape d. Rectangular shape
54. The minimum length of shelterbelts should be about its height-
    a. 10 times b. 20 times
    c. **25 times** d. 50 times
55. The gestation period of rhinoceros is
    a. 240 c. 280
    b. 400 d. **450**
56. *Hipophae rhamnoides* is found in the region of
    a. **Cold desert** c. Tropical Humid
    b. Arid region d. Hot desert
57. Bakarwal tribes is residing in which state of India
    a. **Jammu & Kashmir** c. Rajasthan
    b. Uttarakhand d. Assam
58. Local control is not used in
    a. RBD c. **CRD**
    b. LSD d. SPT

Answers

48. a	49. a	50. c	51. b	52. a	53. b
54. c	55. d	56. a	57. a	58. c	

59. Scientific name of Red panda is
    a. *Panthera uncia* c. ***Ailurus fulgens***
    b. *Panthera leo* d. *Gazella gazella*

60. Partial root parasite-
    a. **Santalum** c. Rafflesia
    b. Viscum d. Cuscuta

61. Partial stem parasite-
    a. Viscum c. Laranhus
    b. Santalum d. **both a and c**

62. Organism eaten by own species called-
    a. **Cannabilism** c. Macrophagy
    b. Commensalism d. Autophagy

63. Transition zone between two communities having greater number of species and diversity-
    a. **Ecotone** c. Ecotype
    b. Estuaries d. Eco cline

64. Species which occur most abundantly and spend their time in ecotone are called-
    a. Key stone species c. Pioneer species
    b. **Edge species** d. Climax species

65. The tendency to increase variety and density of some organism at the community border is known as-
    a. **Edge effect** c. Edge habitat
    b. Edge spectrum d. None

66. Ecological Niche term was given by-
    a. Mohit Husain b. **Grinnel**
    b. Allen d. E. P. Odum

67. The functional role of any species in an ecosystem or community or its functional position or status in an ecosystem is called-
    a. **Niche** b. Habitat.
    b. Strata d. Trophic

68. Physical area where an organism lives called-
    a. Niche c. **Habitat**
    b. Strata d. Trophic

Answers

59. c	60. a	61. d	62. a	63. a	64. b
65. a	66. b	67. a	68. c		

69. This principle state that two closely related species competing for the same resources cannot co-exist indefinitely and the competitively inferior are will be eliminated eventually-
    a. Allen Rule
    b. Mohit's competitive exclusion principle
    c. **Gause's competitive exclusion principle**
    d. Grinnel's competitive exclusion principle
70. Gause's competitive exclusion principle may be true if resources are-
    a. **Limited**
    c. Unlimited
    b. Neither limited nor unlimited
    d. None
71. The increment in growth that takes place in a particular year called-
    a. **Current Annual Increment (CAI)**
    b. Periodic Annual Increment (PAI)
    c. Mean Annual Increment (MAI)
    d. Current Increment Percent
72. The increment in growth that takes place for any short period called-
    a. Current Annual Increment (CAI)
    b. **Periodic Annual Increment (PAI)**
    c. Mean Annual Increment (MAI)
    d. Current Increment Percent
73. The total increment up to a given age divided by that age called-
    a. **Mean Annual Increment (MAI)**
    b. Current Annual Increment (CAI)
    c. Periodic Annual Increment (PAI)
    d. Current Increment Percent
74. The Mean Annual Increment (MAI) at rotation age is called-
    a. Current Annual Increment (CAI)
    b. Current Increment Percent
    c. **Final Mean Annual Increment**
    d. Mean Annual Increment (MAI)

Answers

69. c	70. a	71. a	72. b	73. a	74. c

75. The average annual growth in volume over a specified period expressed as a percentage of the volume either at beginning or, more usually, half way through the period called-
    a. **Increment Percent**
    b. Current Annual Increment (CAI)
    c. Current Increment Percent
    d. Periodic Annual Increment (PAI)

76. In the beginning M.A.I keeps below-
    a. **C.A.I.**
    b. C.I.
    c. P. A. I.
    d. All of the above

77. The C.A.I attains the maximum value before-
    a. P.A.I.
    b. C.I.
    c. **M.A.I**
    d. None

78. Which is never falls up to zero in whole life of a tree-
    a. P.A.I
    b. **M.A.I.**
    c. C.A.I
    d. C.I.

79. Which may falls up to zero in whole life of a tree-
    a. P.A.I.
    b. M.A.I.
    c. C.I.
    d. **C.A.I.**

80. The relation of the increment during a given year to the volume at the beginning of the year called-
    a. **Current Increment Percent (C.I.P)**
    b. Mean Increment Percent (M.I.P)
    c. Periodic Increment Percent (P.I.P)
    d. Annual Increment Percent (A.I.P)

81. The relation of the increment during a given year to a basic volume which may be taken as the mean or average volume for the period, or the volume at the beginning of the period called-
    a. Current Increment Percent
    b. **Periodic Increment Percent**
    c. Mean Increment Percent
    d. Annual Increment Percent

Answers

75. a	76. a	77. c	78. b	79. d	80. a
81. b					

82. The percent ratio which the M.A.I for a given age bears to the total volume at that age called-
    a. **Mean Annual Increment Percent**
    b. Current Increment Percent
    c. Periodic Increment Percent
    d. Mean Increment Percent

83. Pressler's formula for Increment Percent is given as-

    a. Increment Percent (p) = $\frac{V+v}{V-v} \times \frac{100}{n}$

    b. **Increment Percent (p)** = $\frac{V-v}{V+v} \times \frac{200}{n}$

    c. Increment Percent (p) = $\frac{V-v}{V+v} \times \frac{300}{n}$

    d. Increment Percent (p) = $\frac{V-v}{V \times v} \times \frac{400}{n}$

84. Most convenient formula for finding the increment percent (p) of standing trees is that developed by-
    a. **Professor Schneider in 1853**
    b. Professor Schneider in 1753
    c. Professor Schneider in 1953
    d. Professor Schneider in 1653

85. Schneider formula is based on the determination of the-
    a. Diameter at breast height
    b. The number of the rings in the last centimetre of the radius
    c. Tree volume
    d. **Both a and b**

86. Schneider formula for increment percent is-

    a. Increment Percent (p) = $\frac{100}{nD}$

    b. **Increment Percent (p)** = $\frac{400}{nD}$

    c. Increment Percent (p) = $\frac{200}{nD}$

    d. Increment Percent (p) = $\frac{300}{nD}$

Answers

82. a	83. b	84. a	85. d	86. b

87. A fast growing species is one which yields a minimum-
a. **10 m^3 /ha/year**
b. 20 m^3 /ha/year
c. 15 m^3 /ha/year
d. 30 m^3 /ha/year

88. World forest average mean annual increment is-
a. 4.1 m^3/ha.
b. 3.1 m^3/ha.
c. **2.1 m^3/ha.**
d. 1.2 m^3/ha.

89. A fast growing species is one which height increment must not be less than-
a. 40 cm per year
b. **60 cm per year**
c. 50 cm per year
d. 70 cm per year

90. Mean annual increment of Indian forest is-
a. 0.5-0.7 m^3/ha
a. 0.6-0.8 m^3/ha
c. 1.0-1.9 m^3/ha
d. **0.7-0.9 m^3/ha**

91. The increment in the value per unit volume of a tree or a crop, independent of any increase in the price of forest produce resulting from any change in money value in general, or the supply and demand position in particular called-
a. **Quality Increment**
b. Increment Percent
c. Price Increment
d. Increment

92. The increment in price independently of quality increment, resulting from any change in the market on an account of change in money value in general, the demand and supply position in particular called-
a. **Price Increment**
b. Increment Percent
c. Price Increment
d. Quality Increment

93. Price Increment and Quality Increment can be determined and expressed in the same way as volume increment percent by using-
a. Von Mental method
b. **Pressler's method**
c. Schneider method
d. Howard method

94. Increment from C.A.I. can be calculated by using-
a. Area Method
b. Increment Percent method
c. Per Tree Method
d. **All the above**

95. Determination of increment in irregular crop is done by using-
a. **Andre's formula**
b. Huber's formula
c. Behre's formula
d. Von Mental formula

Answers

87. a	88. c	89. b	90. d	91. a	92. a
93. b	94. d	95. a			

96. Biolley's check method is used to determine increment and based on-
    a. Diameter
    b. **Successive inventories**
    c. Tree height
    d. Tree volume

97. According to Biolley the M.A.I can be calculated by using formula-
    a. M.A.I. = $\frac{(V_2 + V_1) + N - P}{n}$
    b. M.A.I. = $\frac{(V_2 - V_1) - N - P}{n}$
    c. **M.A.I. =** $\mathbf{\frac{(V_2 - V_1) + N - P}{n}}$
    d. M.A.I. = $\frac{(V_1 - V_2) + N - P}{n}$

98. The intervals into which the range of age of the trees growing in a forest is divided for classification or use-
    a. Age gradation
    b. **An age-class**
    c. Gradation series
    d. Class interval

99. The unit of Yield Regulation in both regular and irregular forests is a-
    a. **Felling series**
    b. Coupe
    c. Felling cycle
    d. Periodic block

100. The total volume of trees in a fully stocked forest with normal distribution of age-classes for a given rotation called-
    a. Normal yield
    c. **Normal growing stock**
    c. Growing stock
    d. Progressive yield

101. The relationship of the height of the form point above ground level to the total height of the tree is called-
    a. Form coefficient
    c. **Form point ratio**
    b. Form factor ratio
    d. Form class ratio

102. Hojer and Behre formula are used to determine-
    a. Form coefficient
    c. Form quotient
    b. Volume quotient
    d. **Diameter quotient**

103. The ratio of the diameter (d) of a stem at any given height to its breast-height diameter (d.b.h) is called-
    a. **Diameter quotient**
    c. Height quotient
    b. Volume quotient
    d. Form quotient

Answers

96. b	97. c	98. b	99. a	100. c	101. c
102. d	103. a				

104. The area of the bole surface of a tree and log can be calculated by the formula-

a. $S = \frac{g_1 + g_2}{2} \times l$ b. S = g × l

c. $S = g \times l \times$ or $\frac{g_1 + g_2}{2} \times l$ d. **All of the above**

105. The horizontal projection on the ground of the tree crown is called-

a. Crown width b. Crown spreadness

c. Crown height d. **Crown cover**

106. Which formulae is/are used to calculate the volume of logs-

a. Smalain's formula b. Huber' formula

c. Newton's formula d. **All of the above**

107. Smalain's formula for calculation volume of the logs is-

a. $\frac{s_1 - s_2}{2} \times l$ b. $S_m \times l$

c. $\frac{(s_1 + 4Sm + s_2)}{6} \times l$ d. $\mathbf{\frac{s_1 + s_2}{2} \times l}$

108. Huber' formula for calculation volume of the logs is-

a. $\frac{s_1 - s_2}{2} \times l$ b. $\mathbf{S_m \times l}$

c. $\frac{(s_1 + 4Sm + s_2)}{6} \times 6$ d. $\frac{s_1 + s_2}{2} \times l$

109. Newton's formula for calculation volume of the logs is-

a. $\frac{s_1 - s_2}{2} \times l$ b. $S_m \times l$

c. $\mathbf{\frac{(s_1 + 4Sm + s_2)}{6} \times l}$ d. $\frac{s_1 + s_2}{2} \times l$

110. Newton's formula is used to calculate the error in the volume calculated by-

a. Huber' formula b. Smalain's formula

c. Lowry's formula d. **Both a and b**

Answers

104. d 105. d 106. d 107. d 108. b 109. c
110. d

111. Smalain's formula requires the which cross-sections of the log for volume calculation-
a. Side cross section
b. Mid cross-sections
c. **End cross-sections**
d. Corner sections

112. Huber's formula requires the which cross-sections of the log for volume calculation-
a. **Mid cross-sections**
b. Base cross-sections
c. End cross-sections
d. Corner sections

113. The difference between volumes obtained by Smalain's and Newton's formula is-
a. $\Delta = \frac{l}{2} \times (s_1 - s_2 + 2\text{Sm})$
b. $\boldsymbol{\Delta = \frac{l}{3} \times (s_1 + s_2 - 2\text{Sm})}$
c. $\Delta = \frac{l}{2} \times (s_1 \times s_2 - 2\text{Sm})$
d. $\Delta = \frac{l}{4} \times (s_1 - s_2 - 2\text{Sm})$

114. Volumes obtained by Smalain's and Newton's formula in case of cone and neiloid the error was-
a. Negative
b. **Positive**
c. Zero
d. None

115. Volumes obtained by Huber's and Newton's formula in case of cone and neiloid the error was-
a. **Negative**
b. Positive
c. Zero
d. None

116. In India volumes of logs are calculated by using-
a. Half girth formula
b. **Quarter girth formula**
c. Newton formula
d. Smalian formula

117. Quarter girth formula in Britain also known as-
a. Allen's rule
b. Hartig' rule
c. Hojer's rule
d. **Hoppus's rule**

118. Quarter girth formula for calculation volumes of logs is-
a. $V = (g/4)^3 \times l$
b. $\boldsymbol{V = (g/4)^2 \times l}$
c. $V = (g/4) \times l$
d. $V = (g/2)^2 \times l$

Answers

111. c	112. a	113. b	114. b	115. a	116. b
117. d	118. b				

119. Quarter girth formula gives how much percent of the true volume-

a. 75% b. **75.8%**
c. 78.5% d. 85%

120. The space occupied by a stack as distinct from the cubic contents of the wood itself i. e. solid volume called-

a. Stock volume b. Total volume
c. Actual volume d. **Stacked volume**

121. Volume of solid firewood can be determined by using which methods-

a. Xylometric method b. Specific gravity method
c. **Both a and b** d. None

122. Which is/are the methods to determine volume of standing tree-

a. Ocular estimation method
b. Partly ocular and partly by measurement
c. Direct/Indirect measurement
d. **All of the above**

123. Which instruments is/are used to determine the volume of standing tree in indirect measurement method-

a. Spiegel Relaskop b. Tele Relaskop
c. Wheeler Penta Prism d. **All of the above**

124. Which instruments is/are not used to determine the volume of standing tree in indirect measurement method-

a. Tele Relaskop b. **Wooden scale**
c. Stroud Dendrometer d. Callipers

125. A table showing for a species the average contents of trees, logs or sawn timber for one or more given dimensions is called-

a. Yield table b. Money table
c. **Volume table** d. Stand table

126. The volume of a tree depends mainly upon how many variables-

a. Two b. **Three**
c. Four d. Five

127. The volume of a tree does not depends upon which variable-

a. Diameter b. Height
c. Form d. **Area**

Answers

119. b	120. d	121. c	122. d	123. d	124. b
125. c	126. b	127. d			

128. Most important variables for volume table among the three variables is/are-
   a. **Diameter (BH)** b. Height
   c. Form d. Area

129. General volume table is/are based on which variables-
   a. One variables b. **Two variables**
   c. Three variables d. Four variables

130. Local volume table is/are based on which variables-
   a. Area b. Height
   c. **Diameter** d. Form

131. Volume table which is based on one variable-
   a. General volume table
   b. Regional volume table
   c. **Local volume table**
   d. Standard volume table

132. Khus oil is obtained from which species-
   a. *Cymbopogon martini* c. *Cymbopogon Flexosus*
   b. *Cymbopogon nardus* d. ***Vetiver zizaniodes***

133. Ginger oil is obtained from which species-
   a. ***Cymbopogon martini*** c. *Cymbopogon Flexosus*
   b. *Cymbopogon nardus* d. *Vetiver zizaniodes*

134. Geraniol is obtained from which species-
   a. *Cymbopogon nardus* d. *Vetiver zizaniodes*
   b. ***Cymbopogon martini*** c. *Cymbopogon Flexosus*

135. Wood is/are obtained from which species-
   a. *Santalum album* c. *Aquilaria agallocha*
   b. *Cedrus deodara* d. **All of the above**

136. Wood oil is/are obtained from which wood of the tree-
   a. Sapwood c. Softwood
   b. **Heartwood** d. All of the above

137. Santalol is obtained from which species-
   a. ***Santalum album*** c. *Aquilaria agallocha*
   b. *Cymbopogon nardus* d. *Vetiver zizaniodes*

Answers

128. a	129. b	130. c	131. c	132. d	133. a
134. b	135. d	136. b	137. a		

138. Leaf oil is/are obtained from which species-
a. Pinus
b. Cedrus
c. **Eucalyptus**
d. Sandal

139. Which is called as blue gum tree-
a. *Acacia nilotica*
b. *Eucalyptus citriodora*
c. ***Eucalyptus globulus***
d. *Acacia catechu*

140. Citriodora leaf oil is obtained from which species-
a. *Acacia nilotica*
b. ***Eucalyptus citriodora***
c. *Eucalyptus globulus*
d. *Acacia catechu*

141. Camphor leaf oil is obtained from which species-
a. *Cinnamomum cardamomum*
b. *Gaultheria fragrantissima*
c. *Mentha arvensis*
d. ***Cinnamomum camphora***

142. Mint leaf oil is obtained from which species-
a. *Cinnamomum cardamomum*
b. *Gaultheria fragrantissima*
c. ***Mentha arvensis***
d. *Cinnamomum camphora*

143. Wintergreen oil is obtained from which species-
a. *Cinnamomum cardamomum*
b. ***Gaultheria fragrantissima***
c. *Mentha arvensis*
d. *Cinnamomum camphora*

144. Root oil is/are obtained from which species-
a. ***Sassuria lappa***
b. *Gaultheria fragrantissima*
c. *Santalum album*
d. *Cinnamomum camphora*

145. Valerian oil is obtained from the rhizome of which species-
a. *Sassuria lappa*
b. ***Valerian wallichii***
c. *Gaultheria fragrantissima*
d. *Valerian chinensis*

146. Flower oil is/are obtained from which species-
a. *Pandanus tectorius*
b. *Michelia champa*
c. *Accacia farnesiana*
d. **All of the above**

147. Keora oil is obtained from which species-
a. *Michelia champa*
b. ***Pandanus tectorius***
c. *Accacia farnesiana*
d. *Gaultheria fragrantissima*

148. Linaloe oil is obtained from the heartwood of-
a. ***Bursera delpechiana***
b. *Shorea robusta*
c. *Cedrus deodara*
d. *Bursera wallichii*

Answers

138. c	139. c	140. b	141. d	142. c	143. b
144. a	145. b	146. d	147. b	148. a	

149. The highest concentration of non-biodegradable pollutant occurs in-

a. **Top consumer**
b. Secondary consumer
c. Primary consumer
d. Tertiary consumer

150. Tung oil is obtained from which species-

a. ***Aleurites fordii***
b. *Madhuca butyracea*
c. *Azadirachta indica*
d. *Garcinia indica*

151.

a. Soil testing	a. **Troug**
b. Marble	b. **Limestone**
c. Clay	c. **<0.002 mm**
d. Law of minimum	d. **Leibig**
e. Hue	e. **Dominant spectral colour**

152.

a. Gurgaon Project	a. **F. L Bryne**
b. Sevagram Project	b. **M. K. Gandhi**
c. Sriniketan Project	c. **R. N. Tagore**
d. Marthandam Project	d. **Hatch**
e. Etawah Project	e. **Albert Mayer**

153.

a. Kaziranga National Park	a. **Rhinoceros**
b. Sunderbans tiger reserve	b. **Tiger**
c. Dachigam National Park	c. **Hangul**
d. Gir National Park	d. **Asiatic Lion**
e. Dudhwa National Park	e. **Tiger**

154.

a. Green revolution	a. **Rice and wheat production**
b. White revolution	b. **Milk production**
c. Blue revolution	c. **Fish production**
d. Yellow revolution	d. **Oilseeds production**
e. Red revolution	e. **Tomato/meat production**

155.

a. World Biodiversity Day	a. **May-22**
b. World Environment Day	b. **Jun-05**
c. World Oceans Day	c. **Jun-08**
d. World Population Day	d. **Jul-11**
e. International Tiger Day	e. **Jul-29**

Answers

149. a 150. a

156. a. Very light timber — a. **300 kg/m^3**
b. Light timber — b. **301-450 kg/m^3**
c. Moderately heavy timber — c. **451-600 kg/m^3**
d. Heavy timber — d. **601-800 kg/m^3**
e. Very heavy timber — e. **801-950 kg/m^3**

157. a. *Mallotus philippensis* — a. **Kamela dye**
b. *Bixa orellana* — b. **Annatto dye**
c. *Butea monosperma* — c. **Dhak dye**
d. *Indigofera species* — d. **Blue dye**
e. *Lawsonia inermis* — e. **Heena dye**

158. a. Succession in fresh water — a. **Hydrosere**
b. Succession in salty water — b. **Halosere**
c. Succession in acidic water — c. **Oxalosere**
d. Succession in dry region — d. **Xerosere**
c. Succession on rock — e. **Lithosere**

159. a. Damping-off — a. **Fusarium**
b. Root disease — b. **Ganoderma spp.**
c. Pink disease — c. **Corticium salmonicolor**
d. Spike disease — d. **Mycoplasma**
e. Heart-rot — e. **Fomes fungi**

160. a. Poisson distribution — a. $\mathbf{0 \leq x \leq \infty}$
b. Correlation coefficient — b. **–1 to +1**
c. f-test — c. $\mathbf{0 \leq x \leq \infty}$
d. Regression coefficient — d. **–∞ to +∞**
e. Probability — e. **0 to 1**

Answers

CHAPTER 22

Memory based NET Paper 2014

1. Chilgoza is obtained from-
   a. *Cycas revoluta*
   b. ***Pinus gerardiana***
   c. *Pinus roxburghii*
   d. *Pinus wallichiana*
2. Agricultural and Processed Food Products Export Development Authority (APEDA)-
   a. Kanpur
   b. Bangalore
   c. Madhya Pradesh
   d. **New Delhi**
3. Most predominant soil type in India is-
   a. Sandy loam
   b. Black soil
   c. **Red soil**
   d. Brown soil
4. World wildlife day is celebrated on-
   a. Feb-02
   b. **Mar-03**
   c. Mar-21
   d. Apr-22
5. Indian forest Act was formulated in the year-
   a. 1860
   b. 1886
   c. **1927**
   d. 1976
6. International day of forest is celebrated on-
   a. Mar-02
   b. Mar-03
   c. **Mar-21**
   d. Apr-22
7. The control of timber and other forest-produce in transit is deals in the chapter of IFA, 1927-
   a. *Chapter vi*
   b. *Chapter v*
   c. ***Chapter vii***
   d. *Chapter xi*

Answers

1. b 2. d 3. c 4. b 5. c 6. c
7. c

8. Black revolution is related with-
   a. Onion production
   b. Potato production
   c. Rice and wheat production
   d. **Biofuel/Jatropa revolution**

9. Agro climatic zones of India is-
   a. 10
   b. 12
   c. **15**
   d. 20

10. Silviculture is a-
    a. Greek word
    b. **Latin word**
    c. Arabic word
    d. French word

11. Beef wood botanically know as-
    a. *Ceiba patendra*
    b. *Butea monosperma*
    c. ***Casuarina equisitifolia***
    d. *Cassia fistlua*

12. Butter tree is botanically know as-
    a. *Gmelia arborea*
    b. *Butea monosperma*
    c. *Tecomella undulata*
    d. ***Madhuca indica***

13. Which climatic factor play most important role in vegetation distribution on earth-
    a. Rainfall
    b. **Temperature**
    c. Topography
    d. Light

14. *Pterocapus marsupium* belong to family-
    a. Combretaceae
    b. **Dipterocaarpaceae**
    c. Verbenaceae
    d. Santalaceae

15. First organic state in India is-
    a. Mizoram
    b. Kerala
    c. **Sikkim**
    d. Meghalaya

16. The tendency to increase variety and density of some organism at the community border-
    a. **Edge effect**
    b. Edge habitat
    c. Edge spectrum
    d. None

17. 10% law of energy was given by-
    a. **Lindeman**
    b. Elton
    c. Odum
    d. Clements

Answers

8. d	9. c	10. b	11. c	12. d	13. b
14. b	15. c	16. a	17. a		

18. Management of forest broadly classified into-
    a. Control of composition and structure of the growing stock
    b. Harvesting and marketing of forest produce
    c. Administration of forest property personnel
    d. **All of the above**

19. The concept of Progressive Yield was given by-
    a. **German forester Hartig**
    b. Indian forester Hartig
    c. British Forester Hartig
    d. Dutch forester Hartig

20. The principle of Progressive Yield as against the Sustained Yield principle was discussed-
    a. Fourth Indian Silviculture Conference
    b. Fifth Indian Silviculture Conference
    c. **Sixth Indian Silviculture Conference**
    d. Seventh Indian Silviculture Conference

21. Third World Forestry Conference was held in 1948-
    a. New Delhi
    b. **Helsinki**
    c. Brazil
    d. Geneva

22. Which rotation is useful in forests managed primarily for aesthetic and recreational purpose-
    a. Technical rotation
    b. Financial rotation
    c. Rotation of maximum volume production
    d. **Silviculture Rotation**

23. Rotation under which a species yields the maximum material of a specified size or suitability for economic conversion or for special use called-
    a. Rotation of maximum volume production
    b. **Technical rotation**
    c. Financial rotation
    d. Physical rotation

24. A fast growing species is one which yields a minimum-
    a. **10 m^3 /ha/year**
    b. 15 m^3 /ha/year
    c. 20 m^3 /ha/year
    d. 30 m^3 /ha/year

Answers

18. d 19. a 20. c 21. b 22. d 23. b
24. a

25. World forest average mean annual increment is-
a. 4.1 m^3/ha.
b. **2.1 m^3/ha**
c. 3.1 m^3/ha.
d. 1.2 m^3/ha

26. According to Biolley the M.A.I can be calculated by using formula-
a. $\text{M.A.I.} = \frac{(V_2 + V_1) + N - P}{n}$
b. $\mathbf{M.A.I.} = \frac{(V_2 - V_1) + N - P}{n}$
c. $\text{M.A.I.} = \frac{(V_2 - V_1) - N - P}{n}$
d. $\text{M.A.I.} = \frac{(V_1 - V_2) + N - P}{n}$

27. A term generally applied to the determination of the yield and the prescribed means of releasing it, called-
a. **Yield regulation**
b. Yield capacity
c. Yield determination
d. Total Yield

28. Period of Working Plan in India is-
a. 5 years
b. 8 years
c. **10 years**
d. 15 years

29. The forest survey of India (FSI) conducts forest cover survey once in every-
a. One year
b. **Two** years
c. Five years
d. Ten years

30. Which is based only on area for yield regulation in regular forest-
a. Annual coupe by gross area method (A1)
b. Annual coupe by reduced area method (A2)
c. Annual coupe by total area method (A1)
d. **Both a and b**

31. The ranges of regression coefficient lies between-
a. –1 to +1
b. 0 to 1
c. **–∞ to +∞**
d. 0 to ∞

32. Headquarter of IUCN is situated in-
a. Gland
b. **Morgis**
c. Geneva
d. Nairobi

33. IMD was established in the year-
a. 1875
b. 1938
c. **1961**
d. 1985

Answers

25. b	26. b	27. a	28. c	29. b	30. d
31. c	32. b	33. c			

34. *Gmelia arborea* belong to family-
a. Rubiaceae
b. **Sapindaceae**
c. Meliaceae
d. **Lamiaceae**

35. *Grewia tilifolia* belong to family-
a. Combretaceae
b. **Malvaceae**
c. Verbenaceae
d. Legumniosae

36. Which scale is used for Forest type map & Working Plan map in India-
a. 1 : 25000
b. 1 : 500
c. 1 : 20000
d. **1 : 50000**

37. Colour of tropical semi-evergreen forest group on the forest type map shows-
a. Violet
b. **Purple**
c. Red
d. Pink

38. India's first national park which is situated in Nainital, Uttarakhand-
a. **Jim Corbett National Park**
b. Dachigam national park
c. Bennergutta national park
d. Dudhwa national park

39. Smallest tiger reserve in India is-
a. Tandoba national park
b. Gir national park
c. Kaziranga national park
d. **Ranthambore national park**

40. Largest tiger reserve in India which is located in Andhra Pradesh-
a. **Nagarjuna Sagar sanctuary**
b. Periyar sanctuary
c. Chilka lake Bird sanctuary
d. Annamalai sanctuary

41. Kaziranga National Park is situated in-
a. Tamil Nadu
b. Gujarat
c. Kerala
d. **Assam**

42. Hunting of Wild Animals is deal in which chapter of WPA-
a. *Chapter vii*
b. *Chapter iv*
c. *Chapter vi*
d. ***Chapter iii***

43. Which schedule of WPA deals with Big game animals-
a. Schedule *i*
b. Schedule *ii*
c. **Schedule *iii***
d. Schedule *iv*

Answers

34. d	35. b	36. d	37. b	38. a	39. d
40. a	41. d	42. d	43. c		

44. Project Lion was started in the year-
a. **1972** b. 1973
c. 1974 d. 1977

45. Gestation period of Rhinoceros and Kangaroo in days are respectively-
a. **450 and 42** b. 550 and 90
c. 450 and 120 d. 650 and 92

46. Edge of the tooth which faces the cutting direction called-
a. Space b. Back
c. **Face** d. Gullet

47. Distance between two adjacent teeth called-
a. Face b. Gullet
c. Pitch d. **Space**

48. Entire opening between two adjacent teeth is called-
a. Gauge b. **Gullet**
c. Pitch d. Kerf

49. Angle tween the face of a toot and the line passing through points of the teeth called-
a. **Pitch** b. Gullet
c. Gauge d. Kerf

50. Standard sleeper size of Broad Gauge is-
a. **275 cm × 26 cm × 13 cm** b. 183 cm × 21 cm × 12 cm
c. 153 cm × 18 cm × 12 cm d. 153 cm × 16 cm × 11 cm

51. Square or rectangular shape having 15 cm under section called-
a. Hakries b. Squares
c. **Scantlings** d. Poles

52. Kheersal is obtained from the tree-
a. *Shorea robusta* b. *Madhuca latifolia*
c. ***Acacia catechu*** d. *Butea monosperma*

53. Which is/are not the extraction method of cutch and katha-
a. Country method b. Modified method
c. Factory method d. **All of the above**

Answers

44. a 45. a 46. c 47. d 48. b 49. a
50. a 51. c 52. c 53. d

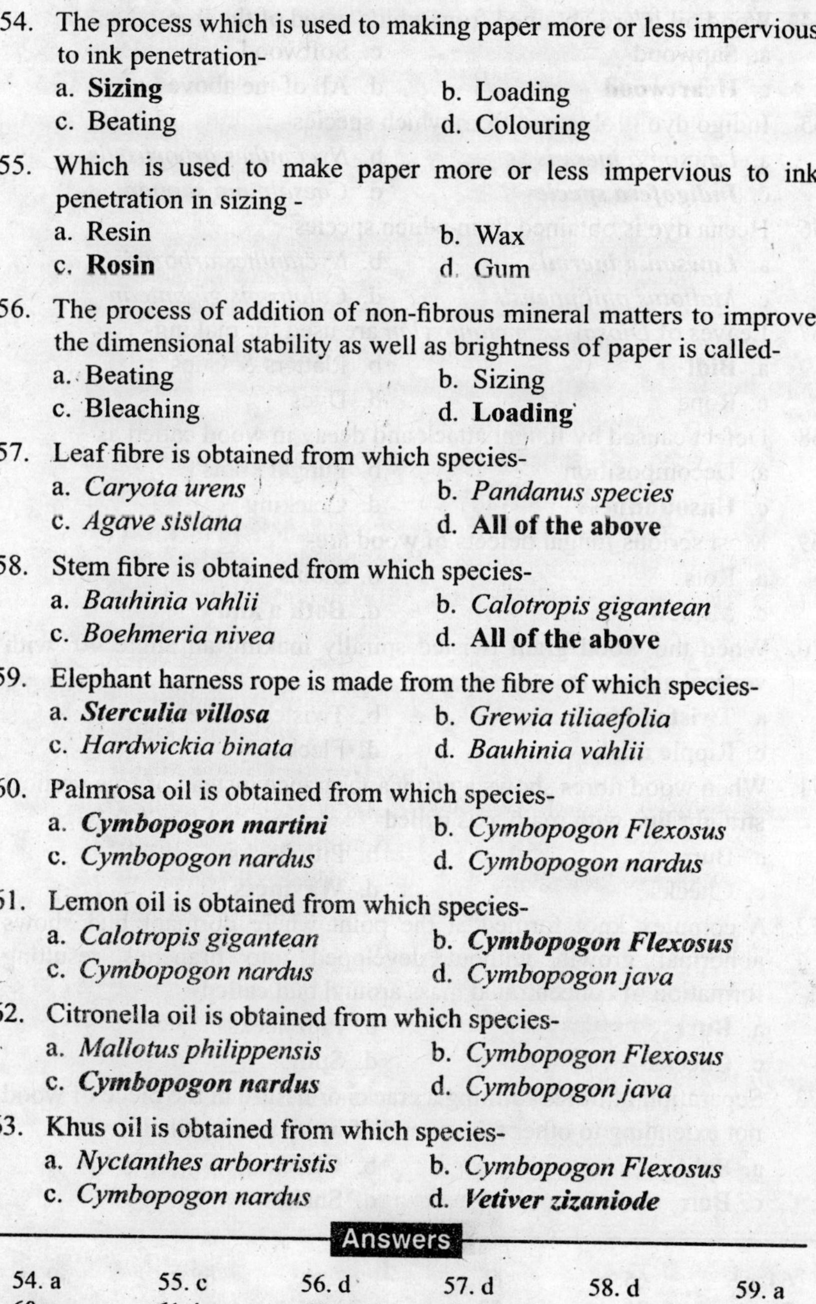

54. The process which is used to making paper more or less impervious to ink penetration-
    a. **Sizing** b. Loading
    c. Beating d. Colouring

55. Which is used to make paper more or less impervious to ink penetration in sizing -
    a. Resin b. Wax
    c. **Rosin** d. Gum

56. The process of addition of non-fibrous mineral matters to improve the dimensional stability as well as brightness of paper is called-
    a. Beating b. Sizing
    c. Bleaching d. **Loading**

57. Leaf fibre is obtained from which species-
    a. *Caryota urens* b. *Pandanus species*
    c. *Agave sislana* d. **All of the above**

58. Stem fibre is obtained from which species-
    a. *Bauhinia vahlii* b. *Calotropis gigantean*
    c. *Boehmeria nivea* d. **All of the above**

59. Elephant harness rope is made from the fibre of which species-
    a. ***Sterculia villosa*** b. *Grewia tiliaefolia*
    c. *Hardwickia binata* d. *Bauhinia vahlii*

60. Palmrosa oil is obtained from which species-
    a. ***Cymbopogon martini*** b. *Cymbopogon Flexosus*
    c. *Cymbopogon nardus* d. *Cymbopogon nardus*

61. Lemon oil is obtained from which species-
    a. *Calotropis gigantean* b. ***Cymbopogon Flexosus***
    c. *Cymbopogon nardus* d. *Cymbopogon java*

62. Citronella oil is obtained from which species-
    a. *Mallotus philippensis* b. *Cymbopogon Flexosus*
    c. ***Cymbopogon nardus*** d. *Cymbopogon java*

63. Khus oil is obtained from which species-
    a. *Nyctanthes arbortristis* b. *Cymbopogon Flexosus*
    c. *Cymbopogon nardus* d. ***Vetiver zizaniode***

Answers

54. a	55. c	56. d	57. d	58. d	59. a
60. a	61. b	62. c	63. d		

64. Wood oil is/are obtained from which wood of the tree-
a. Sapwood c. Softwood
c. **Heartwood** d. All of the above

65. Indigo dye is obtained from which species-
a. *Lawsonia inermis* b. *Nyctanthes arbortristis*
c. ***Indigofera species*** d. *Caesalpinia sappan*

66. Heena dye is obtained from which species-
a. ***Lawsonia inermis*** b. *Nyctanthes arbortristis*
c. *Mallotus philippensis* d. *Calotropis gigantean*

67. Leaves of *Diospyros melanoxylon* are used for making-
a. **Bidi** b. Platters & Cups
c. Rope d. Dyes

68. Defect caused by fungal attack and decay in wood called as-
a. Decomposition b. Fungal knots
c. **Unsoundness** d. Cracking

69. Most serious fungal defects of wood are-
a. Rots b. Stains
c. Mildew d. **Both a and c**

70. When the wood grain twisted spirally making an angle 40^0 with vertical axis called-
a. **Twisted fibre** b. Twisted shake
c. Ripple marks d. Flecks

71. When wood fibres shows a wavy appearance instead of true vertical straight line with main axis called-
a. Burr b. Pith flecks
c. Checks d. **Waviness**

72. A complex knot formed at the point where dormant bud shows abnormal growth without developed into branches resulting formation of concentrated mass around bud called-
a. **Burr** b. Pith flecks
c. Checks d. Split

73. Separation of fibres forming a cracks or fissure in the piece of wood not extending to other face or end of wood piece called-
a. Split b. **Checks**
c. Burr d. Shakes

Answers

64. c	65. c	66. a	67. a	68. c	69. d
70. a	71. d	72. a	73. b		

74. Separation of fibres forming a cracks or fissure in the piece of wood and extending to other face or end of wood piece called-

a. Waviness b. **Split**

c. Shakes d. Checks

75. Density of the moderately heavy timber is-

a. 201-350 kg/m^3 b. **451-600 kg/m^3**

c. 301-450 kg/m^3 d. 401-650 kg/m^3

76. Woods which require little protection against during rapid seasoning called-

a. Highly refractory wood b. **Moderately refractory wood**

c. Non refractory wood d. Refractory wood

77. Which types of wood preservatives is not suitable for indoor works-

a. **Oil type** b. Water soluble type

c. Organic solvent types d. All of the above

78. Which is/are not the constituents of water soluble preservatives-

a. As_2O_5 b. $CuSO_4$

c. Na/K Cr_2O_7 d. **All of the above**

79. Which is/are not the constituents of ASCU wood preservatives-

a. As_2O_5 b. $CuSO_4$

c. Na/K Cr_2O_7 d. **All of the above**

80. Which method is suitable for preservation of dry as well as green woods-

a. **Soaking method** b. Heating and cooling method

c. Sap displacement method d. None

81. Full cell process of wood preservation is applied at a pressure-

a. 1.50 to 10 kg/cm^2 b. 2.50 to 15 kg/cm^2

c. **3.50 to 12 kg/cm^2** d. 5.0 to 10 kg/cm^2

82. In which pressure method of wood preservation maximum penetration is obtained with minimum absorption of preservatives-

a. Rueping process b. **Empty cell process**

c. Lowry process d. Full cell process

83. The average number of bamboos in a bamboo raft is-

a. 2000 b. 5000

c. 1000 d. **10000**

Answers

74. b	75. b	76. b	77. a	78. d	79. d
80. a	81. c	82. b	83. d		

84. The centre face of plywood is called-
a. Cross bands b. Back
c. **Core** d. Face

85. The numbers of veneers in a plywood is always-
a. Even numbers b. **Odd numbers**
c. Both d. None

86. The maximum bordered piths are found in the tracheids of-
a. Angiosperms b. Bryophytes
c. **Gymnosperms** d. Pteridophytes

87. Closed vascular bundles are found in-
a. **Monocotyledon** b. Dicotyledons
c. Bryophytes d. Pteridophytes

88. Girth of tree is measured in-
a. Centimetre b. Nearest centimetre
c. Metre d. **Metre and centimetre**

89. In India Breast-Height is taken as above the ground level -
a. 1.30 m b. 1.37 cm
c. 1.30 cm d. **1.37 m**

90. On slopping ground diameter at Breast-Height should be measured on the-
a. Low hill side b. **Uphill side**
c. Aspect side d. Topography side

91. Which bands are used in diameter measurement using Spiegel Relaskop-
a. Band 1 and 2 narrow bands b. **Band 1 and 4 narrow bands**
c. Band 2 and 4 narrow bands d. Band 2 and 3 narrow bands

92. Which is the correct equation of Metzger theory-
a. $d^2 = \frac{32 \times W \times F \times L}{2\pi \times S}$ b. $\mathbf{d^3 = \frac{32 \times W \times F \times L}{\pi \times S}}$

c. $d^2 = \frac{32 \times W \times F}{\pi \times S \times L}$ d. $d^3 = \frac{32 \times W \times F \times L}{2\pi \times S}$

Answers

84. c	85. c	86. c	87. a	88. d	89. d
90. b	91. b	92. b			

93. Breast-height form factor also known as-
  a. **Artificial form factor** b. Absolute form factor
  c. Normal form factor d. True form factor

94. The concept of form quotient was given by-
  a. German forester A. Schiffel **b. Austrian forester A. Schiffel**
  c. Indian forester Mohit Husain d. German forester Hartig

95. Quarter girth formula gives how much percent of the true volume-
  a. 75% b. **75.8%**
  c. 78.5% d. 85%

96. Most important variables for volume table among the three variables is/are-
  a. **Diameter (BH)** b. Height
  c. Form d. Area

97. General volume table is/are based on which variables-
  a. One variables b. **Two variables**
  c. Three variables d. Four variables

98. Pressler's formula for volume increment percent is-
  a. $p = \frac{100(D-d)}{n} \times \frac{(V+v)}{n(V-v)}$ b. $p = \frac{200(D+d)}{n} \times \frac{(V-v)}{n(V+v)}$
  c. $\boldsymbol{p = \frac{200(D-d)}{n} \times \frac{(V-v)}{n(V+v)}}$ d. $p = \frac{400(D-d)}{n} \times \frac{(V+v)}{n(V-v)}$

99. Two phase sampling or double sampling is commonly used in-
  a. Bamboo enumeration
  b. Aerial photograph interpretation
  c. Forest inventories
  d. **All of the above**

100. The Bitterlich or variable plot or point sampling or probability proportional to size (PPS) and PPP or sampling with probability proportional to prediction are based on-
  a. Multi-stage sampling
  b. Stratified random sampling
  c. Multiphase sampling
  d. Sampling with varying probability

Answers

93. a	94. b	95. b	96. a	97. b	98. c
99. d	100. d				

101. The formula of basal area factor of wedge-prism is-

a. $BAF = \dfrac{10,00}{1+3\left(\dfrac{L}{D}\right)^2}$

b. $BAF = \dfrac{10,000}{1+2\left(\dfrac{L}{D}\right)^2}$

c. $BAF = \dfrac{10,00}{1+4\left(\dfrac{L}{D}\right)^2}$

d. $\mathbf{BAF = \dfrac{10,000}{1+4\left(\dfrac{L}{D}\right)^2}}$

102. Tiger Grass botanically known as-
a. *Dichanthium annulatum*
b. *Panicum antidotale*
c. *Sorghum sudanense*
d. ***Thysanolaena maxima***

103. A forest fire that burns ground cover (herbaceous plants and shrubs) only is called-
a. **Ground fire**
b. Intentional fires
c. Surface fire
d. Deliberate fire

104. Full abbreviation of CVP index is-
a. Climate, Vegetation and Precipitation
b. Climate, Vegetation and Productivity site
c. **Climate, Vegetation and Productivity index**
d. Climate, Vegetation and Potential index

105. Which is/are moderately drought hardy species-
a. *Acacia nilotica*
b. ***Acacia catechu***
c. *Madhuca indica*
d. *Toona ciliata*

106. Which is/are a strong coppicer-
a. *Dalbergia sissoo*
b. *Diospyrus melanoxylon*
c. *Ougenia oojensis*
d. **All of the above**

107. During tetrazolium test live seed takes a colour-
a. White
b. Green
c. **Red**
d. Black

Answers

101. d 102. d 103. a 104. c 105. b 106. d
107. c

108. Which formula is used to calculate number of plant (N) in square planting plantation-

a. $N = \dfrac{100\times100}{\text{planting distance (X)}}$

b. $N = \dfrac{\mathbf{100\times100}}{\textbf{Square of planting distance (X)}}$

c. $N = \dfrac{2\times100\times100}{\text{Square of planting distance (X)}}$

d. $N = \dfrac{100\times100\times1.155}{\text{Square of planting distance (X)}}$

109. Which is/are not a types groups of montane sub-tropical forests major group-
   a. Subtropical broad leaved forests
   b. Subtropical pine forests
   c. **Littoral and swampy forests**
   d. Subtropical dry evergreen forests

110. Time of flowering and seed collection in *Ailanthus excelsa* respectively-
   a. June-September and Dec-March
   b. Sept-November and Feb-March
   c. **Feb-March and April-May**
   d. Sept-November and Jan – February

111. Time of flowering and seed collection in *Albizia procera* respectively-
   a. June-September and Dec-March
   b. **Sept-November and Feb-March**
   c. May and June-December
   d. Sept-November and Jan – February

112. Time of flowering and seed collection in *Anogeissus latifolia* respectively-
   a. Feb-April and May-June
   b. Jan-March and April-May
   c. March-May and June-July
   d. **June-September and Dec-March**

Answers

108. b 109. c 110. c 111. b 112. d

113. Time of flowering and seed collection in *Azadirachta indica* respectively-
   a. Jan-March and April-May
   b. **March-May and June-July**
   c. Feb-April and May-June
   d. May and June-December
114. Time of flowering and seed collection in *Boswellia serrata* respectively-
   a. **Feb-April and May-June**
   b. May and June-December
   c. August-September and Jan-February
   d. March and April-May
115. Time of flowering and seed collection in *Casuarina equisetifolia* respectively-
   a. August-September and Jan-February
   b. Jan-March and April-May
   c. **May and June-December**
   d. Feb-April and May-June
116. Wind-breaks should be planted at an angles to the wind from which protection is needed-
   a. 30° b. 45°
   c. 60° d. **90°**
117. For better protection against environmental factors wind break should be planted in which direction-
   a. East-west b. South-east
   c. North-west d. **North-south**
118. The first taungya plantation in India was raised in which region-
   a. Gujarat b. West Bengal
   c. **North Bengal** d. Meghalaya
119. Shifting cultivation is practised extensively in which region-
   a. **North-eastern** b. Southern
   c. Western d. Northern
120. The selection of parent trees on the basis of performance of their progeny is called-
   a. **Progeny testing** b. Within family selection
   c. Sib selection d. Mass selection

Answers

113. b 114. a 115. c 116. d 117. d 118. c
119. a 120. a

121. Heritability value can be-
a. Positive
b. Negative
c. Zero
d. **Both a and c**

122. The ratio of total genetic variation (V_G) in a population to the phenotypic variation (V_P)-
a. Broad-sense heritability (H^2)
b. Narrow-sense heritability (h^2)
c. Specific combining ability (SCA)
d. General combining ability (GCA)

123. Breeder seed is the progeny of-
a. Breeder seed
b. **Nucleus seed**
c. Foundation seed
d. Certified seed

124. Tag colour of foundation seed is-
a. Azar blue
b. **White**
c. Purple
d. Pink

125. Bulk density has value-
a. **1.65 gm/c.c**
b. 1.85 gm/c.c
c. 2.65 gm/c.c
d. 2.85 gm/c.c

126. Land capability soil class suitable for wildlife & watershed-
a. Class IV
b. Class VI
c. Class VII
d. **Class VIII**

127. Percent of innovators in a society to adopt a new research or innovation is-
a. **2.5**
b. 13.5
c. 34
d. 16

128. Sewagram project was started in the year-
a. 1920
b. 1921
c. **1929**
d. 1948

129. Local control is not applied for the design-
a. Randomized Block Design
b. Completely Randomized Block Design
c. Latin Square Design
d. **Completely Randomized Design**

Answers

121. d	122. a	123. b	124. b	125. a	126. d
127. a	128. c	129. d			

130. In hypothesis testing the most common level of significance that is used by statisticians is-
a. 0.5 b. **0.05**
c. 1.0 d. 0.10

131. Which theory is most acceptable theory for stomatal movement-
a. **Active potassium (K^+) and hydrogen (H^+)**
b. Transpiration pull & cohesion force theory
c. Mass flow hypothesis
d. Passive potassium (K^+) and hydrogen (H^+)

132. Plant hormone which helps in closing of stomata is-
a. GA_3 b. Auxin
c. Cytokinin d. **ABA**

133. Effect of low temperature on the initiation and development of flower called-
a. Cryopreservation b. Photoperiodism
c. **Vernalization** d. Anthesis

134. World environment day is celebrated on-
a. **Jul-11** b. May-22
c. Mar-21 d. Jun-22

135. Economics is a-
a. **Greek word** b. Latin word
c. Arabic word d. French word

136. Young ones of tiger is called-
a. Calf b. Fawn
c. **Cub** d. Lamb

137. Group of camel is called-
a. Herd b. Flock
c. **Caravan** d. Pride

138. Young ones of rhinoceros is called-
a. **Calf** b. Fawn
c. Cub d. Lamb

139. Meat of cattle is called-
a. Chevon b. Mutton
c. **Beef** d. Venison

Answers

130. b	131. a	132. d	133. c	134. a	135. a
136. c	137. c	138. a	139. c		

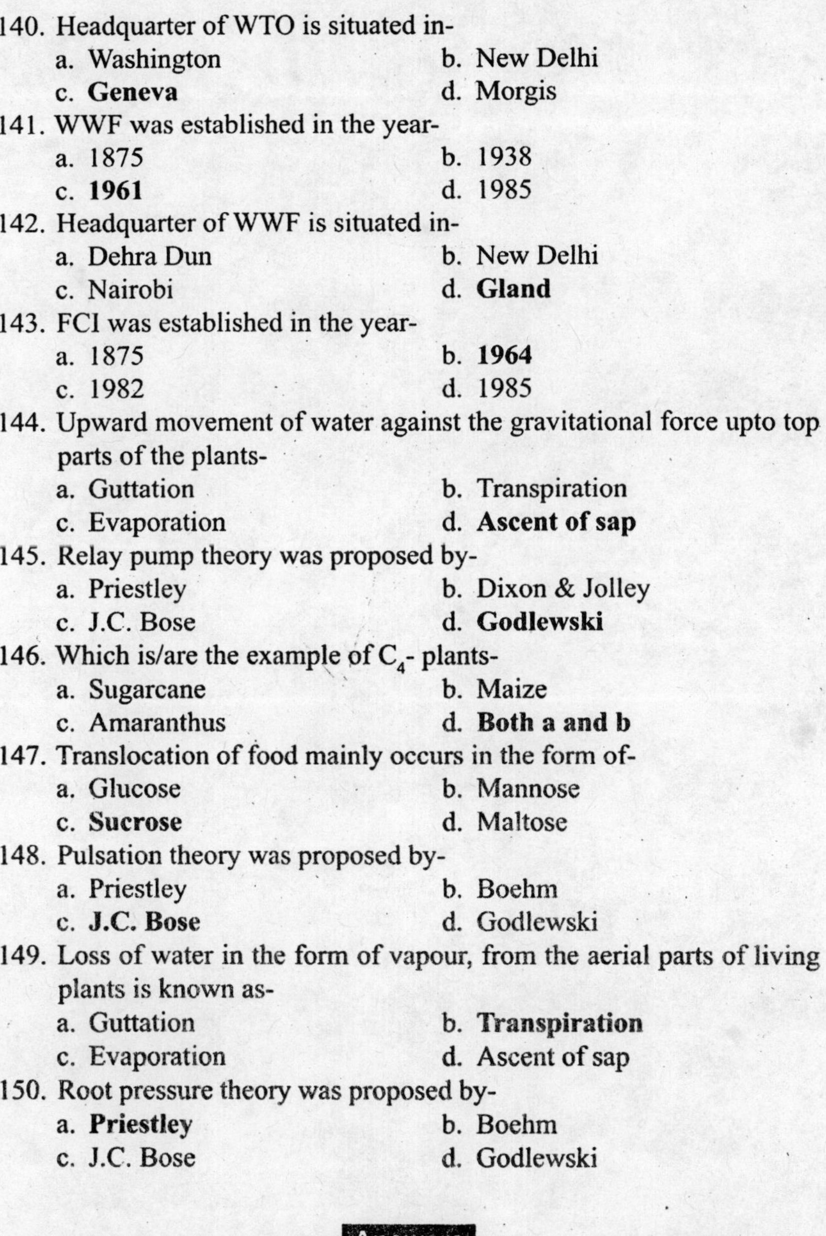

140. Headquarter of WTO is situated in-
   a. Washington   b. New Delhi
   c. **Geneva**   d. Morgis
141. WWF was established in the year-
   a. 1875   b. 1938
   c. **1961**   d. 1985
142. Headquarter of WWF is situated in-
   a. Dehra Dun   b. New Delhi
   c. Nairobi   d. **Gland**
143. FCI was established in the year-
   a. 1875   b. **1964**
   c. 1982   d. 1985
144. Upward movement of water against the gravitational force upto top parts of the plants-
   a. Guttation   b. Transpiration
   c. Evaporation   d. **Ascent of sap**
145. Relay pump theory was proposed by-
   a. Priestley   b. Dixon & Jolley
   c. J.C. Bose   d. **Godlewski**
146. Which is/are the example of C_4- plants-
   a. Sugarcane   b. Maize
   c. Amaranthus   d. **Both a and b**
147. Translocation of food mainly occurs in the form of-
   a. Glucose   b. Mannose
   c. **Sucrose**   d. Maltose
148. Pulsation theory was proposed by-
   a. Priestley   b. Boehm
   c. **J.C. Bose**   d. Godlewski
149. Loss of water in the form of vapour, from the aerial parts of living plants is known as-
   a. Guttation   b. **Transpiration**
   c. Evaporation   d. Ascent of sap
150. Root pressure theory was proposed by-
   a. **Priestley**   b. Boehm
   c. J.C. Bose   d. Godlewski

Answers

140. c	141. c	142. d	143. b	144. d	145. d
146. d147. c	148. c	149. b	150. a		

CHAPTER 23

Memory based SRF Paper 2014

1. Term foris is derived from-
   a. Greek word
   b. **Latin word**
   c. English word
   d. French word
2. State having highest forest cover as % of its geographical area-
   a. **Mizoram**
   b. Uttarakhand
   c. Madhya Pradesh
   d. Assam
3. Currently Mangrove cover in India is-
   a. 1740 km²
   b. 2740 km²
   c. 3740 km²
   d. **4740 km²**
4. The total carbon stock in the country's forest is around 7, 044 million tonnes.
   a. 4, 044 mil tonnes
   b. 5, 044 mil tonnes
   c. **7, 044 mil tonnes**
   d. 9, 044 mil tonnes
5. Center for Tropical Forest Science is located in-
   a. Morgis
   b. Geneva
   c. Glans
   d. **Panama**
6. Central Drug Research Institute is located in-
   a. New Delhi
   b. Dehra Dun
   c. Kolkata
   d. **Lucknow**
7. Centre for Forest-based Livelihoods and Extension is situated in-
   a. Morgis
   b. Geneva
   c. Glans
   d. **Agartala**

Answers

1. b 2. a 3. d 4. c 5. d 6. d
7. d

8. Centre for Social Forestry and Eco-Rehabilitation is located in-
   a. Jaipur b. **Allahabad**
   c. New Delhi d. Bombay

9. Indian Grassland and fodder research institute (IGFRI) located in-
   a. **Uttar Pradesh** b. Rajasthan
   c. Odisha d. Kerala

10. Indian Institute of Forest Management (IIFM) is located in-
    a. Aizawl b. **Bhopal**
    c. Kerala d. New Delhi

11. Indian Institute of Plantation Management (IIPM) is situated-
    a. New Delhi b. Kolkata
    c. Mumbai d. **Bangalore**

12. Headquarter of CITES is situated in-
    a. Geneva b. Rome
    c. **Washington** d. Switzerland

13. Headquarter of IUCN is situated in-
    a. America b. Rome
    c. Washington d. **Switzerland**

14. Indian forest service was started in the year-
    a. 1893 b. 1948
    c. **1966** d. 1983

15. International day of forest is celebrated on-
    a. Mar-02 b. Mar-03
    c. **Mar-21** d. Apr-22

16. Which chapter of Indian Forest Act (IFA-1927) deals with village forests-
    a. *Chapter i* b. *Chapter ii*
    c. ***Chapter iii*** d. *Chapter iv*

17. Rainbow revolution is related with-
    a. Onion production b. Potato production
    c. Rice and wheat production d. **All sector of agriculture**

18. ICRAF was established in the year-
    a. 1875 b. 1956
    c. 1961 d. **1977**

Answers

8. b	9. a	10. b	11. d	12. c	13. d
14. c	15. c	16. c	17. d	18. d	

19. Forest flame tree is botanically know as-
    a. *Ceiba patendra* b. ***Butea monosperma***
    c. *Bauhania variegata* d. *Ashoka indica*

20. Red cedar botanically know as-
    a. ***Toona ciliata*** b. *Cupresses torulosa*
    c. *Dalbergia latifolia* d. *Cassia fistlua*

21. Mysoore gum belong to family-
    a. **Myrtaceae** b. Iridaceae
    c. Ebenaceae d. Bixaceae

22. Study of the relationship of the group of the different species with their environment called-
    a. **Synecology** b. Morphology
    c. Ecology d. Autecology

23. The species which have great influence on the community's characteristics relative to their low abundance and biomass are called-
    a. **key stone species** b. Dominant species
    c. Apex species d. None

24. Under unfavourable conditions many zooplankton species in lakes and ponds are known to enter –
    a. Mitigation b. Hibernation
    c. Aestivation d. **Diapause**

25. Association between two organism, when one behave as master and another as slave called-
    a. Commensalism b. Symbiosis
    c. Mutualism d. **Helotism**

26. Succession in acidic water-
    a. Hydrosere b. Halosere
    c. **Oxalosere** d. Xerosere

27. Total stem parasite-
    a. Viscum b. Rafflesia
    c. **Cuscuta** d. Santalum

Answers

19. b	20. a	21. a	22. a	23. a	24. d
25. d	26. c	27. c			

28. Organism who are obtained their food directly from producers or plants-
    a. **Primary consumer** b. Secondary consumer
    c. Tertiary consumer d. Top consumer

29. Total amount of living organic matter present in particular area in particular time in an ecosystem is known as-
    a. Standing form b. Standing state
    c. **Standing crop** d. standing matter

30. The total amount of energy fixed in an ecosystem in unit time including the organic matter used up in respiration during the measurement is called-
    a. Primary productivity b. Secondary Productivity
    c. Net Primary Productivity d. **Gross Primary Productivity**

31. The pH of acid rain water is lesser than-
    a. **5.6** b. 7.0
    c. 8.6 d. 6.6

32. Pampas grassland found in-
    a. North America b. **South America**
    c. Europe and Asia d. New Zealand

33. India's first forest policy was enunciated in
    a. **1894** b. 1952
    c. 1988 d. 2011

34. Which is the unit of forest management-
    a. Coupe b. Beat
    c. **Working plan** d. Working circle

35. A forest area forming the whole or part of a Working circle for distribute felling and regeneration to maintain or create a normal distribution of age-classes is called-
    a. Felling cycle b. Regeneration cycle
    c. **Felling Series** d. Seeding felling

36. Third World Forestry Conference was held in 1948-
    a. New Delhi b. **Helsinki**
    c. Brazil d. Geneva

Answers

28. a	29. c	30. d	31. a	32. b	33. a
34. c	35. c	36. b			

37. Rotation which is applicable in case of protection and amenity forest, park lands, roadside avenues-
    a. Silviculture rotation
    b. Technical rotation
    c. **Physical rotation**
    d. Financial rotation
38. Soil Expectation Value (S_e) is-
    a. S_e = x/0.0k
    b. S_e = x/0.0h
    c. S_c = x/0.0d
    d. **S_e = x/0.0p**
39. Conversion period is usually-
    a. More than rotation
    b. **Less than rotation period**
    c. Equal to rotation
    d. Always greater than rotation period
40. In the beginning M.A.I keeps below-
    a. **C.A.I.**
    b. P. A. I.
    c. C.I.
    d. All of the above
41. Mean annual increment of Indian forest is-
    a. 0.5-0.7 m^3/ha
    b. 1.0-1.9 m^3/ha
    c. 0.6-0.8 m^3/ha
    d. **0.7-0.9 m^3/ha**
42. Judeich's Stand Selection Method for yield regulation in regular forest is based on-
    a. Volume
    b. Area and Increment
    c. Volume and Increment
    d. **Area and volume**
43. Von Mental's Formula (based on volume) for the determination of annual yield is-
    a. $Y_a = \frac{N.G.S}{r}$
    b. $\mathbf{Y_a = \frac{N.G.S}{2r}}$
    c. $Y_a = \frac{2N.G.S}{r}$
    d. $Y_a = \frac{2N.G.S}{4r}$
44. Which scale is used for management map in India-
    a. 1 : 5000
    b. 1 : 15000
    c. **1 : 50000**
    d. 1 : 40000

Answers

37. c	38. d	39. b	40. a	41. d	42. d
43. c	44. c				

45. Colour of littoral & swamp forest group on the forest type map shows-
a. Purple b. Green
c. Red d. **Prussian blue**

46. Smythies' Safe-guarding formula is also known as U.P Safe-guarding formula, It was formed mainly for sal forest in U.P. is-
a. $X = \frac{t}{f}(II + Z\%\, II)$ b. $X = \frac{f}{t}(II \times Z\%\, II)$
c. $X = \frac{t}{f}(II - Z\%\, II)$ d. $\boldsymbol{X = \frac{f}{t}(II - Z\%\, II)}$

47. Formula of Duchafour's Crown Cover which is used to calculate the modified or reduced area (C) is given below-
a. $C = K^2g^2n^2$ b. $C = Kg^2n^2$
c. $C = K^2gn^2$ d. $\boldsymbol{C = K^2g^2n}$

48. Rewa national park in Madhya Pradesh is famous for the-
a. **White tiger** b. Tigers
c. Asiatic Lion d. Chinkara

49. India's first and only butterfly park was established in 1992-
a. Gujarat b. Kerala
c. Uttarakhand d. **Sikkim**

50. The species which are in danger of extinction and whose survival is unlikely if the causal factors continue to be operating called-
a. **Endangered species** b. Extinct species
c. Threatened species d. Rare species

51. World Wildlife Day is celebrated on-
a. 21 March b. **3 March**
c. 21 May d. 8 October

52. Width of the saw cut is called-
a. Gauge b. Gullet
c. **Kerf** d. Pitch

Answers

45. d	46. d	47. d	48. a	49. d	50. a
51. b	52. c				

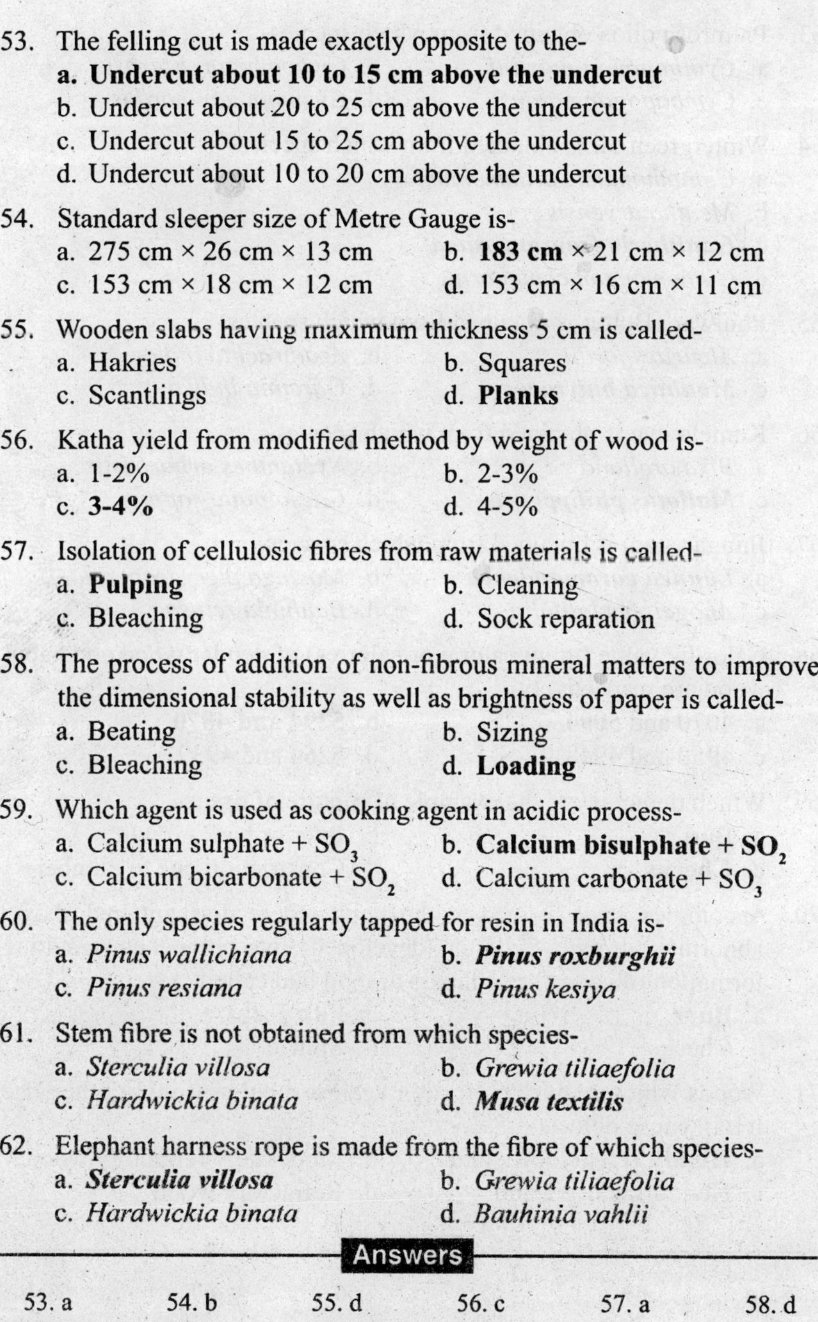

53. The felling cut is made exactly opposite to the-
    a. Undercut about 10 to 15 cm above the undercut
    b. Undercut about 20 to 25 cm above the undercut
    c. Undercut about 15 to 25 cm above the undercut
    d. Undercut about 10 to 20 cm above the undercut

54. Standard sleeper size of Metre Gauge is-
    a. 275 cm × 26 cm × 13 cm
    b. **183 cm** × 21 cm × 12 cm
    c. 153 cm × 18 cm × 12 cm
    d. 153 cm × 16 cm × 11 cm

55. Wooden slabs having maximum thickness 5 cm is called-
    a. Hakries
    b. Squares
    c. Scantlings
    d. **Planks**

56. Katha yield from modified method by weight of wood is-
    a. 1-2%
    b. 2-3%
    c. **3-4%**
    d. 4-5%

57. Isolation of cellulosic fibres from raw materials is called-
    a. **Pulping**
    b. Cleaning
    c. Bleaching
    d. Sock reparation

58. The process of addition of non-fibrous mineral matters to improve the dimensional stability as well as brightness of paper is called-
    a. Beating
    b. Sizing
    c. Bleaching
    d. **Loading**

59. Which agent is used as cooking agent in acidic process-
    a. Calcium sulphate + SO_3
    b. **Calcium bisulphate + SO_2**
    c. Calcium bicarbonate + SO_2
    d. Calcium carbonate + SO_3

60. The only species regularly tapped for resin in India is-
    a. *Pinus wallichiana*
    b. ***Pinus roxburghii***
    c. *Pinus resiana*
    d. *Pinus kesiya*

61. Stem fibre is not obtained from which species-
    a. *Sterculia villosa*
    b. *Grewia tiliaefolia*
    c. *Hardwickia binata*
    d. ***Musa textilis***

62. Elephant harness rope is made from the fibre of which species-
    a. ***Sterculia villosa***
    b. *Grewia tiliaefolia*
    c. *Hardwickia binata*
    d. *Bauhinia vahlii*

Answers

53. a	54. b	55. d	56. c	57. a	58. d
59. b	60. b	61. d	62. a		

63. Palmrosa oil is obtained from which species-
a. ***Cymbopogon martini*** b. *Cymbopogon Flexosus*
c. *Cymbopogon nardus* d. *Cymbopogon nardus*

64. Wintergreen oil is obtained from which species-
a. *Cinnamomum cardamomum*
b. *Mentha arvensis*
c. ***Gaultheria fragrantissima***
d. *Cinnamomum camphora*

65. Phulwara Butter is obtained from which species-
a. *Aleurites fordii* b. *Azadirachta indica*
c. ***Madhuca butyracea*** d. *Garcinia indica*

66. Kamela dye is obtained from which species-
a. *Bixa orellana* b. *Nyctanthes arbortristis*
c. ***Mallotus philippensis*** d. *Caesalpinia sappan*

67. Jhingan gum is obtained from which species-
a. ***Lannea coromandelica*** b. *Moringa pterygosperma*
c. *Anogeissus latifolia* d. *Bauhinia retusa*

68. Calorific value (gram calorie or calories) of deodar wood and babul wood are respectively-
a. 4070 and 5094 b. **5294 and 4870**
c. 4950 and 4948 d. 5264 and 4939

69. Which defect is/are the example of rupture of tissue-
a. Burr b. Interior bark growth
c. **Checks** d. Constriction due to climber

70. A complex knot formed at the point where dormant bud shows abnormal growth without developed into branches resulting formation of concentrated mass around bud called-
a. **Burr** b. Pith flecks
c. Checks d. Split

71. Woods which require protection against rapid seasoning otherwise it will cause defects-
a. **Highly refractory wood** b. Moderately refractory wood
c. Non refractory wood d. Refractory wood

Answers

63. a	64. c	65. c	66. c	67. a	68. b
69. c	70. a	71. a			

72. Which method is suitable for preservation of dry as well as green woods-
a. **Soaking method** b. Heating and cooling method
c. Sap displacement method d. Steeping method

73. Full cell process of wood preservation is applied at a pressure-
a. 1.50 to 10 kg/cm^2 b. 2.50 to 15 kg/cm^2
c. **3.50 to 12 kg/cm^2** d. 5.0 to 10 kg/cm^2

74. Formation of whole individual from a single cell called-
a. Regeneration b. Vegetation
c. Totipotency d. **All of the above**

75. Alburnum and duramen respectively also known as-
a. Heartwood and Sapwood b. **Sapwood and heartwood**
c. Soft wood and hardwood d. Hardwood and softwood

76. In India Breast-Height is taken as above the ground level -
a. 1.30 m b. 1.37 cm
c. 1.30 cm d. **1.37 m**

77. Which is used to measure stem diameter accurately at any point on the tree stem from any convenient distance-
a. Clinometer b. Spiegel Relaskop
c. **Wheeler Penta Prism** d. Digital calliper

78. Height measuring instrument which are based on trigonometric principle are-
a. Brandis Hypsometer
b. Abney's level
c. Blume-Leiss Hypsometer
d. **All of the above**

79. Breast-height form factor also known as-
a. **Artificial form factor** b. Absolute form factor
c. Normal form factor d. True form factor

80. The ratio between the mid-diameter and d.b.h. is known as-
a. Form factor b. **Form quotient**
c. Form height d. Form coefficient

Answers

72. a	73. c	74. d	75. b	76. d	77. c
78. d	79. a	80. b			

81. Newton's formula for calculation volume of the logs is-

a. $\frac{s_1 - s_2}{2} \times l$ b. $S_m \times l$

c. $\frac{(s_1 + 4Sm + s_2)}{6} \times 6$ d. $\frac{s_1 + s_2}{2} \times l$

82. In India volumes of logs are calculated by using-

a. Half girth formula b. **Quarter girth formula**

c. Newton formula d. Smalian formula

83. Quarter girth formula gives how much percent of the true volume-

a. 75% b. **75.8%**

c. 78.5% d. 85%

84. Local volume table is/are based on which variables-

a. Area b. Height

c. **Diameter** d. Form

85. Term ecosystem was given by-

a. Haeckel b. Odum

c. Reiter d. **Tansley**

86. Pressler's formula for volume increment percent is-

a. $p = \frac{100(D-d)}{n} \times \frac{(V+v)}{n(V-v)}$ b. $p = \frac{200(D+d)}{n} \times \frac{(V-v)}{n(V+v)}$

c. $\boldsymbol{p = \frac{200(D-d)}{n} \times \frac{(V-v)}{n(V+v)}}$ d. $p = \frac{400(D-d)}{n} \times \frac{(V+v)}{n(V-v)}$

87. Compound interest formula for diameter increment percent is-

a. $P = 100\left[\left(\frac{D}{d}\right)^{1/n} + 1\right]$ b. $P = 200\left[\left(\frac{D}{d}\right)^{1/n} - 1\right]$

c. $\boldsymbol{P = 100\left[\left(\frac{D}{d}\right)^{1/n} - 1\right]}$ d. $P = 200\left[\left(\frac{D}{d}\right)^{1/n} - 1\right]$

88. A sampling in which sampling units are selected in such a manner that all possible units of the same size have equal chance of being chosen is called-

a. **Random sampling** b. Point sampling

c. Non- random sampling d. Partial sampling

Answers

81. c	82. b	83. b	84. c	85. d	86. c
87. c	88. a				

89. The number of trees counted by horizontal point sampling and multiplied by a constant factor gives-
a. **Basal area per hectare (BA/ha)**
b. Number of trees per hectare
c. Diameter per hectare
d. Tree height per hectare

90. The ratio between the tree basal area & area of circular plot is-
a. $5k^2/2$ b. $2k^2/7$
c. $3k^2/4$ d. **$k^2/4$**

91. Which bands is/are not used in spiegel relaskop-
a. Band 1 b. Band 2
c. **Band 3** d. Band 4

92. The ratio between the volume of a tree to the product of basal area and height is called-
a. Form height b. Form quotient
c. Form value d. **Form factor**

93. Buffalo grass or paragrass is botanically known as-
a. ***Brachiaria mutica*** b. *Cenchrus ciliaris*
c. *Sehima nervosum* d. *Panicum maximum*

94. Who has submitted his report on "Improvement of Indian Agriculture" in 1893-
a. Dr. Swaminathan b. Dr. B.P Pal
c. **Dr. Voelcker** d. Dr. N. E. Barlog

95. Which forest fire most commonly occurs in plains-
a. Ground fire b. **Surface fire**
c. Crown fire d. Creeping fire

96. Which is teak defoliator-
a. *Hyblaea tectonae* b. ***Hyblaea puera***
c. *Plecoptera reflexa* d. *Tonica niviferana*

97. Which is deodar defoliator-
a. *Hyblaea cedrae* b. *Hyblaea puera*
c. *Plecoptera reflexa* d. ***Ectropis deodarae***

Answers

89. a	90. d	91. c	92. d	93. a	94. c
95. b	96. b	97. d			

98. A forest which is regenerated from vegetative parts of tree either naturally or artificially is called as-
   a. Production forest
   b. **Coppice forest**
   c. Pollard forest
   d. High forest
99. An area with complete protection, constituted according to chapter □ of Indian forest Act-1927 is called as-
   a. **Reserved forest**
   b. Protected forest
   c. Village forest
   d. Protection forest
100. The foundation of scientific forestry was laid during the year-
   a. 1825
   b. 1847
   c. 1856
   d. **1864**
101. Who is the first Inspector General (IGF) of Indian forest-
   a. L. S. Khanna
   b. **Dr. Dietrich Brandis**
   c. Dr. Dietrich Hokip
   d. Ram Prakash
102. A day in which 2.5 mm or more rain falls called-
   a. **Rainy day**
   b. Dry day
   c. Wet day
   d. All of the above
103. Which is/are light demander species-
   a. Dipterocarpus
   b. Calophylum
   c. Shorea
   d. **All of the above**
104. CVP index was given by-Paterson-
   a. Dietrich Brandis
   b. Backer
   c. **Paterson**
   d. Bruce Zobel
105. Accumulation of heavy cold air into natural depression is called-
   a. Radiation frost
   b. **Pool frost**
   c. Advective frost
   d. Ground frost
106. When a stem shows irregular involution and swellings called-
   a. Forking
   b. Buttressing
   c. **Fluting**
   d. Canker
107. Lantana camera is a notorious weed controlled by-
   a. ***Orthezia insignis***
   b. *Sporobolus tecti*
   c. *Pasturella conyza*
   d. *Bacillus lantanae*
108. Which is/are drought hardy species-
   a. ***Acacia nilotica***
   b. *Acacia catechu*
   c. *Madhuca indica*
   d. *Toona ciliata*

Answers

98. b	99. a	100. d	101. b	102. a	103. d
104. c	105. b	106. c	107. a	108. a	

109. Which is/are moderately drought hardy species-
   a. *Acacia nilotica*   b. ***Acacia catechu***
   c. *Madhuca indica*   d. *Toona ciliata*
110. Ordinary thinning also called-
   a. German thinning   b. Low thinning
   c. Thinning from the below   d. **All of the above**
111. Which is/are a non coppicer-
   a. *Butea monosperma*   b. *Hardwickia binata*
   c. *Bombax ceiba*   d. ***Picea smithiana***
112. The percentage of seed in a sample that have germinated in a test upto the time when the rate of germination reaches its peak is called-
   a. **Germination energy**   b. Germination potential
   c. Germinative capacity   d. Germination
113. Which formula is used to calculate number of plant (N) in line planting plantation-
   a. $N = \dfrac{\mathbf{100 \times 100}}{\textbf{distance of plants in lines (X)} \times \textbf{distance between the lines (Y)}}$
   b. $N = \dfrac{100 \times 100}{\text{Square of planting distance (X)}}$
   c. $N = \dfrac{100 \times 100 \times 1.155}{\text{Square of planting distance (X)}}$
   d. $N = \dfrac{2 \times 100 \times 100}{\text{Square of planting distance (X)}}$
114. Master seed year in *Shorea robusta*-
   a. **3-5 year**   b. 4-5 year
   c. 7-8 year   d. 5 year
115. Time of flowering and seed collection in *Casuarina equisetifolia* respectively-
   a. August-September and Jan-February
   b. Jan-March and April-May
   c. **May and June-December**
   d. Feb-April and May-June

Answers

109. b	110. d	111. d	112. a	113. a	114. a
115. c					

116. Which is/are frost tender species-
   a. ***Bombax ceiba*** b. *Dalbergia sissoo*
   c. *Schleichera oleosa* d. *Diospyros melanoxylon*
117. First agroforestry seminar in India was organised at-
   a. **Imphal** b. Pantanagar
   c. New Delhi d. Gujarat
118. The process of taking nutrients from deeper soil profile and depositing them on the surface layer is referred-
   a. Leaching b. Nutrient immobilization
   c. **Nutrient pumping** d. Mineralization
119. Spacing between trees within rows in alley cropping-
   a. **0.25-2.0 meter** b. 0.5-2.0 meter
   c. 1-2.0 meter d. 1.5-2.5 meter
120. A typical shelter belt has a shape-
   a. Quadrangles shape b. **Triangular shape**
   c. Square shape d. Rectangular shape
121. Home garden is most commonly practiced in which region-
   a. Temperate b. **Humid tropical**
   c. Tropical d. Arid
122. Which is/are a common example of the zonal pattern-
   a. Home garden b. **Alley cropping**
   c. Kitchen garden d. Silvopastural system
123. Shifting cultivation is practised extensively in which region-
   a. **North-eastern** b. Southern
   c. Western d. Northern
124. Shifting cultivation in Philippines is known as-
   a. Karen b. Ray
   c. **Hanumo** d. Chena
125. Joint forest management (JFM) was firstly started in which state-
   a. Odisha b. Rajasthan
   c. **West Bengal** d. Gujarat
126. Segment of DNA which moves from one chromosome to another chromosome within the genome of an individual called-
   a. Royal gene b. **Jumping gene**
   c. Transgene d. Operator gene

Answers

116. a	117. a	118. c	119. a	120. b	121. b
122. b	123. a	124. c	125. c	126. b	

127. Knife of DNA is-
   a. Exonuclease
   b. **Endonuclease**
   c. RNA polymerase
   d. Both b and c

128. The difference between the mean of the selected individuals (X_s) and the population mean (X) is called-
   a. **Selection differential**
   b. Genetic gain
   c. Gene load
   d. Heritability

129. The selection procedure that involves many cycles of selection and breeding is known as-
   a. Mass selection
   b. Current selection
   c. **Recurrent selection**
   d. Progeny selection

130. Foundation seed is the progeny of-
   a. Nucleus seed
   b. **Breeder seed**
   c. Foundation seed
   d. Registered seed

131. Tag colour of breeder seed is-
   a. **Yellow**
   b. White
   c. Purple
   d. Green

132. Montmorillonite is an example of-
   a. 2:1 non-expanding types
   b. **2:1 expanding types**
   c. 1:1 expanding types
   d. 1:2 expanding types

133. Land capability soil class suitable for wildlife & watershed-
   a. Class □
   b. Class □
   c. Class □
   d. **Class □**

134. T and V programme was introduced in the year-
   a. 1943
   b. **1974**
   c. 1986
   d. 2002

135. Lab to Land programme was started by ICAR on-
   a. **Silver jubilee**
   b. Golden jubilee
   c. Diamond jubilee
   d. Platinum jubilee

136. Panchayati Raj was first started in-
   a. Bihar
   b. **Rajasthan**
   c. Pantanagar
   d. Pondicherry

Answers

127. b	128. a	129. c	130. b	131. a	132. b
133. d	134. b	135. a	136. b		

137. z-test is applicable when the sample size is/are-

a. ≥ 50 b. **> 30**

c. < 50 d. < 30

138. Degree of freedom is-

a. (n + k) b. **(n – k)**

c. (n/k) d. (n × k)

139. The ranges of t-test lies between-

a. –1 to +1 b. 0 to 1

c. **–∞ to +∞** d. 0 to ∞

140. The ranges of probability lies between-

a. –1 to +1 b. **0 to 1**

c. –∞ to +∞ d. 0 to ∞

141. Plant hormone which helps in closing of stomata is-

a. GA_3 b. Auxin

c. Cytokinin d. **ABA**

142. Effect of low temperature on the initiation and development of flower called-

a. Cryopreservation b. Photoperiodism

c. **Vernalization** d. Anthesis

143. First carboxylation in C_4-cycle occurs by-

a. PGA b. RNA polymerase

c. Rubisco d. **PEP case**

144. Concept of photorespiration was given by -

a. **Decker & Tio** b. Calvin & Benson

c. Haberlandt d. Emerson & Arnold

145. Which is/are micronutrients-

a. Ca b. Mg

c. **Zn** d. P

146. Which is natural auxin hormone-

a. NAA b. IBA

c. **IAA** d. IPA

Answers

137. b	138. b	139. c	140. b	141. d	142. c
143. d	144. a	145. c	146. c		

CHAPTER 24

UPPSC ACF Examination 2015

1. Who was the ruler of Vatsa during the time of Buddha-
   a. **Bodhi** b. Udayana
   c. Satanika d. Nichakshu
2. Which one countries has not yielded any Ashoka edict-
   a. **Afghanistan** b. India
   c. Sri Lanka d. Pakistan
3. The first Gupta ruler to isssue silver coins was-
   a. Chandragupta □ b. Samundragupta
   c. **Chandragupta** □ d. Kumargupta □
4. Who was the founder of Vikramshila monastery-
   a. Gopal b. Devpal
   c. Mahipal d. **Dharampal**
5. In Vedic times the basis of untouchability was-
   a. Impurity b. **Occupation**
   c. Poverty d. None
6. Who has written *Taj-ul-Maasir*-
   a. Isami b. **Hasan Nizami**
   c. Al-Beruni d. Amir Khusrau
7. Who among the following Sultans renamed 'Devagiri' as 'Qubbatul Islam'-
   a. Ala-ud-din Khalji b. Firoz Shah Tughluq
   c. **Mohd Bin Tughluq** d. Abu Bakr Shah

Answers

1\. a 2. a 3. c 4. d 5. b 6. b
7\. c

8. Who among the following emperor called as 'Qalander'-
   a. Akbar b. Jahangir
   c. Shah Jahan d. **Babur**

9. Who among the following was disciple of Baba Farid-
   a. **Shaikh Nazamuddin Auliya** b. Amir Khusrau
   c. Shaikh Salim Chisti d. None

10. Who among the following had annexed the independent kingdom of Golkunda in the Mughal empire-
    a. Akbar b. Jahangir
    c. **Aurangzeb** d. Shah Jahan

11. Who is called 'Parrot of India'-
    a. Jaisi b. Amir Hasan
    c. Faizi d. **Amir Khusrau**

12. Who is called 'Shahe Bekhabar'-
    a. Farrukhsiyar b. Jahandar
    c. **Muhammad Shah** d. Bahadur Shah

13. Who among the following sultans ordered that in his dominion foreigners should be addressed as 'aizza;-
    a. Ala-ud-din Khalji b. **Firoz Shah Tughluq**
    c. Mohd Bin Tughluq d. Ghiyas-ud-din

14. Who among the following sultans first introduced token currency in India-
    a. Iltutmish b. **Mohd Bin Tughluq**
    c. Khizr Khan d. Bahlol Lodi

15. Who among the following sultans shifted the capital of the Sultanat from Delhi to Agra-
    a. Ibrahim Lodi b. Sher Shah
    c. **Sikander Lodi** d. Akbar

16. The Bombay Plan prepared in post-independence period intended to focus on-
    a. Political development b. **Social development**
    c. Educational development d. Economic development

Answers

8. d	9. a	10. c	11. d	12. c	13. b
14. b	15. c	16. b			

17. Who among the following was the first Indian to go to jail for performing his duty as a journalist-
    a. Surendranath Banerjee   b. Bal Gangadhar Tilak
    c. **Rashbehari Bose**   d. Pheroz Shah Mehta
18. Who among the following had declared that the real revolutionary armies are in the villages and factories
    a. Mahatma Gandhi   b. **Bhagat Singh**
    c. Karl Marx   d. Jai Prakash Narayan
19. Who among the following was the first to bring a printing press in India-
    a. Dutch   b. Briton
    c. **Portuguese**   d. French
20. During the early phase of resistance to British rule, the newspaper Indian Mirror was edited by-
    a. Bal Gangadhar Tilak   b. Surendranath Banerjee
    c. **Manmohan Ghosh**   d. N.N Sen
21. Which one of the following river is known as Ganga of South-
    a. Kaveri   b. Tungabhadra
    c. Krishna   d. **Godavari**
22. Selvas in Brazil are-
    a. Coffee producing areas   b. **Dense equatorial forest**
    c. Tropical grasslands   d. High mountain pastures
23. Periyar wildlife sanctuary is located in-
    a. **Kerala**   b. Karnataka
    c. Tamil Nadu   d. Arunchal Pradesh
24. Which Sea is surrounded by three continents-
    a. Red Sea   b. Arabian Sea
    c. Bering Sea   d. **Mediterranean Sea**
25. Sun appear earliest in India-
    a. Nagaland   b. Mizoram
    c. Assam   d. **Arunchal Pradesh**
26. Which is caused by rotation of the earth-
    a. Tides   b. **Day & Night**
    c. Change of season   d. Eclipse

Answers

17. c	18. b	19. c	20. c	21. d	22. b
23. a	24. d	25. d	26. b		

27. 'Ring of fire' is associated with the distribution of-
    a. **Volcanoes** b. Warm sea currents
    c. Petroleum d. Global warming
28. How many high pressure belts are found over the globe-
    a. 5 b. 4
    c. 3 d. **2**
29. Cocoa triangle is located in-
    a. Togo b. Ethiopia
    c. Brazil d. **Ghana**
30. Golden triangle in south South-East Asia is known in the world for-
    a. Oil seeds production b. Golden fleece production
    c. **Opium production** d. Gold mining
31. The preamble of the Indian Constitution was amended in the year-
    a. 1975 b. **1976**
    c. 1978 d. 1979
32. Articles of Indian Constitution is related to Indian Foreign Policy-
    a. Article 51 b. Article 60
    c. Article 50 d. **Article 380**
33. The minimum age of Indian voters was reduced from 21 to 18 years in-
    a. 62nd amendment b. 60th amendments
    c. 52 amendments d. **61st amendments**
34. Panchayati Raj is included in the-
    a. Union list b. Concurrent list
    c. **State list** d. Residuary list
35. Articles never implemented into action of Indian Constitution-
    a. Article 356 b. **Article 360**
    c. Article 352 d. Article 60
36. In which schedules of Indian Constitution anti defection law is placed-
    a. **10th** b. 9th
    c. 6th d. None

Answers

27. a	28. d	29. d	30. c	31. b	32. d
33. d	34. c	35. b	36. a		

37. Which can not be dissolved but can be abolished-
    a. Lok Sabha b. Rajya Sabha
    c. State Legislative Assembly d. **State Legislative Council**
38. The concept of concurrent list was borrowed from the Constitution of-
    a. USA b. UK
    c. **Australia** d. Canada
39. How many cadre in the all India services-
    a. **3** b. 4
    c. 5 d. 6
40. Which word is not a part of Indian Constitution-
    a. Democratic b. **Political**
    c. Republic d. Socialist
41. Who is the Head of the executive-
    a. **Prime Minister** b. President
    c. Cabinet d. None
42. The members of Rajya Sabha are elected for a term of-
    a. 4 years b. 5 years
    c. **6 years** d. 8 years
43. In India right to property was taken away by which one of the constitutional amendments-
    a. 43rd b. **44th**
    c. 49th d. None
44. Legislative powers of the Indian Union are vested in-
    a. **Parliament** b. President
    c. Prime minister d. Supreme court
45. The maximum gap between two successive Sessions of the Indian Parliament can be-
    a. 12 months b. 9 months
    c. **6 months** d. 4 months
46. The parliament of India consists of-
    a. Both the House of Parliament
    b. Lok Sabha only
    c. Rajya Sabha only
    d. **President, Lok Sabha and Lok Sabha**

Answers

37. d	38. c	39. a	40. b	41. a	42. c
43. b	44. a	45. c	46. d		

47. Who among the following presides over the joint sitting of both the Houses of the Parliament-
a. **Speaker of Lok Sabha** b. President
c. Vice President d. None

48. Schedule-xi of Indian Constitution deals with-
a. Municipality b. **Panchayati Raj**
c. Central State relation d. Anti-defection

49. Which part of Indian Constitution deals with the citizenship-
a. Part iv b. Part iii
c. **Part ii** d. Part i

50. Founder of 'Swatantra Party'-
a. **C. Rajagopalchari** b. C. R. Das
c. K. M. Munshi d. K. Kamraj

51. Which country supported the membership of India in Nuclear Suppliers Group-
a. **China** b. Brazil
c. Switzerland d. France

52. Brand ambassador of Beti Bachao, Beti Padhao scheme-
a. Aishwarya Bachchan b. Rekha
c. **Madhuri Dixit** d. Shabana Azmi

53. 2018 FIFA world cup will be held at-
a. **Russia** b. Qatar
c. France d. Netherlands

54. The lonely Planet group has given the 'Best Indian Destination for Wildlife' award in May 2016-
a. Kerala tourism b. **Madhya Pradesh tourism**
c. Gujarat tourism d. Assam tourism

55. The female tribal of a district have first declared to impose social boycott on those who cut trees-
a. Chhattisgarh b. Uttarakhand
c. **Himanchal Pradesh** d. Rajasthan

56. In whose memory a commemorative coin o Rs. 100 and a circulation coin of Rs. 10 have been released by te union govt. in May 2016-
a. Dr. Ambedkar b. Swami Chinmayananda
c. **Maharana Pratap** d. Jamshedji Tata

Answers

47. a	48. b	49. c	50. a	51. a	52. c
53. a	54. b	55. c	56. c		

57. 'Swabhiman Yojna' is related to-
a. **Spreading rural banking**
b. Women empowerment
c. Scheme of pension for landless
d. Adolescence girl education

58. The Govt. of India has banned which food additives in June 2016 as the study reports found that its presence causes cancer-
a. Potasium chromate
b. Potassium bromate
c. Potassium sulphate
d. **Potassium gluconate**

59. Which state declared 35 per cent quota for females on non-gazetted posts-
a. Jharkhand
b. West Bengal
c. **Bihar**
d. Maharashtra

60. Which state become first to provide free treatment to confirmed cases of Hepatitis C-
a. Rajasthan
b. Gujarat
c. Bihar
d. **Punjab**

61. Global e-company, eBay has entered into an agreement with Handloom Corporation with which state-
a. Rajasthan
b. Gujarat
c. Haryana
d. **Uttar Pradesh**

62. 2017 FIFA under-17 World cup will be held-
a. Spain
b. **India**
c. Argentina
d. Brazil

63. Which country receive maximum funds from India for promotion of technical co-operation-
a. **Nepal**
b. Afghanistan
c. Bhutan
d. Bangladesh

64. Green Passage scheme for orphan students was released in May 2016 by the state-
a. Bihar
b. Kerala
c. Madhya Pradesh
d. **Odisha**

Answers

57. a	58. d	59. c	60. d	61. d	62. b
63. a	64. d				

65. Theme of 'International Day against Drug Abuse and Illicit Trafficking' observed on the 26th June, 2016-
   a. Drug-stop it
   b. **Listen first**
   c. Save health-say no to drugs
   d. Let's develop without Drugs

66. World longest and deepest rail tunnel has been officially opened in June 2016-
   a. **Switzerland**
   b. China
   c. Japan
   d. Germany

67. First country in the world to ban deforestation-
   a. New Zealand
   b. Germany
   c. **Norway**
   d. Poland

68. Afghanistan-India Friendship Dam inaugurated in the June 2016 is situated in which provenance of Afghanistan-
   a. Helmand
   b. **Herat**
   c. Farah
   d. Nimruz

69. According to 'Good Country Index, 2015' report released in June 2016, which among the following countries has got the highest positions-
   a. Iceland
   b. Netherlands
   c. Ireland
   d. **Sweden**

70. China has recently launched its first 'Dark Sky Reserve' in which area of Tibet-
   a. **Ngari Prefecture**
   b. Lhasa
   c. Shannam
   d. Ahasa

71. India Stonemart 2015 was organized in-
   a. Hyderabad
   b. Mumbai
   c. **Jaipur**
   d. Pune

72. According to WHO, first country to eliminate 'Mother to Baby' HIV-
   a. Malaysia
   b. **Thailand**
   c. South Korea
   d. Myanmar

73. Fat tax introducing state of India is-
   a. Karnataka
   b. Tamil Nadu
   c. Maharashtra
   d. **Kerala**

Answers

65. b	66. a	67. c	68. b	69. d	70. a
71. c	72. b	73. d			

74. Which committee has submitted its reports on New Education Policy on 27th May, 2016-
   a. **T.S.R Subramanian Committee**
   b. J. S. Rajput Committee
   c. R. M. Lodha Committee
   d. Ashok Dalwai Committee

75. Which one of the following pairs is not correctly matched-
   a. Vinesh Phogat - Wrestling
   b. Kothgai - Hockcy
   c. **Vijendra Singh - Football**
   d. Pintala - Chess

76. 'Nanga Parbat' is located in-
   a. Uttarakhand b. Himanchal Pradesh
   c. **Jammu & Kashmir** d. Meghalaya

77. Which district have largest area in U.P-
   a. Allahabad b. **Lakhimpur Khiri**
   c. Lalitpur d. Sonbhadra

78. The longest river in the world-
   a. **Nile** b. Amazon
   c. Brahmaputra d. Mississippi

79. The longest mountain range in the world-
   a. Himalayas b. Rocky
   c. Alps d. **Andes**

80. Nasik is located along the river-
   a. Narmada b. Penganga
   c. **Godavari** d. Mahi

81. Wildlife Institute of India is located in-
   a. Almora b. **Dehra Dun**
   c. Bhopal d. Guwahati

82. Number of National Parks notified upto 2014 is-
   a. 103 b. 98
   c. **106** d. 94

Answers

74. a	75. c	76. c	77. b	78. a	79. d
80. c	81. b	82. c			

83. Project tiger was launched in the year-

a. 1965 b. **1973**

c. 1983 d. 1978

84. World wildlife day is observed on-

a. **3rd March** b. 22nd March

c. 28th March d. 7th April

85. Green revolution is related with-

a. **Food grain** b. Oil seeds

c. Fish d. Potato

86. Which one of the following States was not among the three new States included in 'National Dairy Plan (Plane □)' in June 2015-

a. Chhattisgarh b. **Uttar Pradesh**

c. Jharkhand d. Uttarakhand

87. Who among the following has won 'Mr. World 2016' crown at a contest held in Great Britain on July 9, 2016-

a. Fernandez b. Aldo Esparza

c. Abrahim Alexander d. **Rohit Khandelwal**

88. The UP Govt. has celebrated 2015-2016 as which one of the following-

a. **Kisan Year**

b. Yuva Year

c. Mahila Suraksha Year

d. Environmental Conservation Year

89. Which one of the following States had the highest production of onion in India in the year 2015-16-

a. Gujarat b. Karnataka

c. **Maharashtra** d. West Bengal

90. The maximum speed of Talgo train run on 13th July, 2016-

a. 200 km/hr. b. **180 km/hr.**

c. 150 km/hr. d. 120 km/hr.

91. The scientists of which one of the following countries have recently claimed to discover a new state of matter 'Jahn-Teller metal-

a. Germany b. China

c. USA d. **Japan**

Answers

83. b 84. a 85. a 86. b 87. d 88. a
89. c 90. b 91. d

92. Who among the following inaugurated 'Rio 2016 Olympic Torch' unveiling ceremony in the first week of July, 2015-
a. **Dilma Rousseff** b. Jaqueline Mourao
c. Carlos Arther Nuzman d. Beth Lula

93. The meeting of Inter-State Council held on 16th July, 2016 in New Delhi was-
a. 9th b. 10th
c. **11th** d. 12th

94. Recently a robot took life of a man, In which one of the following countries this incident occurred-
a. **Germany** b. France
c. Japan d. USA

95. What was the theme of 6th Jagran Film Festival started on 1st July, 2015 at Siri Fort Auditorium, New Delhi-
a. Cinema for 'Better India' b. **Cinema for 'Happy Lives'**
c. Promotion of 'Art Film' d. None

96. The largest uranium producing country of the world-
a. France b. China
c. USA d. **Kazakhstan**

97. The number of Islands in the South China Ocean is-
a. 350 b. 300
c. **250** d. 200

98. GAGAN system launched in July 2015 will facilitate-
a. Launching missiles b. **Swift civil aviation**
c. Guiding of submarines d. None

99. The advised to replacement of the word 'Adhinayak' by 'Mangal' in the National song of India was given-
a. Smiriti Irani b. Sushma Swaraj
c. Kesari Nath Tripathi d. **Kalyan Singh**

100. The Islamic proclaimer associated with Islamic Research Foundation-
a. **Zakir Naik** b. Zahir Naik
c. Salman Naik d. None

Answers

92. a	93. c	94. a	95. b	96. d	97. c
98. b	99. d	100. a			

101. The number of the member countries in the Nuclear Suppliers Group till July, 2016-
a. 38 b. **48**
c. 58 d. 68

102. CNG train was started for the first time in-
a. Madhya Pradesh b. Maharashtra
c. Rajasthan d. **Haryana**

103. Which Indian state declared 'Blue Mormon' as the state butterfly-
a. Uttarakhand b. Tamil Nadu
c. **Maharashtra** d. Gujarat

104. The Commonwealth Games 2018 will be held in-
a. Australia b. India
c. **South Africa** d. Pakistan

105. Artificial island in South China Ocean was buit by-
a. Japan b. **China**
c. Indonesia d. Malaysia

106. Which Chief Minister demanded scrapping of Governer's post in the Inter-state Council held on 16th July, 2016 in New Delhi-
a. Mamata Banerjee b. Harish Rawath
c. Akhilesh Yadav d. **Nitish Kumar**

107. The high speed Talgo train was run on 13th July, 2016 between-
a. Agra-Gwalior b. New Delhi-Rewari
c. **Mathura-Palwal** d. New Delhi-Mathura

108. The number of countries in European Union from which UK is likely to be out is-
a. **28** b. 38
c. 18 d. 08

109. The Prime Minister of UK in July, 2016 after resignation of David Cameron was-
a. Boris Johnson b. **Theresa May**
c. Priti Patel d. Dominic Raab

110. 'Water is not any substance but it is a human rights' has been said by-
a. Nitisk Kumar b. Manohar Lal Khattar
c. **Arvind Kejriwal** d. Mamata Banerjee

Answers

101. b	102. d	103. c	104. c	105. b	106. d
107. c	108. a	109. b	110. c		

111. Which one of the following pairs is not correctly matched-
   a. Decibel-unit of sound intensity
   b. Horse power- unit of power
   c. Nautical mile- Unit of Distance
   d. **Celsius- unit of heat**

112. Kilowatt-hour is unit-
   a. Mass b. time
   c. Electrical energy d. **Electric power**

113. The objects are visible in light due to-
   a. **Scattering** b. Refraction
   c. Absorption d. Fluorescence

114. Rainbow is seen due to-
   a. Interference by thin film b. **Dispersion of sunlight**
   c. Scattering of sunlight d. Diffraction of sunlight

115. One gram of ice is kept with one gram of stream in a thermal chamber. After thermal equilibrium, the temperature of this mixture will be-
   a. 20 ^{0}C b. 50 ^{0}C
   c. 80 ^{0}C d. **100 ^{0}C**

116. A 100-watt bulb illuminates for eight hours daily. The electrical energy consumed in the entire month of April will be-
   a. 12 units b. 36 units
   c. **24 units** d. 48 units

117. A green-coloured glass plate when seen through red light appears-
   a. **Black** b. Red
   c. Green d. Blue

118. A massless rubber ballon is filled with 200 gram water. Its weight in water will be-
   a. 100 grams b. **zero**
   c. 200 grams d. 400 grams

119. Which one of the following is not the unit of time-
   a. Year b. Day
   c. **Light-year** d. Hour

Answers

111. d	112. d	113. a	114. b	115. d	116. c
117. a	118. b	119. c			

120. Which of the following substances is also called 'liquid gold'-
a. **Petroleum** b. Platinum
c. Aqua regia d. Mercury

121. Which one of the following pairs is not correctly matched-
a. Chloromycetin- Antityphoid b. Crystal violet- Antiseptic
c. Quinine- Antimalarial d. **Cinchonine- Anaesthetic**

122. The light from the sun reaches the earth in about
a. 0.5 second b. 5 second
c. 50 second d. **500 second**

123. German silver is an alloy of-
a. **Copper, zinc and nickel** b. Zinc, nickel and cobalt
c. Nickel, cobalt and copper d. Silver, zinc and copper

124. Which one of the following compounds is known as gammexane-
a. Trinitrobenzene b. Trinitrotoluene
c. **Benzene hexachloride** d. Benzene hexabromide

125. Which one of the following gases absorbs ultraviolet radiation-
a. Oxygen b. **Ozone**
c. Argon d. Sulphur dioxide

126. Which one of the following compounds is called 'Chinese white'-
a. Lead oxide b. Mercuric oxide
c. Lead peroxide d. **Zinc oxide**

127. Which one of the following pairs is not correctly matched-
a. **Vitamin A- colour blindness**
b. VitaminB_1- beriberi
c. VitaminB_{12}- anemia
d. Vitamin-C- scurvy

128. Which one among the following substances is sweetest-
a. **Sucronic acid** b. Sucralose
c. Saccharin d. Aspartame

129. Lead pencil contains-
a. Lead (Pb) b. **Graphite**
c. Iron sulphide d. Lead sulphide

Answers

120. a	121. d	122. d	123. a	124. c	125. b
126. d	127. a	128. a	129. b		

130. Baking soda is the commercial name of-
a. Sodium carbonate
b. Sodium borate
c. **Sodium bicarbonate**
d. Sodium thiosulphate

131. CO_2 is transported in the human blood as-
a. Carbonic acid
b. Sodium carbonate
c. **Sodium bicarbonate**
d. Potassium carbonate

132. In human body, urea is synthesized in-
a. **Liver**
b. Kidney
c. Uninary bladder
d. Spleen

133. Which statement is true about bile juice-
a. Synthesized in the bile juice
b. **Contains salts which emulsify fats**
c. Contains enzymes which digest fats
d. Acidic in nature

134. The main nitrogenous excretory product of human is-
a. Ammonia
b. Uric acid
c. Guanine
d. **Urea**

135. Which one of the following is not an essential fatty acids-
a. Linolenic acid
b. Linoleic acid
c. Stearic acid
d. **Arachidonic acid**

136. During respiration, the conc. Of which one of the following gases remains unchanged in the inspired and expired air-
a. **N_2**
b. O_2
c. CO_2
d. None

137. Human teeth appears only in permanent dentition-
a. Incisor
b. **Molar**
c. Canine
d. Premolar

138. Bird produces milk is-
a. Sparrow
b. Dove
c. Crow
d. **Pigeon**

139. The fastest running bird is-
a. Kiwi
b. **Emu**
c. Ostrich
d. Cassowary

Answers

130. c	131. c	132. a	133. b	134. d	135. d
136. a	137. b	138. d	139. b		

140. Albinism is characterized by the complete absence of which of the following-
a. **Melanin** b. Albumin
c. Cholesterol d. Bilirubin

141. Chilgoza is obtained from-
a. *Taxus baccata* b. *Cedrus deodara*
c. *Cycas revoluta* d. ***Pinus gerardiana***

142. Which is used for manuring rice field-
a. Pteris b. ***Azolla***
c. Seleginella d. Equisetum

143. Most poisonous fungi-
a. *Agaricus bisporus* b. *A. campestris*
c. ***Amantia phalloides*** d. *Podaxon podaxis*

144. Binomial system of nomenclature was given by-
a. **C. Linnaeus** b. A. P. Candolle
c. Darwin d. Mendel

145. Which is not a modified root-
a. Carrot b. **Potato**
c. Asparagus d. Sweet potato

146. Intraxylary phloem is found in-
a. *Achyranthes* b. *Boerhavvia*
c. ***Dracaena*** d. *Strychnos*

147. Which does not fix nitrogen-
a. Rhizobium b. Cyanobacteria
c. **Mycorrhiza** d. Frankia

148. C_4 cycle was first reported in-
a. **Maize** b. Sugarcane
c. Wheat d. Carrot

149. Pyramid of energy in an ecosystem is always-
a. Inverted b. Spindle shape
c. Curved d. **Upright**

150. The initiation codon in protein synthesis is-
a. **AUG** b. GUA
c. GCA d. CCA

Answers

140. a	141. d	142. b	143. c	144. a	145. b
146. c	147. c	148. a	149. d	150. a	

INDIAN COUNCIL OF FORESTRY RESEARCH AND EDUCATION

POST: SCIENTIST-B (FORESTRY)

1. 2Gs/r formula for yield regulation proposed by
   a. Masson b. Von mantel
   c. Howard d. Simmon
2. Growing Grasses in rotation with field crops
   a. Dry farming b. Contour farming
   c. Lay farming d. Relay cropping
3. Growth rings are formed by activities of
   a. Xylem b. Cambium
   c. Xylem & Phloem d. Phloem
4. In the Raunkier's life forms trees are categorized under
   a. Chamaephytes b. Phanerophytes
   c. Hemicryptophytes d. Cryptophytes
5. A simple mechanical tissue devoid of lignin is
   a. Collenchyma b. Sclerenchyma
   c. Parenchyma d. None of these
6. A Distinct and unique forest type found in carnatic coast from Tirunelveli to Nellore
   a. Tropical dry evergreen forest
   b. Littoral and swamp forest
   c. Red sanders beuring forest
   d. Southern tropical semi evergreen
7. State which statement is correct
   a. The rate of nutrient cycling is high in coniferous forests
   b. Nutrient cycling is low in northern latitude compared to tropical region
   c. Nutrient cycling between regions are same
   d. No difference in nutrient cycling between Silvicultural system

Answers

1. b 2. c 3. b 4. b 5. c 6. a
7. b

8. Sclerophyllus natural vegetation is found in
   a. Tundra regions
   b. Equatorial regions
   c. Monsoon regions
   d. Mediterranean regions

9. Gugul is obtained from
   a. *Pinus kesiya*
   b. *Kingliodendron pinnatum*
   c. *Dipterocarpus turbinatus*
   d. *Boswellia serrate*

10. Concept of progressive yield attributable to
    a. Von montel
    b. Brandis
    c. Smith
    d. Harting

11. Shingles are called as
    a. Bobbin
    b. Sleepers
    c. Wood tiles
    d. Picker arms

12. Which is a population of individuals which has become adapted to a specific environment in which it has been planted
    a. Provenance
    b. Ecotype
    c. Seed Source
    d. Land race

13. National Green Tribunal came into existence in the year
    a. 2010
    b. 2012
    c. 2011
    d. 2015

14. Class I Lands having slope of
    a. Upto 8%
    b. 1-33 %
    c. 33% to 100%
    d. 8% to 33%

15. Spike disease of *Santalum album* caused.
    a. Mycoplasam like bodies
    b. *Fusarium solani*
    c. *Fomes badius*
    d. *Ganoderma lucidum*

16. *Gulatheria fragrantissima* yields
    a. Winter green oil
    b. Citriodra oil
    c. Camphor oil
    d. Costus oil

17. The Colour of mammals and birds of warm region is darker than those living in cold areas called as.
    a. Gloger's rule
    b. Jordan's rule
    c. Allen's rule
    d. Bergmann's rule

Answers

8. d	9. d	10. d	11. c	12. d	13. a
14. b	15. a	16. a	17. a		

18. Protein banks are form of
a. Agri Silvicultural system
b. Alley cropping
c. Agro Silvopastoral system
d. Silvopastoral system

19. $\frac{V-V}{V+V} \times \frac{200}{N}$ V-V × 200 increment percent formula proposed by
a. Pressler
b. Schneider
c. Biolley
d. Von mantel

20. Structural material resulting from glued lamination called as
a. Plywood
b. Glulam
c. Core board
d. Particle board

21. Method of yield regulation suitable for extensive tropical forests and where selection fellings are the rule.
a. Volume unit method
b. French method
c. Brandis method
d. Hufnagl method

22. *Iosoma beesoni* used for control of
a. Defoliator of sal
b. Defoliator of the toon
c. Defoliator of deodar
d. Defoliator of shisham

23. Kokam butter obtained from
a. *Madhuca indica*
b. *Garcinia indica*
c. *Aleurites fordi*
d. *Shorea robusta*

24. One metric tonnes of Carbon equals the following metric tones of CO_2 equivalent?
a. 3.76
b. 3.06
c. 3.02
d. 3.66

25. The average growing stock in recorded forest area per hectare in India is
a. 25.2 m^3
b. 121.8 m^3
c. 59.79 m^3
d. 95.4m^3

26. The Scheduled Tribes and Other Traditional Forest Dwellers (Recognition of Forest Rights) Act was enacted in
a. 2004
b. 2006
c. 2005
d. 2007

Answers

18. d	19. a	20. b	21. c	22. c	23. b
24. d	25. c	26. b			

27. One of the following project does not require Environmental Impact Assessment
    a. Chemical fertilizer industry
    b. Nuclear Power Project
    c. Petrolium refining industry
    d. Micro-hydel Projects

28. Institute of Wood Science and Technology located at
    a. Mumbai
    b. Bengaluru
    c. Itanagar
    d. Chennai

29. The Forest Conservation Act was enacted in the year
    a. 1980
    b. 1978
    c. 1988
    d. 2000

30. *Chi* square test was developed by
    a. Fisher
    b. Mahalanobis
    c. Galton
    d. Karl pearson

31. Agro-climatic zones of our country recognized by planning commission of India are
    a. 11
    b. 13
    c. 15
    d. 17

32. Correlation coefficient ranges between
    a. 0 to 1
    b. –1 to +1
    c. 1 to 0
    d. 0 to –1

33. Normal distribution curve is exactly
    a. Modal shape
    b. Sphere shape
    c. Bell shape
    d. All of these

34. Asia's first marine biosphere reserve is
    a. Gulf of Kutch
    b. Gulf of Cambay
    c. Sunderbans
    d. Gulf of Mannar

35. Method of handling of dormant seeds, in which the imbibed seeds are subjected to a period of chilling after-ripen embryo is called as,
    a. Scarification
    b. Caprification
    c. Fructification
    d. Stratification

36. Which of these is a defoliator
    a. *Calandra glandium*
    b. *Eutectona machaeralis*
    c. *Tonica niviferana*
    d. *Hyblaea puera*

Answers

27. d 28. b 29. a 30. d 31. c 32. b
33. c 34. d 35. d 36. d

37. Resin canals located in plants fibro-vascular system is characteristic feature of *Anacardiacea* family, which one of the following species do not exhibit it

a. *Pistachio vera* b. *Anacardium occidentale*
c. *Semecarpus anacardium* d. *Mangifera indica*

38. A Sub-division of a felling series formed with the object of regulating felling in some special manner

a. Coupe b. Beat
c. Cutting section d. Block

39. State of Forest Report published by FSI, Dehradun is once in

a. Two years b. Three years
c. Annually d. Quinquennially

40. This deer species Endemic to Manipur

a. Dancing deer b. Mouse deer
c. Swamp deer d. Spotted deer

41. Largest area covered by this forest type in India

a. Tropical Dry Evergreen b. Tropical Wet Evergreen
c. Tropical dry Deciduous d. Tropical Moist Deciduous

42. Indicator plant for lime rich soils in the peninsular India

a. *Ixora parviflora* b. *Acaia leucophloea*
c. *Chloroxylon swietenia* d. *Cupressus torulosa*

43. A nail was driven into the trunk of a tree at a point 1.5 m from the soil level. After 3 years, the nail will

a. Move up b. Move sideways
c. Remain where it was d. Move down

44. Cork cambium is also called as

a. Phelloderm b. Phellogen
c. Phellem d. Vascular cambium

45. Factor- product relationship is also known as

a. Law of equi-marginal returns
b. Least cost combination
c. Law of returns
d. All of these

Answers

37. a	38. c	39. a	40. a	41. c	42. a
43. c	44. c	45. a			

46. Tyloses occur in
    a. Heart wood
    b. Sap wood
    c. Hard wood
    d. Soft wood
47. Incomplete designs refers to
    a. Lattice design
    b. RBD
    c. Latin square design
    d. Augmented design
48. Rhea fibre obtained form
    a. *Boehmeria nivea*
    b. *Calotropis gigantea*
    c. *Sterculia villosa*
    d. *Cannabis sativa*
49. Burrs of which species yields figured decorative wood
    a. Walnut
    b. Siris
    c. Laurel
    d. All of these
50. Disease free plants can be produced through
    a. Meristem culture
    b. Callus culture
    c. Protoplast culture
    d. Embryo culture
51. How many percent of the incoming solar radiation is reflected back to atmosphere
    a. 10
    b. 46
    c. 28
    d. 42
52. If the level of contour lines increased from outer to inner portion, it indicates
    a. Close loop
    b. Valley
    c. Saddle
    d. Hill
53. Development of an embryo from any cell of the gametophyte other than egg cell called
    a. Apogamy
    b. Parthenogenesis
    c. Sporophytic budding
    d. Conjugation
54. Maximum net return is obtained when marginal cost (MC) is_ the marginal return (MR)
    a. Less than
    b. Equal to
    c. More than
    d. All of these
55. Kranz type of anatomy is found in
    a. C3 plants
    b. C4 plants
    c. Both in C3 & C4
    d. CAM Plants

Answers

46. a	47. a	48. a	49. d	50. A	51. d
52. d	53. a	54. b	55. b		

56. Number of sample plots required for a Species in each thinning regime to prepare a yield table
    a. 400 b. 500
    c. 600 d. 700

57. Ral, an oleoresin extracted from
    a. *Vateria indica* b. *Pinus roxburgii*
    c. *Shorea robusta* d. *Pinus wallichiana*

58. Which one of the following is a species of grasses suitable for soil conservation of ravine lands and the land affected by shifting cultivation
    a. *Pennisetum polystachyon* b. *Cenchurus setigerus*
    c. *Lasiurus sindicus* d. *Penicum antidotale*

59. Good seed year of *Cedrus deodara* is
    a. 10 years b. 1-2 years
    c. 5-6 years d. 4-5 years

60. Indicator plant for favourable conditions for natural regeneration of deodar and kail
    a. *Spirea sorbifolia* b. *Viola canescens*
    c. *Narenga porphyrocoma* d. *Saccharum procerum*

61. In regeneration survey, 'U' symbol represents
    a. Establish regeneration
    b. Sub-whippy seedlings whose height is less than 50 cm
    c. Browsed Whippy
    d. Unbrowsed

62. Whippy How many seedlings per acre will you need for planting trees at a spacing of 9 feet x 9 feet ?
    a. 81 b. 304
    c. 436 d. 538

63. Which of the following is NOT true about the role forests play in protecting water quality?
    a. Filtering pollutants b. Shading water
    c. Stabilizing banks d. Releasing nutrients

Answers

56. c 57. c 58. a 59. d 60. b 61. b
62. d 63. d

64. Which of the following is NOT a common characteristic of invasive plants?
    a. High seed production and good ability of those seeds to germinate
    b. Seed is dispersed by wind, water and/or animals over great distances
    c. Predator avoidance and/or deterrence
    d. Leaf out later in the spring and lose their leaves earlier in the fall
65. Which of the following factors does NOT have a significant effect on the behavior of a forest fire?
    a. Relative humidity b. Topography
    c. Temperature d. Soil texture
66. Which term is used to define the capacity of a tree to develop and grow in the shade while in competition with other trees?
    a. Tracheids b. Reproduction
    c. Toughness d. Tolerance
67. Competition for water, minerals, light and space is most severe between two
    a. Closely related species occupying the same niche
    b. Closely related species occupying different niche
    c. Unrelated species occupying the same niche
    d. Speciesoccupying different overlapping ecosystems
68. A habitat of long severe winters and growing season limited to a few months of summer constitutes:
    a. Savanah ecosystem b. Taiga ecosystem
    c. Tundra ecosystem d. None of these
69. Which one of the following terms does not reflect the idea of ecosystem:
    a. Holocene b. Cherposem
    c. Biosystem d. Geobiocoenosis
70. Which of the following statement is NOT true about mangrove forests?
    a. They grow along thecoast especially in the river deltas
    b. These plants are able to grow in a mix of saline and fresh water
    c. They grow luxuriantly in muddy areas
    d. All of these statements are false

Answers

64. d	65. d	66. d	67. a	68. b	69. b
70. d					

71. According to State of Forest Report, 2017 published by Ministry of Environment, Forest and Climate Change, the total forest cover in India is around:
a. 33% b. 23%
c. 21% d. 28%

72. A tabulation of volume, basal area, number of trees, etc. per acre found in full stands on specified sites at specified ages
a. consistence b. pole stand
c. yield table d. Site index

73. Harvesting or killing trees infected or highly susceptible to insects or diseases to protect the rest of the forest stand.
a. sanitation cut b. consistence
c. site index d. selection thin

74. Which of the following species is endemic to India?
a. King Cobra b. Snow Leopard
c. Bonnet Macauque d. Great Hornbill

75. Which of the following protected areas comprises of montane grasslands interspersed with forest patches called "Sholas"?
a. Kaziranga NP, Assam
b. Eravikulam NP, Kerala
c. Kedarnath WLS, Uttarakhand
d. Kanha NP, MP

76. Which among the following is the first widely used herbicide ?
a. 2,4-D b. TCDD
c. Limonene d. Vinegar

77. The study of the life history and general characteristics of forest tree, with particular reference to environmental factors
a. Silvics b. Silviculture
c. Volume regulation d. Dendrology

78. India's first National Centre for Marine Bio-diversity (NCMB) is located in:
a. Bhavnagar b. Jamnagar
c. Mumbai d. Puducherry

Answers

71. c	72. c	73. a	74. c	75. b	76. a
77. a	78. b				

79. A branch of life sciences that studies the short term and long-term changes in the size and age composition of population, and the biological and environmental processes influencing those changes
    a. population dynamics
    b. conservation biology
    c. mutations
    d. evolution

80. As per the India State of Forest Report, 2017, the three states which have shown increase in the forest cover in the country are:
    a. Tamil Nadu, Karnataka and Kerala
    b. AndhraPradesh, Karnataka and Kerala
    c. Himachal Pradesh, Uttarakhand and Sikkim
    d. Madhya Pradesh, Odisha and Jharkhand

81. Area wise which state has the largest forest cover in the country?
    a. Arunachal Pradesh
    b. Chhattisgarh
    c. Madhya Pradesh
    d. Odisha

82. In which of the following marine protected area would you not find coral reefs?
    a. Gulf of Mannar
    b. Gulf of Kutch
    c. Andaman & Nicobar
    d. Sundarbans

83. Currently Asiatic Lions are found only in the Gir National Park of Gujarat. A major disaster/ disease outbreak is likely to wipe out the entire population in one stroke, In order to have a back up population in case of such eventuality, an alternate home for the Asiatic Lion is being created at:
    a. Kanha Natoinal Park in Madhya Pradesh
    b. Kuno Palpur Sanctuary in Madhya Pradesh
    c. Pench National Park in Maharashtra
    d. Ranthambore National Park, Rajasthan

84. Which of the following is a non-poisnous snake of India?
    a. King Cobra
    b. Saw scaled viper
    c. Common Krait
    d. Indian Rock Python

85. Organisms which are typically saprophytic, but under certain conditions may be parasitic eg. Decay fungi in living trees
    a. facultative saprophytes
    b. characteristics of hyphea
    c. obligate parasites
    d. facultative parasites

Answers

79. a	80. b	81. c	82. d	83. b	84. d
85. d					

86. The swollen tip of a hyphae or germ tube that facilitates attachment and penetration of the host by the fungi

a. appressorium
b. haustorium
c. witches broom
d. proboscis

87. Which is not a part of National Action Plan on Climate Change?

a. National Water Mission
b. National Mission on Pollution Control
c. National Mission for Sustainable Agriculture
d. National Mission on Sustainable Habitat

88. Which common names of insects is associated with the order Hymenoptera?

a. Wasps and ants
b. Beetles and weevils
c. Flies and mosquitoes
d. Grasshoppers and crickets

89. Which of the following engineering structures is not a part of Soil & Moisture control measures?

a. Gully Plugs
b. Check dams
c. Causeway
d. Loose boulder structure

90. Which of the following is the right chronology for various forest related legislation of our country?

a. Wildlife Protection Act, Forest Conservation Act, Indian Forest Act
b. Indian Forest Act, Forest Conservation Act, Wildlife Protection Act
c. Indian Forest Act, Wildlife Protection Act, Forest Conservation Act
d. None of these

91. Which among the following multilateral convention seeks to protect the human health and environment from Persistant Organic Pollutants (POPs)?

a. Bonn Convention
b. Stockholm Convention
c. Rotterdam Convention
d. Basel Convention

Answers

86. a 87. b 88. a 89. c 90. c 91. b

92. Which among the following multilateral environment agreements (MEAs) is not correctly paired with the respective issue it deals with?
   a. Montreal Protocol of 1987 – Ozone Depleting Substances
   b. Bonn Convention of 1979 – The conservation of Migratory Species
   c. Basel Convention of 1989 – Regulation of transboundary movement, transit, handling and use of Living Modified Organisms.
   d. Rotterdam Convention of 1998 – Consensual International Trade in certain Hazardous Chemicals and Pesticides.

93. Which one of the following is a useful biological indicator of Sulphur-dioxide pollution ?
   a. Bryophytes
   b. Algal blooms
   c. Pseudomonas
   d. Lichens

94. Identify the non Green-House Gas(GHG) from the following :
   a. Methane
   b. Nitrous oxide
   c. Sulphur Hexafluoride
   d. Carbon Monoxide

95. Which of the following statements is NOT true?
   a. Photochemical smog always contains Ozone.
   b. The toxic effect of Carbon Monoxide is due to its greater affinity for haemoglobin as compared to oxygen.
   c. Lead is the most hazardous metal pollutant of automobile exhaust.
   d. None of these

96. The First Chairperson of the National Green Tribunal (NGT) was ?
   a. Justice A.S. Naidu
   b. Justice Lokeshwar Singh Panta
   c. Justice Markandey Katju
   d. Justice A K Ganguly

97. Green Economy is led by?
   a. United NationsEnvironment Programme
   b. InternationalHydrological Organization
   c. Intergovernmental Panel for Climate Change
   d. European Union

Answers

92. c	93. d	94. d	95. d	96. b	97. a

98. Which among the following protocols is also related to the Access and Benefit Sharing (ABS) mechanism?
    a. Kyoto protocol
    b. Nagoya Protocol
    c. Geneva Protocol
    d. Cartagena Protocol

99. Which among the following city was declared as "Tiger Gateway of India"?
    a. Bhopal
    b. Nagpur
    c. Mysore
    d. Lucknow

100. Which among the following will emit maximum mass of carbon dioxide, when the same quantity is burnt of all of them?
    a. Kerosene
    b. Natural Gas
    c. Gasoline
    d. Propane

Answers

98. b 99. b 100. a